Algal and cyanobacterial biotechnology

Algal and Cyanobacterial Biotechnology

Editors

R. C. Cresswell
Wolfson Institute of Biotechnology,
Division of Molecular Biology and Biotechnology,
University of Sheffield,
Sheffield S1O 2TN, UK

T. A. V. Rees
Leigh Marine Laboratory,
University of Auckland,
R.D.,
Leigh, New Zealand

N. Shah
Division of Biosphere Sciences,
King's College London,
Campden Hill Road,
London W8 7AH, UK

Longman
Scientific &
Technical

Copublished in the United States with
John Wiley & Sons, Inc., New York

Longman Scientific & Technical
Longman Group UK Limited
Longman House, Burnt Mill, Harlow
Essex CM20 2JE, England
and Associated Companies throughout the world.

Copublished in the United States with
John Wiley & Sons Inc., 605 Third Avenue, New York, NY 10158

© Longman Group UK Limited 1989

First published in 1989

British Library Cataloguing in Publication Data
Algal and cyanobacterial biotechnology.
1. Algae. Biotechnology
I. Cresswell, R.C.   II. Rees, T.A.V.
III. Shah, N.   IV. Series
660′.62

ISBN 0-582-49730-2

Library of Congress Cataloging in Publication Data
Algal and cyanobacterial biotechnology/editors, R.C. Cresswell,
  T.A.V. Rees, N. Shah.
      p.   cm.
    Bibliography: p.
    Includes index.
    ISBN 0-470-21477-5
    1. Algae – Biotechnology.   2. Cyanobacteria – Biotechnology.
I. Cresswell, R.C., 1956–   . II. Rees, T.A.V., 1953–   .
III. Shah, N. (Nishith), 1957–   .
TP248.27.A46A44   1989   89-12587
660′.62 – dc20   CIP

Printed and Bound in Great Britain
at the Bath Press, Avon

Preface

For the purpose of this book we have chosen to define algal and cyano-bacterial biotechnology as the commercial exploitation of algae and cyanobacteria or their products. For biotechnologists, algae and cyanobac-teria offer a number of distinct advantages including their diversity and their combination of important characteristics of both plants and microbes. The degree of algal and cyanobacterial diversity is exhibited in their habitats, sizes and levels of organization, life histories, photosynthetic pigments and metabolism. Habitats include the sea, freshwater, soil, snow and ice, hot springs and animal cells. They range in size and level of organization from unicells less than 1 μm in diameter (picophytoplankters) to large (up to 50 m long) seaweeds with parenchymatous tissue and translocation systems. Life histories vary from the simple cell division of microalgae to the baroque complexity of red algal alternation of generation. Algae and cyanobacteria possess more accessory pigments than any other group of plants and their classification is, in part, based on this diversity. Indeed a group of pigments found only in cyanobacteria, cryptomonads and the red algae, the phyco-biliproteins, have recently attracted attention because of their ability to form highly fluorescent conjugates with molecules which possess binding specifi-city (e.g. antibodies) (Glazer and Stryer, 1984). These phycofluors are potentially of great importance in both immunoassays and fluorescence-activated cell sorting. Finally, and perhaps most importantly, is their metabolic diversity. Examples include photosynthetic nitrogen-fixation in certain cyanobacteria, osmoregulatory metabolites and the release of specific metabolites by symbiotic algae and cyanobacteria. A principal aim of this book is to highlight this diversity and its role, both real and potential, in biotechnology.

In common with other plants, algae and cyanobacteria possess chlorophyll *a* and are photosynthetic (with a few exceptions). In principle, therefore, they are cheap to grow requiring only light, CO_2, and inorganic nutrients. Moreover, they will often grow in environments (e.g. brine lakes) which are largely inhospitable to other plants. In addition to being photosynthetic autotrophs, unicellular algae and cyanobacteria possess a number of microbial characteristics including rapid growth rate, high protein content and a variable metabolism which responds rapidly to environmental changes (Myers, 1980). Again the latter is perhaps the most important biotechnologically. A specific example would be those unicellular algae which alter their biochemical composition from predominantly protein to predominantly lipid during nitrogen-deficiency. However, it is the high protein content of unicellular algae and cyanobacteria which makes them attractive as food sources both in the Third World and on spacecrafts.

Perhaps the ultimate biotechnological property of unicellular algae is to be found in certain strains of *Chlorella* which are able to bind gold to their cell walls – the basis, one presumes, for a multimillion dollar industry.

The foregoing has tended to emphasize the value of unicells, but seaweeds are also important biotechnologically. Indeed the history of algal biotechnology is dominated by the uses of seaweeds. A considerable proportion of the medical uses of algae, including the recent application (literally!) of alginate gels as wound dressing, are derived from seaweed products.

The uses of algae and cyanobacteria in both biotechnology and the laboratory are considerable, but these uses have been restricted to relatively few species. A vital component of the future use of algae and cyanobacteria in biotechnology will depend on extending laboratory work to a greater range of species. Given the diversity of algae and cyanobacteria there are probably many new applications to be derived from such investigations.

Finally we would like to thank the authors for the high standard of their contributions. In general, they made our editorial task straightforward and pleasurable.

R. C. Cresswell,
T. A. V. Rees,
N. Shah

References

Glazer A N, Stryer L 1984 Phycofluor probes. *Trends Biochem. Sci.* **9**: 423–7

Myers J 1980 On the algae: thoughts about physiology and measurements of efficiency. In Falkowski P G (ed) *Primary Productivity in the Sea*, Plenum Press, pp 1–15

Contents

3. **The transport, assimilation and production of nitrogenous compounds by cyanobacteria and microalgae**

N. W. Kerby, P. Rowell and W. D. P. Stewart

4. The biotechnology of mass culturing *Dunaliella* for products of commercial interest

A. Ben-Amotz and M. Avron

5. The biotechnological potential of symbiotic algae and cyanobacteria

T. A. V. Rees

6. Fuels from algae

M. Calvin and S. E. Taylor

7. Secondary metabolites of pharmaceutical potential

K.-W. Glombitza and M. Koch

8. The genetic manipulation of cyanobacteria and its potential uses

O. Ciferri, O. Tiboni and A. M. Sanangelantoni

9. Immobilized cells: An appraisal of the methods and applications of cell immobilization techniques

M. Brouers, H. deJong, D. J. Shi and D. O. Hall

10. Industrial production: methods and economics

L. J. Borowitzka and M. A. Borowitzka

11. The future of microalgal biotechnology

J. R. Benemann

Contributors

M. Avron,
Biochemistry Department,
Weizmann Institute of Science,
Rehovot, Israel.

A. Ben-Amotz.
The National Institute of
Oceanography,
Israel Oceanographic and
Limnological Research,
Tel-Shikmona, P.O.B. 8030,
Haifa 31080, Israel.

J. R. Benemann,
Sea Ag, Inc,
5220 Old Dixie Hwy,
Fort Pierce, Florida, 34946, USA

L. J. Borowitzka,
Western Biotechnology Ltd,
2–6 Railway Parade,
Bayswater, WA 6053, Australia.

M. A. Borowitzka,
Algal Biotechnology Laboratory,
School of Biological and
Environmental Sciences,
Murdoch University,
Murdoch, WA 6150, Australia.

M. Brouers,
Department of Botany,
University of Liège,
Sart Tilman,
4000 Liège, Belgium.

M. Calvin,
Department of Chemistry and
Lawrence Berkeley Laboratory,
Berkeley, California 94720, USA

D. J. Chapman,
Department of Biology,
University of California,
Los Angeles, California 90024,
USA

O. Ciferri,
Dipartimento di Genetica e
Microbiologia,
'A Buzzati Traverso',
Via S. Epifaniq 14, 27100 Pavia,
Italy.

G. E. Fogg,
School of Ocean Sciences,
University College of North Wales,
Marine Science Laboratories,
Menai Bridge, Anglesey,
Gwynedd LL59 5EH, U.K

K. W. Gellenbeck,
Microbio Resources Inc.,
6150 Lusk Boulevard, Suite B-105,
San Diego, California 92121,
USA

K.-W. Glombitza,
Institute fur Pharmazeutische
Biologie,
Rheinische Friedrich-Wilhelms-
Universitat,
Nussallee 6, D-5300 Bonn 1, West
Germany.

D. O. Hall,
King's College,
University of London,
Campden Hill Road,
London W8 7AH, UK

H. deJong,
King's College,
University of London,
Campden Hill Road,
London W8 7AH, UK

N. W. Kerby,
The AFRC Research Group on
Cyanobacteria,
Department of Biological Sciences,
The University of Dundee,
Dundee DD1 4HN, UK

M. Koch,
Institute fur Pharmazeutische
Biologie,
Rheinische Friedrich-Wilhelms-
Universitat,
Nussallee 6, D-5300 Bonn 1, West
Germany.

T. A. V. Rees,
Leigh Marine Laboratory,
University of Auckland,
RD, Leigh, New Zealand.

P. Rowell,
The AFRC Research Group on
Cyanobacteria,
Department of Biological Sciences,
The University of Dundee,
Dundee DD1 4HN, UK

A. M. Sanangelantoni,
Dipartimento di Genetica e
Microbiologia,
'A Buzzati Traverso',
Via S. Epifanio 14, 27100 Pavia,
Italy.

D. J. Shi,
Institute of Botany,
Academia Sinica,
Beijing, China.

W. D. P. Stewart,
The AFRC Research Group on
Cyanobacteria,
Department of Biological Sciences,
The University of Dundee,
Dundee DD1 4HN, UK

S. E. Taylor,
Chemical Biodynamics Division,
Lawrence Berkeley Laboratory,
1 Cyclotron Road,
University of California,
Berkeley,
California 94720, USA.

O. Tiboni,
Dipartimento di Genetica e
Microbiologia,
'A Buzzati Traverso',
Via S. Epifanio 14, 27100 Pavia, Italy.

1

An historical perspective of algal biotechnology

D. J. Chapman and K. W. Gellenbeck

Introduction

What is algal biotechnology? The purist – and narrow view – would argue that algal biotechnology is the science of genetic manipulation of algae with the techniques of modern molecular biology: mapping, cloning and splicing of specific genes to produce a 'better' alga. That view would limit this contribution to a very few pages as this area of study is only recently receiving broad attention (Craig and Reichelt, 1986). The broad view would equate algal biotechnology with algal technology, but here the problem is that many such aspects are simply plant or animal technology in which the source biomass is an alga rather than some other organism. Methane production from algal biomass, algae as cattle feed and, to a certain extent, the use of algal extractives as fertilizer would fall into this category. We have taken a middle ground and, in providing an historical perspective, regard the topic as algal technology in which the operative factor is some property or attribute unique to algae. This could be extractive chemistry and utilization of the unique phycocolloids, the large-scale production of microalgal biomass in open ponds, marine mariculture in either open or semi-closed systems, or extractive chemistry and utilization of chemical compounds unique to algae or which could be extracted most economically from algae. For convenience we have divided this historical overview into a brief introduction into what algae are, one section discussing products from algae and another that covers the techniques used in growing these organisms.

The nature of algae

Though most people are not able to look at a terrestrial plant, for instance an apple tree, and put a scientific name to it or give details about its physi-

Table 1.1   Chemical characteristics of the major classes of algae with industrial interest.

Class	Common name(s)	Chlorophyll
Cyanophyceae	Cyanobacteria Blue–green algae	a
Rhodophyceae	Red algae	a
Phaeophyceae	Brown algae	a and c_1c_2
Chrysophyceae	Golden algae	a and c_1c_2
Bacillariophyceae	Diatoms	a and c_1c_2
Chlorophyceae	Green algae	a and b
Ulvaphyceae	Green algae	a and b
	Terrestrial plants	a and b

ology, they are none the less able to tell you that it requires water, fertilizer in the soil, sunshine and that it will bloom and bear fruit. Many people would even tell you that they would love to have a good one in their back yard! Algae, on the other hand, remain relatively unknown. In fact, when algae are recognized it is usually in a negative sense as a troublesome aquatic growth. Hence the regrettably common and oft-repeated term 'pond scum' indicative of the bad reputation gained by these organisms.

The reality of the algae is quite different. As a group they defy a strict taxonomic definition and display a tremendous diversity in form ranging from microscopic prokaryotic cells (the cyanobacteria) to huge seaweeds such as the giant kelp *Macrocystis* that, up to 70 m, are among the largest plants on the planet. Within these extremes are unicellular forms from microns to millimetres in diameter, colonies of cells, filaments, sheets and complex thalli with differentiation into leaf-like blades and stem-like stipes. Only a brief summary of the classification of algae will be given here; for more complete treatments the reader is referred to works such as Trainor, 1978; Lee, 1980; Lobban and Wynne, 1981; Bold and Wynne, 1986.

Since the early beginnings of phycology, or the study of algae, colour has

Other pigments	Metabolic storage product	Cell wall components
Phycobiliproteins, carotenoids, xanthophylls	Glycogen, others	Bacterial (sugars, muramic and glutamic acids, etc.)
Phycobiliproteins, carotenoids, xanthophylls	Floridean starch	Cellulose, xylans, sulphated polysaccharides, some calcified
Carotenoids, fucoxanthin, other xanthophylls	Laminarin, mannitol	Cellulose, alginic acid
Carotenoids, fucoxanthin, other xanthophylls	Chryso-laminarin	Cellulose, silica, some calcified
Carotenoids, fucoxanthin, other xanthophylls	Chryso-laminarin	Silica
Carotenoids, xanthophylls	Starch, glycerol	Cellulose, xylans, mannans, some calcified
Carotenoids, siphonaxanthin, other xanthophylls	Starch	Cellulose, xylans, mannans, some calcified
Carotenoids, xanthophylls	Starch	Cellulose, lignin

been the most outstanding characteristic of these plants and has been relied on to separate them into categories. This led to the common names such as the green, red or brown algae that are still used. Perhaps surprisingly, modern systematics and evolutionary theory have maintained many of these common groups, for the most part with a reliance on pigment composition as a major separating character. A caution should be made that visual colour (phenotypic) can often be unreliable; for example, a green alga can appear reddish due to its complement of pigments other than chlorophyll and many red algae, particularly in the intertidal, appear green, due to the very low levels of red photosynthetic pigments, biliproteins, in their cells. Table 1.1 lists the major groupings of algae that have been of industrial interest together with their pigment content and other characters that are used to separate them taxonomically (storage product, flagellar make-up and cell wall composition). The characteristics of terrestrial plants are also listed for comparison.

A special note should be made for the cyanobacteria which for many years were known as the blue–green algae. These organisms are prokaryotic (which allies them with the bacteria) while all remaining algae are eukaryotic. Why then are they included and very often studied by phycologists as well as bacteriologists? This is mostly due to the ecological niches occupied by the cyanobacteria which, on a macroscopic level, make them indistinguishable from other eukaryotic algae that grow in the same areas. Out of convenience then, the techniques of phycology have been frequently applied to the cyanobacteria and make them a part of the algae. This anomaly points out what is perhaps the single most outstanding attribute of the artificial grouping 'algae' in relation to their utilization in biotechnology. This is the vast genetic diversity to be found in the many evolutionary lines. In addition, many of these lines are thought to be quite ancient, most notably the cyanobacteria that are some of the first organisms to have evolved and have survived for more than 3 billion years (Schopf and Packer, 1987). As displayed by the variety of metabolic products in Table 1.1, the biochemical pathways found in the algae are already quite diverse, and the majority of species have not been studied in any detail at all. In comparison, the terrestrial plants have a shorter evolutionary history and are dominated in both an ecological and technological sense by a single evolutionary line, the angiosperms, that have existed for some 200 million years.

Products from algae

The early history of the varied uses of algae has been documented by Chapman and Chapman (1980) and Soeder (1986). A short summary of identified, commercially practical, uses must suffice for this discussion. The use of algae for products other than food can be traced back at least to the seventeenth century when French peasants recovered soda ash from brown algae (kelps) collected from wild populations. The soda ash industry was followed by the development in the mid-nineteenth century of an iodine industry that used the same algal sources. The success and decline and dating of these industries in various countries is mirrored by the discovery of non-biological sources such as Chilean mineral deposits, and the constraints on availability brought about by the upheavals of war.

The alginic acids were discovered in the latter part of the nineteenth century and became a commercial product in the early twentieth century and are now the dominant product industry based on brown algae.

The other industrial developments that paralleled this sequence were the phycocolloid industries centred on red algae. The agar industry can trace its origin to China and Japan in the mid-seventeenth century, while the carrageenan industry originated in Ireland, followed by the United States in the 1830s.

As other industries declined and economics became a driving force, the algal products industry of the last sixty years has been dominated by the big three phycocolloid polysaccharides: alginates, carrageenans and agars. It is easy to understand this dominance. These three families of polysaccharides are present in high yields in their respective groups of algae (Table 1.2) and

Table 1.2 Principal sources of algal phycocolloids

ALGINIC ACID: Phaeophyceae – Brown algae
Laminariales
  Ecklonia
  Eisenia bicyclis
  Macrocystis pyrifera
  Laminaria digitata
  Laminaria hyperborea
Fucales
  Ascophyllum nodosum
  Sargassum spp.

AGAR: Rhodophyceae – Red algae
Gelidiales
  Gelidium spp.
  Gracilaria spp.
  Pterocladia spp.
  Suhria vittata

CARRAGEENANS: Rhodophyceae – Red algae
Gigartinales
  Ahnfeltia plicata
  Chondrus crispus
  Eucheuma spp.
  Furcellaria fastigiata
  Gigartina spp.
  Gymnogongrus spp.
  Hypnea spp.
  Iridaea spp.
  Phyllophora nervosa

within these groups of algae are individual genera or species that are region-
ally very abundant, often occurring in predominantly unialgal stands that
lend themselves to mechanical or manual harvesting. Their widespread uses,
the lack of synthetic substitutes and, with very few exceptions (e.g. animal
gelatins and bacterial exudates), the lack of alternative non-algal sources
make these attractive economic propositions.

The reverse, and negative side, is the problem of availability. Natural
populations mean a limited standing crop, with the very real possibility of
demand outstripping supply. This concern has been the driving force behind
research into marine algal mariculture; efforts to increase standing crops and
bring other algae into economic use are described in the next section of this
chapter.

In the last fifteen years attention has turned to other areas with particular
focus on pharmacological potential, energy biomass, agricultural fertilizer,
and utilization of unicellular algae whose biological properties lend themselves to large-scale production in regions unfavourable for other more
traditional uses such as terrestrial agriculture. Before turning to these
aspects, a brief synopsis of the current state of the colloid future is provided.

Algal colloids

Alginates

The current status of alginate usage has been reviewed by Glicksman (1987) and King (1983). Supply still meets demand, and this has been achieved through a combination of management of harvesting techniques and development of new sources. The usual sources have been *Macrocystis, Ascophyllum, Eisenia, Laminaria digitata* and *Laminaria hyperborea*, the latter producing an alginate quite different from the others (King, 1983). Although we are beginning to understand the mechanism of alginate biosynthesis, the determinants controlling the ratio of guluronic acid to mannuronic acid that make up the alginates and their block combinations are not well understood, and the type of alginate recovered cannot be controlled *de novo*. Additional sources of alginates are now being sought. This expansion takes on additional importance when one remembers that nearly all present day alginate sources are cold temperature algae. Although alginates are common to brown algae, those species growing in sufficiently abundant biomass are, with few exceptions, the fucoids (Fucales) and kelps (Laminariales) of colder water. A cosmopolitan alga, particularly common in the tropics, and with a life history and growth pattern well suited for both open system and closed system mariculture and capable of producing sufficient biomass is *Sargassum* (Gellenbeck and Chapman, 1986; Mshigeni and Chapman, 1990). On a worldwide scale the harvesting of *Sargassum* is a fraction of the others (e.g. some 8000 tons in India in 1978). One can expect an increasing interest in this fucoid brown alga particularly in the tropics. A possibility away from algae, however, is the application of modern genetic technology to alginate biosynthesis. Two bacteria, *Pseuedomonas aeruginosa* and *Azotobacter vinelandii* are alginate producers. *Pseudomonas* is currently the target of much research in alginate biosynthesis, and the regulation of gene expression (Deretic *et al.*, 1987). As these authors have pointed out (p. 477) we may yet see the time when, 'Understanding . . . may also allow manufacturing this important polysaccharide from stable high yielding bacterial strains rather than its extraction from seaweeds, where yield and quantity are extremely variable'.

Carrageenans

These sulphated polysaccharides are found in members of the red algal order Gigartinales. While a number of forms are known (Glicksman, 1983), each with different properties, the commercial utilization is handicapped by the fact that the particular carrageenan present depends upon the individual species, life history stage and location. These smaller red algae, frequently intertidal in nature, do not lend themselves to the harvesting protocol used for large brown algae.

There are no substitute compounds in sight and the lack of abundant sources of these algae prompted the major research thrust on *Chondrus* and

Eucheuma mariculture described in the next section. Other gigartinalean red algae, e.g. *Hypnea, Iridaea* and *Gigartina* are potential sources (Hansen *et al.*, 1981), but only if mariculture becomes economically feasible. They are not abundant in unialgal stands for 'farming', and there is little information to suggest that they are good candidates for open bay farming.

Agars

This polysaccharide, of a family similar to carrageenans, can be regarded as the smallest industry of the big three, both in terms of annual production (Glicksman, 1983) and extent of source material. The red algae *Gelidium* and *Pterocladia* were for years the only viable source. There are few agarophytes and their scattered and limited distribution, both geographically and locally, has worked against anything but principally small collections for the larger processor. Neither of these two algae seem suited to mariculture manipulation of any type. In recent years attention has turned to species of *Gracilaria*. This alga, in contrast to the other two, does frequently occur in readily collectable unialgal stands, and has been the focus of numerous large-scale culturing investigations and industries (Shang, 1976; LaPointe and Ryther, 1978; Hansen, 1984).

Pharmacological use

The phycocolloid industry has represented the backbone of algal utilization for many years, but with very little input from modern biotechnology. In the last twenty years there has been a major drive to search for pharmacologically useful compounds from marine organisms, including algae. During the 1950s and 1960s bacteria and fungi were subjected to extensive chemical surveys for pharmacologically useful compounds. Surveys of higher plants followed and it was inevitable that marine organisms (especially invertebrates and algae) would be the next frontier. The usual research plan (e.g. Fenical and Paul, 1984; Reichelt and Borowitzka, 1984) has followed the sequence: (1) collect what is available or collectable. There is usually no *a priori* reason to suggest screening one particular group; (2) isolate the principal components of a particular chemical class; (3) screen the collected examples for bioactive properties such as cytotoxicity, antiviral, anti-inflammatory, antitumour or antimicrobial activity; and (4) analyse the 'hit' for the exact compound responsible, the quantity produced, etc. In terms of increasing our knowledge of algal chemistry, this approach has proved a bonanza. The publications in such journals as *Journal of the American Chemical Society, Tetrahedron, Tetrahedron Letters, Phytochemistry* and specialized publications, e.g. *Natural Products Reports*; and review articles in series such as *Pure and Applied Chemistry, Marine Natural Products* (Scheuer, 1978–1984) and *International Seaweed Symposia* (Bird and Ragan, 1984; Ragan and Bird, 1987) reveal a wealth of novel compounds isolated from algae, together quite frequently with the results of basic bioactivity testing. More often than not, that is where the investigation ends, and much of this has to do with the economics of pharmaceutical development. In a very enlightening article Baker (1984) has outlined the rise and fall of the marine pharmacology programme at the Roche Institute of Marine Pharmacology in Sydney, Australia. Considerable effort has gone into such phycochemical endeavours, but the payoff has not come, and some less than optimistic outlooks have been expressed. Blunden and Gordon (1986) have commented that the combination of the difficulties of culture and collection, lack of success to date and the less than favourable economic climate have led to a decline in pharmaceutical research. They note that a major break-through is needed to reverse the trend. A similar viewpoint was expressed by D. Renn, quoted in Curtin (1985). Indeed it is pertinent to ask if algae would ever become an economically viable, commercial source of phar-maceuticals. Very many of the compounds isolated are present in very small amounts in algae of very limited distribution and abundance. Their growth patterns and our capabilities with mariculture suggest that large-scale culturing is not presently feasible. Indeed, one suspects that a compound identified as pharmacologically useful would be turned over to the synthetic chemists to provide an alternate route of supply. A case in point is the cyclic peptide didemnin. This natural compound isolated from the tunicate *Trididemnum* (25 g pure per 270 kg of tunicate) is undergoing clinical trials as a potential anticancer agent. Very recently (Rinehart *et al.*, 1987) a total synthesis has been achieved and although no economic figures have been provided one must ask whether large-scale culture or collection of *Trididemnum* would be the ultimate commercial source. By analogy, would pharmaceutically active compounds from algal sources fare any differently once they are discovered?

Other extractives

Outside of the phycocolloid industry there are very few other algal extrac-tives that currently show economic potential. Two examples will be mentioned.

β-*Carotene*

This compound, ubiquitous in green plants, is produced in very high yield by the halotolerant *Dunaliella salina*. This Chlorophycean (green alga) flag-ellate is ideally suited to mass culture in evaporative shallow brackish ponds. It is a fast grower with a wide tolerance to high temperature, salinity and solar irradiation. Major efforts at industrial production (e.g. Ben-Amotz and Avron, 1980; Borowitzka *et al.*, 1984; Ben-Amotz and Avron, this volume) are underway in California and Hawaii in the United States, in Israel, and in western Australia. The extractive chemistry and biological manipulation is straightforward and well understood, as is the laboratory cultivation. The technology has been developed but the question remains: can this approach compete economically with long established chemical syntheses? (See Ch. 10.)

Glycerol

This is another principal osmoregulatory product of *Dunaliella* (Ben-Amotz and Avron, 1980). In this case the economic question does not concern competition with chemical syntheses, but rather recovery during the petroleum extraction process. The vagaries and economic fluctuations, as well as the current low cost of glycerol in the petroleum industry, would seem to preclude the algal industry.

Algal technology has developed the capabilities for large-scale culture of microalgae. It is ideally suited for arid zone agronomy, where there are few demands on the land for other uses. The future lies in applying the technology suited for that environment, and producing a compound of marketable potential. Three examples will suffice, and interestingly enough they are all derived from the one alga *Porphyridium*, a unicellular red alga, that appears readily adaptable to outdoor pond culture (Arad *et al.*, 1985).

Arachidonic acid

This is an essential dietary fatty acid and the natural precursor of such compounds as prostaglandins and thromboxanes. Good natural sources are very few, the principal ones being heterotrophic microorganisms. Since *Porphyridium* (a red alga) probably has one of the highest natural concentrations of arachidonic acid, up to 36% of the total fatty acids, it is thus a candidate for supply. There are both freshwater (*P. aerugineum*) and saline water (*P. cruentum*) species. *Porphyridium* is an adaptable alga, thus an ideal candidate for open pond cultivation and as a source of this fatty acid (Arad, 1986).

Sulphated polysaccharide

Cells of the various species of *Porphyridium* secrete a sulphated polysaccharide into the medium. Arad *et al.* (1985) have commented on their particular resistance to extreme conditions, noting its use in tertiary oil recovery and its potential as a food thickener.

Phycofluors

A third more intriguing use of *Porphyridium* would be as a source of phycofluors (Glazer and Stryer, 1984). Phycofluors are the phycobiliproteins bound to biologically active molecules such as immunoglobulin, biotin and protein A. The phycobiliproteins are photosynthetically active pigments, abundant in red algae, cryptomonads and cyanobacteria. Yield is not a problem, but the choice of alga is. Cryptomonads are not viable choices for open pond culture. Many abundant macrophytic red algae do not lend themselves to extraction of phycobiliproteins, and the few that do are not readily abundant. *Porphyridium*, however, as an organism that is culturable and readily extractable, is an ideal source for the phycoerythrins. Cyanobacteria are organisms that lend themselves to pond culture (e.g. *Spirulina*) and would be excellent sources of phycocyanin phycofluors. *Porphyridium* may be the example where engineering technology, algal biology and the product have come together in an economically marketable situation, not subject to economic pressure and competition from non-algal sources.

Are there other similar algae for the future? *Dunaliella*, with its capacity for β-carotene and glycerol, is a possibility, depending upon market economics (see earlier). There are other microalgae with unique products, e.g. *Botryococcus* with its unusual hydrocarbons. Whether the right combination of product(s) use, cheap and simple culturability, recovery and market economics can come together remains to be seen. Certainly there are many microalgae, for which simple, large-scale and cheap outdoor culturing is available. Cyanobacteria, most notably *Spirulina*, and certain green algae such as *Scenedesmus* and *Chlorella* are possibilities. However, in the absence of an identifiable high value, high yield product, markets such as poultry feed and human protein supplement will continue to be the prime use. The reader is referred to articles in Shelef and Soeder (1980), Ragan and Bird (1987) and a comprehensive review by Venkataraman and Becker (1985).

Genetics

Molecular genetics of algae is still in its infancy. Accordingly, algal biotechnology, defined in the narrow sense, has yet to really begin. However, strain and genetic selection of macroalgae have been practised for some time (Cheney, 1984; van der Meer, 1987). A very well-known example is the selection of the T4 strain of *Chondrus crispus* (Neish, 1976) and the selection programme in China for *Laminaria* (Wu and Lin, 1987).

Strain selection for microalgae (Vonshak, 1987), grown in open ponds, has also been carried out, but its effectiveness is limited by the air and soil borne contaminants that can invade the cultures of selected strains. Elaborate protection against contamination of batch mode culture will usually be needed to allow use of derived strains in open ponds.

Summary

There is insufficient space in this chapter to consider every possible class of compounds with potential marketable value, or with a practical commercial use, or every alga that has been investigated as a potential economic entity. The literature is abundant and widely scattered, with the state of marketable and practical commercial use running the full range from high potential to very low. If there has been a weakness in developing algal biotechnology (or technology), it would stem from the fact that too often phycologists have isolated and grown organisms, without regard for any use; and the non-phycologists have frequently identified a use without concern for the production capabilities and supply of the alga. The future is there for algal (bio)technology, but it will come slowly and haphazardly unless phycologists (researchers in universities), applied phycologists (researchers in industries), engineers and economists come together to pool their talents.

The history of algal culture

As a means of organizing the wide variety of culturing techniques, we will divide them into those applied to macroalgae, mostly the seaweeds, and those applied to microalgae. Within these areas, we hope to describe the major advances and achievements that have led to wide application and operations at an industrial level, and compare these advances to those made in the history of terrestrial agriculture.

Macroalgae

Collections of wild stocks of seaweeds have been made for food and other purposes since as early as 900 BC (Porterfield, 1922). This method of obtaining algal material continues today on a small scale throughout the world for food, phycocolloids, and medicinal purposes (Chapman and Chapman, 1980). Two large-scale industrial operations, that still rely on the collection of wild stocks of seaweeds, are the extraction of alginic acid from the brown algae, e.g. *Macrocystis* along the California coast in the US (Sleeper, 1980), and the extraction of carrageenan from *Chondrus crispus* in the eastern US and Canada (Hansen *et al.*, 1981). The success of these industries depends on large quantities of easily harvestable stands in near-coastal waters. The susceptibility of such supplies to the whims of nature is obviously an ever-present problem. An excellent example of this is the occurrence of an El Niño event in the Pacific in 1982–83 which led to a tremendous decrease in the occurrence and extent of kelp beds, due mostly to uncharacteristically high water temperatures (Dayton and Tegner, 1984). This led the major US producer, Kelco Inc., to import seaweeds from other parts of the world until the beds recovered.

An obvious comparison can be made with these wild harvests to terrestrial agriculture. Only extremely limited populations today can survive with gathering from wild stocks as their sole supply. Since prehistoric times societies have rapidly domesticated useful wild plant species to ease harvesting and ensure predictable production. Not surprisingly, a number of macroalgal species have also been successfully domesticated, with Japan and China leading the way in technological advances.

Probably the most extensively cultivated macroalga is the red alga *Porphyra* that has been cultured in Japan since 1570 (Tseng and Chang, 1955). In its earliest form this cultivation consisted of the planting of brush to increase the available surface area for the algal spores to settle. Careful observation of natural populations taught the early marine farmers where to locate these structures and how to care for them. Until the early 1960s the seeding of such brush, and the nets that replaced them in more recent times, relied on spore dispersal from populations of adult plants in the open ocean. This changed after 1949 when the elucidation of the complete life cycle of *Porphyra* by Drew (1949) pointed out a previously unknown microscopic alternating phase of the species found growing on mollusc shells, and termed the 'conchocelis' phase. With this information, controlled seeding of culture nets was achieved and the industry was able to expand rapidly in response to increased demand for the sheets of dried alga referred to as 'nori'. The state of this industry today is truly impressive, with some 50 km^2 in cultivation (Soeder, 1986) and a production of 300 000 freshweight tons valued at $380 million (Kurogi, 1975). Extensive research and development have resulted in both a thorough understanding of the biology of the algae (spore formation, inoculum storage by freezing, nutrient needs and application, strain improvement and genetic manipulation, disease control, etc.), as well as a commercially consistent growth and processing system using nets spread in coastal waters. For a thorough description the reader is referred to Miura (1975).

The lessons to be learned from this most successful of algal industries are perhaps not that different from those that can be gleaned from terrestrial agriculture: (1) a complete and thorough understanding of the biology and ecological needs of the cultivated species is imperative; (2) to allow the application of the techniques needed to improve the efficiency of the process (e.g. strain selection, breeding) tight controls on growth and production are required; and (3) there must be a market of sufficient size and consistency to support continued improvement and advancement. This is the fusion of talent and effort from basic phycology, applied phycology, engineering and economics referred to earlier.

These lessons have not been lost on other workers that have successfully developed a number of domesticated macroalgal species. Market changes, due to political upheaval in Indonesia in the 1960s, led to the infusion of funds from the carrageenan industry for development of a raw material supply from the alga *Eucheuma* grown in the Philippines (Doty, 1973). With the advantage of inexpensive local labour, a network of small family farms using vegetative reproduction methods on nets strung in calm waters was already able to produce 16 500 tons of extractable material in 1979 (Laite and Ricohermoso, 1981). Again, part of the success can be attributed to basic research on the biology of *Eucheuma* that was translated into useful farming practices (e.g. Polne *et al.*, 1980).

Another culturing success in open waters by the Japanese and Chinese has been achieved with the brown algal genera *Undaria* and *Laminaria*. Historically, harvests of wild populations were sufficient to satisfy the market demand, but an infusion of support from the government in the late 1950s, and an increasing demand due to the highly regarded content of vitamins, minerals and special proteins have provided the impetus to develop a successful industry (Saito, 1975). The process is somewhat similar to the techniques used for *Porphyra*. Spores are impregnated on ropes, germling growth is carried out in controlled tanks and growth of the adult plants is completed in coastal waters on a raft apparatus. Total production as of 1985 was approximately 300 000 tons (Nisizawa *et al.*, 1987).

Mention was made above concerning the harvest of the giant kelp *Macrocystis pyrifera* for alginic acid production. This huge plant with a tremendous growth potential of up to 8 dry tons ha^{-1} $year^{-1}$ under optimal conditions (North, 1980) is an obvious choice for domestication. Though there are not yet any large-scale commercial operations for controlled culture of this species, substantial efforts and progress in this direction have been made. Much of this effort has not been driven by the need for increased alginic acid production, but rather for conversion of the plant material into usable energy through the process of anaerobic digestion (Leone, 1980). Neushul (1977), rightly, has pointed out the need to have a complete understanding of the species to be able to control the life history of the plant and make wise choices for culturing techniques and procedures. Substantial laboratory work has led to an extensive test farm using transplanted individual plants laid out in an array quite analogous to a terrestrial farmer's field (Neushul, personal communication). Data from these tests may soon lead to operational farms if market demands will support their establishment.

A final effort at the growth of *Macrocystis* deserves mention since it opens up such vast potentials. All of the open ocean culturing operations described above rely on the availability of coastal areas somewhat protected from the destructive potential of storms. The overall concept in this case, however, was to grow *Macrocystis* on floating rafts away from the coast where the nutrient poor waters were supplemented with nutrient laden water pumped up from greater depths (Leone, 1980). The advantage of utilization of vast areas of the ocean for photosynthetic energy conversion is obvious and led to the input of a high level of governmental funding in the energy conscious 1970s. A 0.1 ha test module was built and operated for two years off the coast of California (North, 1980), but long-term success was prevented by storms that severely damaged both the plants and structure. The failure to obtain reliable yield data, and substantial management difficulties (Budiansky, 1980), ended the project in the early 1980s. Much was learned, however, about the engineering techniques that will be needed to take advantage of open ocean areas in the future, and interest will surely rise when fossil fuel shortages again come to dominate world politics (see the chapter by Calvin and Taylor in this volume).

The major alternative to the open ocean techniques of seaweed culturing, which attempt to turn an area of the ocean into a monoculture of the desired species, is the use of a pond enclosure on land as the farm plot. This is termed a semi-closed culture as seawater is pumped between the pond and the ocean. The obvious advantage in this process is the control of many environmental factors that are impossible in the ocean, most notably the destructive potential of storms. Just as obvious is the increased cost in construction and operation of the system (Huguenin, 1976).

The most extensive effort with this idea has been the culture of *Chondrus crispus*. Pioneering efforts in the 1970s by workers at the Atlantic Regional

Laboratory, Nova Scotia (NRCC) led to rapid improvement of both the strain and the techniques so that pilot-sized farms of various designs were in operation by 1980 (Hansen *et al.*, 1981). The jump to production-size facilities, originally predicted for 1986 (Neish, 1976) has not occurred though. The major reason appears to be the reluctance of investors to put money into processes that are quite expensive when the current collections from nature are of sufficient quantity and low enough cost to meet the demand.

Other efforts at semi-closed culture have been successful at research and pilot-scale. Ryther *et al.* (1979) have grown *Gracilaria* and *Hypnea* in raceway ponds for agar production in Florida. Edelstein *et al.* (1976) have also grown *Gracilaria* in Nova Scotia. Waaland (1977) has expanded such systems to the growth of *Gigartina*, and Gellenbeck and Chapman (1986) to *Sargassum* . In all of these cases the technical feasibility has been shown, but the economic projections have not been sufficient to support sustained development programmes.

So what has history taught us about the culture of macroalgae? We see two main lessons. First, successful efforts have their roots in a precise and reliable understanding of the biological needs and limitations of the cultured species. The culture system can then be tailored to meet those needs. The second lesson, not surprisingly, concerns money. Funding to do the basic research is an obvious need, but like the terrestrial farmer, being able to grow a crop is only part of the battle. Consistently obtaining a price for the product sufficient to cover the cost of production and leave a profit is imperative to the survival of the process (Gellenbeck and Chapman, 1983).

Microalgae

The domestication of microalgae is in many respects quite different than it is with macroalgae. Unlike the macroalgae or even terrestrial plants, there is no longer a large plant that can be held in the hand. Now a microscope is needed to even see the individual plants that make up the crop, and the established base of information from known agricultural practices is even less applicable in many respects than it has been for macroalgal mariculture. This has led to a number of technical difficulties that have historically required innovative solutions.

The history of attempts at large-scale controlled production of microalgae does not extend nearly as far back as it does for macroalgae perhaps due to the difficulties referred to above. The first published work to summarize advances in microalgal culture up to that time is the classic *Algal Culture from Laboratory to Pilot Plant* (Burlew, 1953a). This work chronicles a project spearheaded by the Carnegie Institution of Washington which was an offshoot of its basic research on the nature of photosynthetic processes. The main driving force was the use of the green microalga *Chlorella* for large-scale production of food, with the main question being whether the high productivities measured in the laboratory for this species could be maintained in larger systems. The culmination of the project was a pilot plant study (57 m^2 using enclosed tubes) carried out by Arthur D. Little, Inc. in 1951 with the established aims of demonstrating continuous growth on a larger scale, gaining information to assist economic appraisals, and producing experimental quantities of *Chlorella* for evaluation of possible end uses of the product. In conjunction with the large-scale work, funding was provided for continued study of the basic physiology and growth characteristics of the *Chlorella* cells.

This volume still makes fascinating reading for algal biotechnologists as most of the major problems presented to expansion of this industry were identified by these early workers. Perhaps the most persistent problem has been the maintenance of yields. Burlew (1953b) summarizes the fact that the culture in their work with the highest efficiency of energy utilization amounted to 70 g m^{-2}, which was over five times the best yield obtained in the pilot plant work. Wassink *et al.* (1953) looked at the overall photosynthetic efficiency of *Chlorella* in comparison to terrestrial plants and found it to be quite similar. In low light irradiances the efficiency in relation to incident radiation was 12–20% for both, while under outdoor full sunlight intensities the efficiency dropped to 2–3% for both the algal and terrestrial plant cultures. The main factor identified as producing low efficiency under natural conditions was excessive illumination. Other problems pointed out by this work (Burlew, 1953b) were: the need to recycle the culture medium, compounded by the difficulty of maintaining production rates with 'old' medium; the problems presented by contamination of the cultures by other competing algae, and predators that feed upon the algae; maintaining environmental conditions (temperature, light, etc.) optimal for growth of the algae; the technical problems of efficiently harvesting the culture relatively constantly without affecting growth of the remaining cells, then drying the harvest to a final product; perhaps most important, what use and value was there for the final product? Despite the obstacles presented by these problems many positive and valuable advancements were made.

The most obvious step was the proof that large-scale culturing of microalgae could be successfully carried out, supported by the final production of some 50 kg of dried *Chlorella* product from the pilot plant operation. In an historical context, however, the most lasting impact was perhaps the realization that the special problems associated with handling a microalgal culture went beyond biology and required a close association with engineering disciplines to make laboratory results scale-up to production facilities. Engineering analysis and questions moved to the forefront of developments in microalgal culture.

The two major obstacles to direct industrial application of the early culture work with *Chlorella* for use as food were the cost of production and the marketing and use of the final product. Tamiya (1957) analysed the results of the early studies and estimated that the cost of production would be about $520 kg^{-1} which was too high to compete with the least expensive protein sources such as soybean meal. This conclusion was still true some twenty-five years later (Behr and Soeder, 1981). Consideration of the final product for food was early on perceived to be a relatively easy step that would combine with the high protein content of the algae. However, the questions of toxicological safety and acceptability of algal meal in the diet were quickly raised and are far from settled even today. Becker (1986) extensively reviews the entire topic of the nutritional aspects of microalgae as feed, and the reader is referred to that volume for details. However, one of the main conclusions to be reached is that at present it is impossible to make any broad generalizations or recommendations. There are often as many studies to support the use of a given microalgal meal as there are against it. The reasons for this are many fold: production systems are usually not consistent and therefore the characteristics of the algal meal are not consistent; the processing of the meal has a great effect on the digestibility of the material and has varied greatly from one study to the next; and different societies (e.g. the US and Japan) react quite differently to the introduction of a new food in the diet. Much progress has been made with the testing of some algal meals as sources of protein (in accordance with guidelines set by the Protein–Calorie Advisory Group of the United Nations System, PAG (1972)). However, a conclusion reached by Becker (1986) summarizes the extent of the problem: 'The use of algae in human feeding must be preceded by substantial and long-term studies with animals and humans. Such testing is likely to be difficult and the unanimous acceptance of the results both by consumers and the authorities, may take a long time.' Some of the difficulty can be overcome by using algal meal as a feed for animals rather than humans though toxicological considerations still remain, especially toxic substances, such as heavy metals, that are passed along in the food chain (Payer et al., 1976; Soeder, 1980). Additional problems to be considered with animal feed, from a production standpoint, are the much lower value of the product (minimal cost is an overriding consideration in the composition of many animal feeds) and the tremendous quantities that must be available to these industries.

Though all of the reasons cited above have made the original goal of the earliest large-scale microalgal culture efforts – low cost, high protein food production – continually elusive, a variety of ingenious alternatives have led to variously successful operations. A few of these will be described below.

One of the direct outgrowths of the early *Chlorella* work is production of this genus in Japan and Taiwan. H. Tamiya was associated with the early work of the Carnegie Institution (Tamiya *et al.*, 1953) and returned to Japan to guide work at the Tokugawa Institute in Tokyo with *Chlorella* culture. This work was so successful that the institute was unfortunately dissolved in 1965 since the authorities felt that the degree of industrialization that had been reached was sufficient for research tasks to be taken over by profitable *Chlorella* companies (Soeder, 1986). The magnitude of the success can be seen in production numbers given by Kawaguchi (1980). As of 1977 it was stated that 48 large-scale *Chlorella* factories were operating in Asia, the majority in Taiwan, with a combined production in excess of 1000 kg of dried algae per month. With the limitations for production identified in the early work outlined above, how has this been possible? The first perceived obstacles of yield and cost of production have been overcome with meticulous culture strategies that consistently move to larger and larger scale to minimize the problem of contamination. Kawaguchi (1980) outlines a process from laboratory flask culture moving to greenhouse covered ponds of 4 m^2 then 20 m^2 then to outdoor ponds of 200 m^2 and finally 500 m^2. In addition, many of the *Chlorella* operations utilize not only the autotrophic growth of the algae using light as the energy source, but they also use additional carbon sources (e.g. acetate or glucose) to support mixotrophic or heterotrophic growth of the cells. This approach supports increased growth rates and, importantly, the opportunity to more completely enclose the cultures (since decreases in light have less effect) to prevent contamination. Production costs are variously reported around $11 kg^{-1} in 1977 dollars (Kawaguchi, 1980). Obviously, there must be a substantial market to support such costs (in comparison, current prices for soybeans are approximately $0.20 kg^{-1}). The answer has been found in the health food market where most of the algal meal is sold in tablet form, some as an extract referred to as 'Chlorella Growth Factor'. Recent prices in the US are around $200–250 kg^{-1} (from local retail outlets), a level that is quite sufficient to support production costs. However, the health food market is limited and can change quite rapidly. The medicinal effect of such preparations is still a point of contention with many claims of beneficial effects (Hills and Nakamura, 1978; Tyml, 1982) as well as many pointing out constraints (Becker, 1986). Economically, however, the final test is the perception of the public, which is still generally favourable (to our knowledge) despite an outbreak of skin irritation caused by a batch of *Chlorella* contaminated with a chlorophyll breakdown product, phaeophorbide-a. Regulations on the amount of this compound have since been put in place by the Japanese Ministry of Health at 10 mg per 100 mg of product (Becker, 1986). Understandably, much of the information concerning *Chlorella* processes is proprietary, but what is available can be a model for further development of microalgal production with other species that produce a relatively high value product.

Another successful application of the early work was made by a group headed by W. J. Oswald at the University of California, Berkeley in the area of sanitary engineering (Oswald *et al.*, 1957). Oswald realized that algal culture could be used to solve two of the major problems of biological waste treatment (based almost entirely on bacterial metabolism), specifically the high cost of supplying oxygen to maintain aerobic conditions and the removal of nutrients remaining after organic wastes are broken down. With the necessary engineering perspective, what have been termed 'high rate algal ponds', which are paddle-wheel mixed and elongate oval in design, have been constructed in many locations from California to Thailand to Israel (Goldman, 1979; Shelef *et al.*, 1980). The water treatment efficiency of these systems has, in general, been quite comparable to conventional secondary treatment with final BODs (biological oxygen demand) of less than 26 mg/l. However, they have presented at least two formidable problems (Oswald and Benemann, 1977). Firstly, it is quite difficult to control the species composition of the culture with a variety of species coming to dominate at different times of the year. Secondly, harvesting of the algal cells from the final effluent has only been possible at a substantial cost. The control of species composition, dominated frequently by *Scenedesmus*, has never been satisfactorily achieved. The need for a species specific media composition does not meld well with the constant variation in chemical composition of waste waters. The area of harvesting has seen some improvement, with processes such as chemical flocculation, dissolved air flotation, etc. making the harvesting step technically possible but still expensive. A successful harvest leads to the final problem: of what use is the final product which is obviously not only algal material, but includes bacterial material, zooplankton and detritus [the term albazod has been coined (Soeder, 1984) to more precisely describe this material]? This has been approached in two major ways. One is conversion of the biomass to methane through the process of anaerobic digestion. However, best estimates (Oswald and Benemann, 1977) indicate that the amounts of methane that could be produced are not substantially greater than the energy needed to operate the system. Another approach has been utilization of the material as a feed for animals. This has been somewhat successful in an integrated system where animal waste is what is treated while the resultant biomass is used as a feed for the same animals (Lincoln and Hill, 1980) though the price competition with existing feed components (e.g. soybeans) is still difficult to beat. With municipal waste as the starting material this approach is complicated by the problems of toxicity arising from various components in the waste and has not had large-scale success. In summary, the use of microalgal culture as a means to treat waste has been quite successful and is expanding slowly, held in check mostly by the social and political problems associated with replacing the entrenched conventional secondary treatment systems. The use of the resulting biomass is limited by the uncontrollable composition and toxicological considerations. However, continued research could lead to much more substantial returns from what, in some ways, can be considered a microalgal resource.

An interesting sidelight to the idea of using microalgae as a means to produce oxygen for waste treatment is their use in enclosed environments to sustain human life. This has been investigated in association with the space programmes in both the US and USSR. Work in Oswald's group showed the theoretical feasibility of such a system, which was expanded by Lachance (1968) to show that approximately 100 l of continuous culture were needed to sustain one man. Parallel work in the USSR resulted in a 30-day experiment to simulate a space capsule environment (Kirensky *et al.*, 1967) between man and algae. Work in this area has slowed in the US, but is probably continuing in the USSR, and may become a major component of large installations, such as space stations, where chemical oxygen systems are more difficult and expensive.

Most of the microalgal culture history described above has been related to a very limited number of species, for the most part the green algae *Chlorella* and *Scenedesmus*. However, to move ahead the culturing technology needs to expand and accommodate other species and tap into the vast genetic diversity available. Back in the Carnegie work, Burlew (1953b) acknowledged this and speculated that 'Once a large-scale process is operating smoothly with *Chlorella*, it will be relatively easy to modify it to grow some other species which small-scale experiments in the meantime will have shown to be preferable.' Obviously, this has not been the case, and if we continue with our analogy to terrestrial agriculture, there is no reason to expect different algal species to require similar culture conditions any more than a farmer would treat apple trees in the same way as he did a lettuce crop. The reactions of different species to environmental conditions, their nutrient needs, their rate of growth have all proven to be quite variable. To some extent the explosion of research effort by phycologists since the 1950s, that has helped to elucidate much of this basic biological information for a variety of species, was driven by questions from the engineers involved in the design of culture systems. Their need to break down the complicated process of growth to its basic elements has led to a greater understanding of light requirements and effects, nutrient uptake kinetics, reproductive processes, genetic expression and many other characteristics. This broad spectrum of ideas cannot be reviewed here and the reader is referred to volumes such as Rosowski and Parker (1971, 1982) for introductions to pertinent literature. We will concentrate on the application of these results to the large-scale culture of other microalgal species.

Another prominent culturing and marketing success has been achieved with the cyanobacterium *Spirulina*. The recorded history of use of this species as food extends all the way back to 1524 in Mexico when Fray Toribio do Benavente reached the area and reported the consumption of tecuitlatl, a preparation of *Spirulina*, by the indians living there (Richmond, 1986). A similar history is available for consumption of this alga by native populations around Lake Chad in central West Africa. The push for inexpensive and novel sources of protein, that hastened the development of *Chlorella* culture, also encouraged work with *Spirulina* once its applicability was understood in the 1960s (Durand-Chastel, 1980). Some key characteristics of *Spirulina* offer the potential for substantial improvement over *Chlorella* for food production. First, the protein content of the cells can constitute approximately 70% of the dry weight, with many other desirable components in small amounts (complete analysis in Richmond, 1986). Second, being a cyanobacterium the cell wall components are more easily digested than the walls of most eukaryotic algae. Third, the alga is filamentous in nature which makes its harvesting somewhat easier than unicellular forms (however, the need to remove other contaminants, often unicellular, can eliminate this benefit). Perhaps most importantly, the environmental conditions preferred by *Spirulina*, high pH, high salinity and high temperature, are not conducive to the growth of too many other organisms and therefore ease the maintenance of a monoculture in open ponds. This is most evident at the largest commercial *Spirulina* production facility at Lake Texcoco near Mexico City and operated by Sosa Texcoco Ltd. The lake supports an extremely pure culture of the alga along its periphery, as well as portions of a huge (900 ha, 3 km diameter) solar evaporator that extracts soda from the lake (Durand-Chastel, 1980). Other installations in Taiwan, the US, Thailand, Japan and Israel use conventional paddle-wheel pond technology to produce this organism, for a total production of some 850 tons annually (Richmond, 1986). The current market for this alga is mostly in health foods, fish feeds for coloration and special purposes in Japanese cuisine. The current price is about $10 kg^{-1} though it peaked at around $150 kg^{-1} in 1983 (Soeder, 1986). The dramatic price fluctuations were based again on popular perceptions of the therapeutic value of the algal meal. The various reports supporting these contentions were not always well controlled and at times somewhat exaggerated, which unfortunately clouded some genuine benefits of this food source (Richmond, 1986). The health food market for this product is fairly settled now and will probably remain at the current size and price levels for at least a while. However, what of the promise of its widespread use as an inexpensive protein source. The cost of production (predominantly pond construction costs), variously estimated at $2–10 kg^{-1}, is still not competitive with traditional sources such as soybeans ($0.20 kg^{-1}). In less developed parts of the world where the protein need is great and arable soils are scarce, low technology production may change this situation and make *Spirulina* culturing for basic food needs an attractive alternative.

Though the volumes of production are not as vast as many of the other efforts described here, mention should be made of microalgal culture aimed at the production of feed for animal culture, typically shellfish culture. Its importance was stated by Epifano (in Persoone and Claus, 1980), 'the major bottleneck to commercialization of intensive bivalve culture systems lies in the inability to economically culture massive quantities of these suitable algal species'. A wide variety of small-scale culture systems from indoor enclosed tanks and clear plastic bag culture to outdoor ponds have been operated to grow a variety of marine microalgal species (see Persoone and Claus, 1980 for detailed descriptions). The main challenge to these systems is the maintenance of a unialgal culture, specifically an alga shown to support optimal growth of the product animals.

One of the experimental investigations aimed at producing shellfish is of special interest because it combined a number of other goals of algal culture described above into an integrated system. Ryther *et al.* (1976) used the effluent from a domestic secondary sewage treatment plant as a nutrient source to mix with seawater and enhance the growth of the natural assemblage of unicellular marine microalgae (most often diatoms). The algae, in turn, were used as a feed to oysters, clams and other bivalve molluscs that removed at least 80% of the algae. A final culture of seaweeds (e.g. *Chondrus* and *Gracilaria*) completed removal of nitrogen from the water and therefore completed a tertiary treatment of the sewage water. Three 'products' are then obtainable from such a system: edible molluscs, seaweed biomass and highly treated wastewater. Though much more research is needed to perfect such a system and guarantee public health considerations (the safety of shellfish that have been in contact with treated wastewater would require many toxicological studies and the onus would be on the purveyor to prove that there are no viral contaminants) there is much promise in such a mimicking of a natural ecosystem. If the balance of a natural system can be approximated in a man-made culture, with production focused on valuable products, many of the recurring problems of monoculture could perhaps be minimized or eliminated.

The early examples of microalgal culture described above have mostly been aimed at the production of food (with the exception of water treatment). As outlined above, these efforts continually run up against the more refined techniques of terrestrial agriculture in the battle of the marketplace, where the ultimate decision on the success of large-scale operations is made. Can the gap in the cost of production be narrowed to allow profitable enterprises? There is no definite answer, but it is certain that answers to some basic questions concerning culture operations are needed. In a volume resulting from an international meeting on algal biomass in 1978 (Shelef and Soeder, 1980), that summarizes much of the work up to that time, Becker and Venkataraman (1980) list what, at that time, were considered some of the major limitations to algal cultivation. A more detailed discussion based upon the Indian experience with *Spirulina* and *Scenedesmus* has been provided (Venkataraman and Becker, 1985). Of particular historical interest are the following four items: infections with other microorganisms, low growth rate, expensive harvesting techniques, and high temperatures required for drying and processing. The reason these are of interest is that they are some of the same problems pointed out in the Carnegie Institution work some twenty-five years earlier (Burlew, 1953a). Perhaps history has a lesson to teach us. Oswald (1980) points out that the greatest barrier to progress towards solving these and other basic problems in algal culture is the lack of consistent support, governmental or otherwise, for algal research and development. Unfortunately, for the most part this is still the case.

Beginning in the 1970s a different approach to the economic difficulties of large-scale microalgal production began to gain widespread attention. Rather than attempting to produce relatively low value commodities, the focus was moved to the utilization of the unique metabolic abilities of the algae to produce high value chemicals and products. Some of these efforts are the subject of later chapters in this volume (see Ben-Amotz and Avron, Calvin and Taylor, Glombitza and Koch, Ciferri *et al*).

All of the above historical examples of algal mass culture have used autotrophic metabolism, photosynthesis, to take advantage of the 'free' energy to be gained from sunlight. However, this advantage is gained at a cost, specifically the loss of control of environmental parameters and the introduction of contaminating organisms. When high value products are to be produced, a heterotrophic or mixotrophic growth mode in enclosed bioreactors can be a more cost effective process (Lee, 1986). Many microalgae are capable of using organic carbon sources as an energy source and therefore can conceivably be grown in a fermenter culture – a highly advanced biotechnology used for a wide variety of microbes. In a fermenter all of the environmental conditions can be maintained very near the optimum of the culture species, and all competing organisms can be excluded. The wide scope of heterotropically capable species of algae has been presented by Droop (1974), and this is far from a complete listing since many species have never been tested for this characteristic. One commercial venture has been reported for this type of growth. Workers with the Grain Processing Corp. (Hanson, 1967) used the green alga *Spongiococcum* as a poultry feed component, due to its content of lutein, a carotenoid pigment known to colour the flesh and egg yolks of chickens (Marusich and Bauernfiend, 1981). Large-scale fermentations were carried out for approximately a year, then ended, presumably for economic reasons. There are numerous other investigations into fermentative production of microalgae that have been undertaken recently, but understandably much of this information is proprietary. Hopefully, breakthroughs in this area will soon be forthcoming and open up tremendous new opportunities for algal biotechnology.

This brief overview of the history of microalgal culture has left out a great many workers who have made significant contributions to the field (e.g. work in Germany and Czechoslovakia with *Scenedesmus*). For further information the reader is referred to other recent historical reviews (Goldman, 1979; Soeder, 1986).

Where has this historical review of microalgal culture brought us? Basically it seems it has come to the marketplace where the over-riding consideration must be economics. The successful operations described above have helped to point out the important parameters to be considered and as stated by Benemann *et al.* (1987), 'only well designed and researched processes are likely to meet the test of the marketplace'. The structural simplicity, yet incredible metabolic complexity, of the microalgae make them very promising candidates for utilization in the expanding world of biotechnology.

Conclusion

So how does the advance of algal biotechnology or culture match up to terrestrial agriculture? Overall, the answer must be that phycology is way behind. In culturing, the pioneering efforts of the Japanese in macroalgae (open ocean net culture) and the Carnegie Institution and Oswald in microalgae (pond systems) have given us the starting tools, perhaps analogous to early implements such as the plough for agriculture. In the area of products from algae, there is some food production, centred in the Orient, and there are only a few chemicals that have reached larger-scale success. The great promise lies in the vast number of species that have yet to be examined for the ability to produce specific products. What is needed now is much more biological and engineering knowledge that can lead to the domestication of wild species. This will in turn allow problems to be identified that can best be solved through application of the techniques of genetic manipulation that fall into the purest definition of 'biotechnology'. Then the wild species of promise need to be improved in a number of characteristics, as has been done with breeding of terrestrial crop plants. This area has seen little work so far and could be a major lesson for algal biotechnologists who have an advantage over terrestrial plant technologists in the simpler structure of algae that allow the use of more rapid micro-biological techniques for experimental manipulation. The information needed must come both as precise detailed laboratory investigations as well as from empirical data that can only be obtained from testing of larger-scale systems. A major question is whether support will be available to provide the pilot scale data that are absolutely essential to investment decisions for production scale operations, which can then lead to further advancements and breakthroughs in technology. The majority of this volume outlines the efforts to date in a wide-range of investigations in both products from and culturing techniques for algae. The final chapters (Borowitzka and Borowitzka, and Benemann) discuss the industrial future of algal biotechnology. Algal biotechnology has an advantage over early agricultural efforts in the fantastic new biological techniques at its disposal. A consideration of its history ten years from now will surely tell a different story.

References

Arad S 1986 Biochemicals from unicellular red algae. *Int. Indus. Biotechnol.* **7**: 281–3

Arad S, Adda M, Cohen E 1985 The potential of production of sulfated polysac-carides from *Porphyridium*. In Pasternak D, San Pietro A (eds) *Biosalinity in Action: Bioproduction with saline water*, Nijhoff Publishers, pp 117–28

Baker J T 1984 Seaweeds in pharmaceutical studies and applications. *Hydrobiologia* **116/117**: 29–40

Becker E W 1986 Nutritional properties of microalgae: potentials and constraints. In Richmond A (ed) *CRC Handbook of Microalgal Mass Culture*, Press, Inc., pp 339–420

Becker E W, Venkataraman L V 1980 Production and processing of algae in pilot plant scale experiences of the Indo–German project. In Shelef G, Soeder C J (eds) *Algae Biomass Production and Use*, Elsevier/North-Holland Biomedical Press, pp 35–50

Behr W, Soeder C J 1981 Commercial aspects of utilizing microalgae with special reference to animal feeds. UOFS Publ. Ser. C, No. 3, University of the Orange Free State, Bloemfontein, Repub. of South Africa, p 63

Ben-Amotz A, Avron M 1980 Glycerol, β-carotene and dry algal meal production by commercial cultivation of *Dunaliella* In Shelef G, Soeder C J (eds) *Algal Biomass Production and Use*, Elsevier/North-Holland Biomedical Press, pp 602–10

Ben-Amotz A, Avron M 1986 On the factors which determine massive β-carotene accumulation in the halotolerant alga *Dunaliella bardawil*. *Pl. Physiol.* **72**: 693

Benemann J R, Tillett D M, Weissman J C 1987 Microalgae biotechnology. *Trends Biotechnol.* **5**: 47–58

Bird C J, Ragan M A 1984 (eds) Proceedings Eleventh International Seaweed Symposium. *Hydrobiologia* **116/117**: 1–605

Blunden G, Gordon S M 1986 Medicinal and pharmaceutical uses of algae. *Pharmacy Int.* Nov: 287–90

Bold H C, Wynne M J 1986 *Introduction to the Algae*, Prentice-Hall

Borowitzka L J, Borowitzka M A, Moulton T P 1984 The mass culture of *Dunaliella salina* for fine chemicals: from laboratory to pilot plant. *Hydrobiologia* **116/117**: 115–21

Budiansky S 1980 The great kelp controversy. *Env. Sci. Technol.* **10**: 1170–1

Burlew J S 1953a *Algal Culture from Laboratory to Pilot Plant* Carnegie Institution of Washington Pub. No 600

Burlew J S 1953b Current status of the large-scale culture of algae. In *Algal Culture from Laboratory to Pilot Plant*, Carnegie Institution of Washington Pub. No 600, pp 3–23

Chapman V J, Chapman D J 1980 *Seaweeds and their Uses*, 2nd edn, Chapman and Hall

Cheney D 1984 Genetic modification in seaweeds: Application to commercial utilization and cultivation. In Colwell R R, Pariser E R, Sinskey A J (eds), *Biotechnology in the Marine Sciences*, John Wiley, pp 161–75

Craig H, Reichelt B Y 1986 Genetic engineering in algal biotechnology. *Trends Biotechnol.* **4**: 280–5

Curtin M E 1985 Chemicals from the sea. *Biotechnology* **3**: 34–7

Dayton P K, Tegner M J 1984 Catastrophic storms, El Niño, and patch stability in a southern California kelp community. *Science* **224**: 283–6

Deretic V, Gill J F, Chakrabarty A M 1987 Alginate biosynthesis: a model system for gene regulation and function in *Pseudomonas*. *Biotechnology* **5**: 469–77

Doty M S 1973 Farming the red seaweed *Eucheuma*, for carrageenans. *Micronesia* **9**: 59–73

Drew K M 1949 Conchocelis-phase in the life-history of *Porphyra umbilicalis*. *Nature, (Lond.)* **164**: 748–9

Droop M R 1974 Heterotrophy of carbon. In Stewart W D P (ed.) *Algal Physiology and Biochemistry*, Univ. of California Press, pp 530–59

Durand-Chastel H 1980 Production and use of *Spirulina* in Mexico. In Shelef G, Soeder C J (eds) *Algae Biomass Production and Use*, Elsevier/North-Holland Biomedical Press, pp 35–50

Edelstein T, Bird C J, McLachlan J 1976 Studies on *Gracilaria* 2. Growth under greenhouse conditions. *Can. J. Bot.* **54**: 2275–90

Fenical W, Paul V J 1984 Antimicrobial and cytotoxic terpenoids from tropical green algae of the family Udoteaceae. *Hydrobiologia* **116/117**: 135–40

Gellenbeck K W, Chapman D J 1983 Seaweed uses: the outlook for mariculture. *Endeavour* **7**: 31–7

Gellenbeck K W, Chapman D J 1986 Feasibility of mariculture of the brown seaweed, *Sargassum* (Phaeophyta): growth and culture conditions, alginic acid content and conversion to methane. *Beih. Nova Hedwigia* **83**: 107–15

Glazer A N, Stryer L 1984 Phycofluor probes. *Trends Biochem. Sci.* **9**: 423–7

Glicksman M 1983 Red seaweed extracts (agar, carrageenans, furcellaran). In Glicksman M (ed) *Food Hydrocolloids*, Vol 1, CRC Press, pp 73–113

Glicksman M 1987 Utilization of seaweed hydrocolloids in the food indusrty. *Hydrobiologia* **151/152**: 31–47

Goldman J C 1979 Outdoor algal mass cultures – I. applications. *Wat. Res.* **13**: 1–19

Hansen J E 1984 Strain selection and physiology in the development of *Gracilaria* mariculture. *Hydrobiologia* **116/117**: 89–94

Hansen J E, Packard J E, Doyle, W T 1981 *Mariculture of Red Seaweeds*, Report NoT-CSGCP-002, California Sea Grant College Program Publication

Hanson A M 1967 Microbial production of pigments and vitamins. In Peppler H J (ed) *Microbial Technology*, pp 222–50

Hills C, Nakamura H 1978 *Food from Sunlight*, Univ. of the Trees Press, California

Huguenin J E 1976 An examination of problems and potentials for future large-scale intensive seaweed culture systems. *Aquaculture* **9**: 313–42

Kawaguchi K 1980 Microalgae production systems in Asia. In Shelef G, Soeder C J (eds) *Algae Biomass Production and Use*, Elsevier/North-Holland Biomedical Press, pp 25–33

King A H 1983 Brown seaweed extracts (alginates). In Glicksman M (ed) *Food Hydrocolloids*, Vol 2, CRC Press, pp 115–88

Kirensky L V, Terskov I A, Gitel'zon I I, Lisovsky G M,Kovrov B G, Sid'ko F Y, Okladnikov Y N, Antonyuk M P,Belyanin V N, Rerberg M S 1967 Gas exchange between man and a culture of microalgae in a 30 day experiment. *Kosm. Biol. Med.* **1**: 23

Kurogi M 1975 Japanese seaweed catches, prepared from: Annual report of catch statistics of fishery and culture in 1975. Statistics and Survey Division Ministry of Agriculture and Forestry, Japan.

Lachance P A 1968 Single-cell protein in space systems. In Mateles R T, Tannenbaum S R (eds) *Single-Cell Protein*, MIT Press, Cambridge, Mass., p 122

Laite P, Ricohermoso M 1981 Revolutionary impact of *Eucheuma* production in the South China Sea on the carrageenan industry. In Levering T (ed) *Proc. 10th Int. Seaweed Symp.*, Walter de Gruyter, pp 595–600

LaPointe B E, Ryther J H 1978 Some aspects of the growth and yield of *Gracilaria tikvahiae* in culture. *Aquaculture* **15**: 185–93

Lee R E 1980 *Phycology*, Cambridge Univ. Press

Lee Y 1986 Enclosed bioreactors for the mass cultivation of photosynthetic microorganisms: the future trend. *Trends Biotechnol.* **4**: 186–9

Leone J E 1980 Marine biomass energy project. *Marine Technol. Soc. J.* **14**: 12–31

Lincoln E P, Hill D T 1980 An integrated microalgae system. In Shelef G, Soeder C J (eds) *Algae Biomass Production and Use*, Elsevier/North-Holland Biomedical Press, pp 229–44

Lobban C S, Wynne M J 1981 *Biology of the Seaweeds*, Bot. Monogr. 17, Blackwell Scientific

Marusich W L, Bauernfiend J C 1981 Oxycarotenoids in poultry feed. In Bauernfiend J C (ed) *Carotenoids as Colorants and Vitamin Precursors*, Academic Press, pp 320–462

Miura A 1975 *Porphyra* cultivation in Japan. In Tokida J and Hirose H (eds) *Advance of Phycology in Japan*, Dr. W. Junk, pp 273–304

Mshigeni K E, Chapman D J 1990 *Sargassum*: its biology, ecology and economic potential, with special reference to Indo European species. (In Press)

Neish I C 1976 Role of mariculture in the Canadian seaweed industry. *J. Fish. Res. Bd Canada.* **33**: 1007–14

Neushul M 1977 The domestication of the giant kelp, *Macrocystis*, as a marine plant biomass producer. In Krauss R W (ed) *The Marine Plant Biomass of the Pacific Northwest Coast*, Oregon State Univ. Press, pp 163–87

Nisizawa K, Noda H, Kikuchi R, Watanabe T 1987 The main seaweed foods in Japan. *Hydrobiologia* **151/152**: 5–29

North W J 1980 Biomass from marine macroscopic plants. *Solar Energy.* **25**: 387–95

Oswald W J 1980 Algal production – problems, achievements and potential. In Shelef G, Soeder C J (eds) *Algae Biomass Production and Use*, Elsevier/North-Holland Biomedical Press, pp 1–8

Oswald W J, Benemann J R 1977 A critical analysis of bioconversion with microalgae. In Mitsui A, Miyachi S, San Pietro A, Tamura S (eds) *Biological Solar Energy Conversion*, Academic Press, New York, pp 379–96

Oswald W J, Gotaas H B, Golueke C G 1957 Algae in waste treatment. *Sewage Ind. Wastes* **25**: 692–705

PAG – Protein Advisory Group of FAO/WHO/UNICEF 1972 Guidelines on the production of single cell protein for human consumption. *PAG Guideline No. 12,. PAG-Bull,* **2**: 21–3

Payer H D, Runker K H, Schramel P, Stengel E, Bhumiratana A, Soeder C J 1976 Environmental influences on the accumulation of lead, cadmium, mercury, antimony, arsenic, selenium, bromine and tin in unicellular algae cultivated in Thailand and in Germany. *Chemosphere* **6**: 413–6

Persoone G, Claus C 1980 Mass culture of algae: a bottleneck in the nursery culturing of molluscs. In Shelef G, Soeder C J (eds) *Algae Biomass Production and Use*, Elsevier/North-Holland Biomedical Press, pp 9–20

Polne M, Neushul M, Gibor A 1980 Growing *Eucheuma uncinatum* in culture: domestication of a marine crop plant. In Abbott A, Foster M S, Eklund, L F (eds) *Pacific Seaweed Aquaculture*, California Sea Grant College Program, Univ. of California, pp 115–22

Porterfield W M 1922 References to the algae in the Chinese classics. *Bull. Torrey Bot. Club* **49**: 297–300

Ragan M A, Bird C J 1987 Proceedings Twelfth International Seaweed Symposium. *Hydrobiologia* **151/152**: 1–590

Reichelt J L, Borowitzka M A 1984 Antimicrobial activity from marine algae: results of a large scale screening program. *Hydrobiologia* **116/117**: 158–68

Richmond A 1986 Microalgae of economic potential. In Richmond A (ed.) *CRC Handbook of Microalgal Mass Culture*, CRC Press, Florida, pp 199–243

Rinehart K L, Kishore V, Nagarajan S, Lake R J, Gloer J B, Bozich F A, Li K-M, Maleczka Jr R E E., Todsen W L, Munro M H G, Sullins D W, Sakai R 1987 Total synthesis of didemnins A, B and C. *J. Amer. Chem. Soc.* **109**: 6846–8.

Rosowski J H, Parker B C 1971 *Selected Papers in Phycology*, The Dept. of Botany, Univ. of Nebraska

Rosowski J H, Parker B C 1982 *Selected Papers in Phycology II*, Phycological Society of America, Inc., Book Division

Ryther J H, DeBoer J A, LaPointe B E 1979 Cultivation of seaweeds for hydrocolloids, waste treatment and biomass for energy conversion. *Proc. Int. Seaweed Symp.* **9**: 1–17

Ryther J H, Goldman J C, Gifford C E, Hueguenin J E, Wing A S, Clarner J P, Williams L D, LaPointe B E 1976 Physical models of integrated waste recycling–marine polyculture systems. *Aquaculture* **5**: 168–77

Saito Y 1975 *Undaria*. In Tokida J and Hirose H (eds) *Advance of Phycology in Japan*, Dr. W. Junk, pp 304–20

Scheuer P 1978–84 (ed) *Marine Natural Products*, Vols 1–5, Academic Press, New York

Schopf J W, Packer B 1987 Early Archean (3.3-billion to 3.5-billion-year-old) microfossils from Warrawoona group, Australia. *Science* **237**: 70–3.

Shang V C 1976 Economic aspects of *Gracilaria* cultivation in Taiwan. *Aquaculture* **8**: 1–7

Shelef G, Azov Y, Moraine R, Oron G 1980 Algal mass production as an integral part of a wastewater treatment and reclamation system. In Shelef G, Soeder C J (eds) *Algae Biomass Production and Use*, Elsevier/North-Holland Biomedical Press, pp 183–9

Shelef G, Soeder C J 1980 *Algae Biomass Production and Use*, Elsevier/North-Holland Biomedical Press

Sleeper H 1980 Alginates from *Macrocystis*, In Abbott A, Foster M S, Eklund L F (eds) *Pacific Seaweed-Aquaculture*, California Sea Grant College Program, Univ. of California, pp 130–5

Soeder C J 1980 The scope of microalgae for food and feed. In Shelef G, Soeder C J (eds) *Algae Biomass Production and Use*, Elsevier/North-Holland Biomedical Press, pp 9–20

Soeder C J 1984 Aquatic bioconversion of excrements in ponds. In Ketelaars E H, Iwema B, (eds) *Animals as Waste Converters*, Pudoc Wageningen, Wageningen, The Netherlands, p 130

Soeder C J 1986 An historical outline of applied algology. In Richmond A (ed) *CRC Handbook of Microalgal Mass Culture*, CRC Press, Inc., pp 25–44

Tamiya H, Hase E, Shibata K, Mituya A, Iwamura T, Nihei T, Sasa T 1953 Kinetics of growth of *Chlorella* with special reference to its dependence on quantity of available light and on temperature. In Burlew J S (ed) *Algal Culture from Laboratory to Pilot Plant*, Carnegie Institution of Washington Pub. No 600, pp 204–32

Tamiya H 1957 Mass culture of algae. *A. Rev. Pl. Physiol.* **8**: 309–44

Trainor F R 1978 *Introductory Phycology*, John Wiley & Sons

Tseng C K, Chang T J 1955 Studies on the life history of *Porphyra tenera* Kjellm. *Scien. Sin.* **4**: 375–98

Tyml H 1982 Present state and possibilities of the medical use of chlorococcal algae. *Acta Univ. Palacki Olomuc Fac. Med.* **103**: 273–80

van der Meer J P 1987 Using genetic markers in phycological research. *Hydrobiologia* **151/152**: 49–56

Venkataraman L V, Becker E W 1985 *Biotechnology and utilization of algae. The Indian experience*, Dept. Sci. and Technol. New Delhi, India

Vonshak A 1987 Strain selection of *Spirulina* suitable for mass production. *Hydrobiologia* **151/152**: 75–8

Waaland J H 1977 Growth of Pacific Northwest marine algae in semi-closed culture. In Krauss R W (ed) *The Marine Plant Biomass of the Pacific Northwest Coast*, Oregon State Univ. Press, pp 163–87

Wassink E C, Kok B, van Oorshot J L P 1953 The efficiency of light-energy conversion in *Chlorella* cultures as compared to higher plants. In Burlew J S (ed) *Algal Culture Laboratory to Pilot Plant*, Carnegie Institution of Washington Pub. No 600, pp 55–62

Wu C Y, Lin G 1987 Progress in the genetics and breeding of economic seaweeds in China. *Hydrobiologia* **151/152**: 57–62

2

Algae as experimental organisms

G. E. Fogg

Introduction

The term alga, in its widest sense, includes an extraordinary range of different organisms. It runs the gamuts between complete autotrophy and heterotrophy with complex growth factor requirements, picoplankter (0.2–2 μm) and giant kelp (up to 50 m), psychrophile and thermophile, prokaryote and eukaryote. Among this assemblage are forms to suit almost every experimental purpose and physiologists and biochemists have taken advantage of this. In surveying this enormous field only a general sketch is possible, and in trying to bring order to the wealth of information available distinction must be made between algae being used as experimental material because they are particularly suitable for the study of general problems and algae being studied for their own sakes, in which case we have to think of difficulties rather than suitabilities. Firstly, however, we should look at some features which are desirable in organisms used as experimental material.

General characteristics affecting the use of algae in experiments

Availability

Microalgae

For studies of metabolism it is nearly always essential to have the organism in an axenic condition and naturally occurring populations are invariably contaminated with other forms. Following the pioneer work of Pringsheim (1946), methods of isolation and culture have been refined and it is now possible with care and patience to get most physiological types of microalgae into pure culture. Methods have been described by Stein (1980). Culture collections have been set up in several countries and about a thousand different species and strains are available. Many of these are strains or mutants of a few species – *Chlorella, Chlamydomonas* and *Euglena* spp. being particularly abundant – and at present groups such as Dinophyceae and Haptophyceae are poorly represented and soil forms are more abundant than planktonic species. Mostly the algae are maintained by subculture but cryopreservation is being used increasingly, e.g. for *Chlorella* and *Chlamydomonas* (Morris *et al.* 1979), and is of particular advantage with genetically unstable strains. If experiments are conducted with a view to understanding the behaviour of natural populations it should be remembered that cultures in collections are clonal, that is, derived from a single individual which may not have been representative of what is most abundant in nature. Thus Gallagher (1982) found that different clones of the common marine diatom *Skeletonema costatum* not only had different electrophoretic banding patterns but distinctly different physiological properties, clones with higher capacities for photosynthesis and growth being commonest in summer whereas those characterized by low capacities were most abundant in winter.

Macroalgae

A few of the larger algae can be grown satisfactorily in culture and are available from culture collections, *Acetabularia* being one of these, but it is not always easy to get them free from bacteria. Some require rather specialized conditions and even a comparatively simple form such as *Ulva* may not attain its natural form in culture except under as yet ill-defined circumstances in which the presence of specific strains of bacteria seems to be of importance (Buggeln, 1981). Much work with macroalgae has been done with naturally grown material or plants from relatively crude culture. Charophytes can be grown easily in crude culture and clones obtained but these do not yield material any more uniform than that which can be collected in the field (Hope and Walker, 1975). A point to be remembered is that biochemical alternation of generations has been found in both green and red seaweeds (McCandless, 1981) so that superficially similar specimens, if representing different generations, may be biochemically different. Thus the gametophyte of *Chondrus crispus* produces κ-carrageenan whereas the sporophyte produces λ-carrageenan.

Standardization of physiological state

In experimental work it is usually desirable to have a readily available supply of material in a known physiological state. The most commonly used technique for growing microalgae is batch culture, in which growth follows a characteristic pattern with a phase of rapid exponential increase followed by a stationary phase. During this growth the photosynthetic capacity of the cells, their rate of respiration and their metabolic pattern, vary continuously (Fogg and Thake, 1987). It is possible to obtain material in a standard physiological condition by harvesting after a fixed period from cultures prepared by a standard procedure but variation can occur all too easily. For example, photosynthetic capacity peaks sharply for a short time during exponential growth, the time after inoculation at which this peak occurs depending on rate of growth and hence on factors such as light intensity and temperature. Thus samples harvested after the same number of days of growth at slightly differing temperatures may show considerable differences in maximum photosynthetic rate (Morris and Glover, 1974).

Continuous culture avoids this difficulty. By regulated addition of fresh medium a population may be maintained at a fixed cell concentration with constant rate of growth and physiological properties. Control is accomplished either via photometric monitoring of cell concentration with regulated inflow of fresh medium, as in the turbidostat, or by addition at a constant rate of medium in which one nutrient is at a limiting level, as in the chemostat. Eithcr method works well for microalgae (see Fogg and Thake, 1987) and, in fact, the turbidostat was introduced by Myers and Clark (1944) to provide *Chlorella* in a standard state for physiological experiments six years before the chemostat, devised originally for growing bacteria, was described. Once set up, a turbidostat or chemostat can provide a continuous supply of uniform axenic material for a month or more.

For some purposes continuous culture is unsatisfactory in that the uniform properties of the material produced are average properties from a population of cells in all stages of their division cycle. If it is necessary to work with samples in which the cells are all in the same stage of their life cycle then synchronous culture must be used. Again, algal physiologists were first in the field, Tamiya *et al.* (1953) reporting *Chlorella* cultures in which cell division was synchronized, a year before similar work with bacteria was published. Synchronous cultures may be started from an inoculum of cells all of which are in the same stage of the division cycle, obtained by mechanical means such as differential filtration or centrifugation. They may also be achieved by entraining cell division into a light–dark cycle, or by shock treatment such as a temperature change. Material from synchronous cultures has been used to study changes in metabolic pattern during the life cycle, the biochemistry of cell division, endogenous rhythms and similar problems. Continuous and synchronous culturing can be combined as in the so-called cyclostat (see Fogg and Thake, 1987).

Size

Some microalgae, being minute and of simple structure, are ideal for experimental work because not only can they be handled in uniform suspension, almost like a chemical reagent, but the diffusion pathways involved in exchanges between cell and medium are reduced to a minimum. The high surface/volume ratio of small cells combined with the steepening of diffusion gradients around sinks with a small radius of curvature makes for rapid uptake of solutes. Raven (1986) has argued that the minimum cell size in which all the vital functions of a phototrophic organism can be accommodated is one of 0.25 μm diameter. Algae of this size may exist but are not at present available in culture. However, even for a cell 1 μm in diameter, such as the cyanobacterium, *Synechococcus*, supply of nutrients by molecular diffusion is sufficient to maintain rapid growth. Raven (1986) has calculated that the potential flux of phosphate ion by molecular diffusion from a low bulk phase concentration of 10^{-4} mol m^{-3} into a 1 μm diameter cell is more than 400 times greater than the actual flux needed to sustain the most rapid observed growth. Mierle (1985) obtained experimental evidence appearing to show that for *Synechococcus* the diffusion resistance of the unstirred layer around the cell was appreciable but Raven (1986) considers that this was actually the resistance to diffusion presented by the cell wall. Even for a cell of 10 μm diameter the potential flux of phosphate is four times that needed for maximum growth. Thus, provided cells do not settle out and nutrients are not exhausted, multiplication of such cells is as rapid in unstirred as in stirred culture. Stirring, of course, may be necessary to maintain the cells in suspension or to accelerate uptake of carbon dioxide from the gas phase. Fogg (1986) gives a diagram showing the relations of sinking rate, surface/volume ratio and potential flux of phosphate, to cell size.

Compartmentalization

Many kinds of experiment are facilitated if there is good spatial separation or demarcation in an organism between parts carrying out different functions. A prime example of this is the coenocytic alga *Acetabularia* which is sufficiently large (50 mm) to make it easy to remove or transplant the single large nucleus. The large internodal cells of charophytes have been used for some 200 years in studies on protoplasmic streaming (Hope and Walker, 1975). These and other algae such as *Valonia* were important for early investigations of ion uptake and permeability (Davson and Danielli, 1943) since they have vacuoles large enough to provide sap for analysis. More recently, because they are particularly amenable to investigation with microelectrodes they have played a notable part in work on transport across the vacuole membrane; indeed, electrophysiology has been said to have begun with *Valonia* (Hope and Walker, 1975; Spanswick, 1981; Raven 1984). In some cyanobacteria the localization of nitrogen fixation in heterocysts has facilitated study of the biochemistry and differentiation of the nitrogenase system (Bothe *et al.*, 1984). These are all examples in which the intact organism is used but, of course, processes may also be studied in cell fractions. Much is known about the plastids (Bisalputra, 1974) and other organelles of algal cells (Evans, 1974) but most of this knowledge is derived from electron microscopy rather than experiment. On the whole, it has proved less easy to obtain cell-free preparations of organelles such as chloroplasts (Lien, 1978) and mitochondria (Grant, 1978) from algae than from higher plants (see p. 44).

Genetic manipulation

A powerful approach to the understanding of metabolic pathways is by comparative study of mutants in which individual steps in a pathway have been impaired. Mutants of algae may be obtained by similar means to those used with other organisms but two circumstances have restricted their use in biochemical studies: one is that many algae are obligate phototrophs and impairment of any part of the photosynthetic system usually means that growth is not possible; the other is that many algae do not have sexual reproduction so that gene recombination is not readily achieved. Few algae have played much part in biochemical genetics. An abundance of mutant strains has been induced in *Anacystis* (*Synechococcus*) sp. among the cyanobacteria and in *Chlamydomonas* spp., *Chlorella* spp., *Scenedesmus* spp. and *Euglena* spp. among the eukaryotic algae. Of these, only *Chlamydomonas* shows sexual reproduction. The value of *Chlamydomonas*, in which the ordinary vegetative cell is haploid and the products of meiosis can easily be separated on germination of the zygote, for genetical studies was realized by Pascher in 1916 but it was only after Lewin (1949) had again drawn attention to its advantages that it began to be widely used. Sexual reproduction is easily induced, there is a high yield of zygotes and the life cycle can be completed within a week. Part of the genetic information necessary for chloroplast function is contained in the chloroplast and shows non-Mendelian, uniparental, transmission. However, chloroplast gene markers occasionally recombine and this unique property of *Chlamydomonas* has permitted the construction of a map of the chloroplast genome in addition to an extensive nuclear genetic map (Goldschmidt-Clermont *et al.*, 1984). Studies on *Chlamydomonas* and the other algae mentioned above have contributed importantly to our knowledge of photosynthesis (see p. 35), chlorophyll synthesis (Castelfranco and Beale, 1983), plastid inheritance, chloroplast development and flagellar mechanisms (see p. 42, Wiessner *et al.*, 1984). There are many other algae, with different life cycles, that might also be useful in genetical studies (Lewin, 1976). Tandeau de Marsac and Houmard (1987) have emphasized the attractiveness of the cyanobacteria for the genetic engineer.

Algae and investigation of general physiological and biochemical problems

Photosynthesis

Algae have played a key role in the experimental investigation of photosynthesis since Engelmann in 1883 took advantage of the wide separation of colourless protoplasm and chloroplast in *Spirogyra* to demonstrate, by means of bacteria which migrated towards higher concentration of oxygen, the crucial fact that this gas is liberated only from the chloroplast. It was assumed from the beginning that the process of photosynthesis is essentially the same in green algae as in higher plants and the demonstration by

Willstätter and Stoll in 1913 that all the green plants they examined contained chlorophylls *a* and *b* confirmed this. Their further finding that brown algae contain chlorophyll *a* made it reasonable to suppose also that photosynthesis is basically the same in whichever organism it takes place.

At this stage the most promising approach to understanding the mechanism of photosynthesis appeared to be by study of its kinetics. This was taken up by Warburg (1919) who used *Chlorella* as experimental material in which the functioning of the green plant could be examined in its simplest terms. *Chlorella*, which had been isolated in pure culture by Beijerinck in 1890, could be grown easily in the laboratory and was well suited for the manometric technique for studying gas exchanges which Warburg had devised. His finding that the photosynthetic yield from a given amount of light is greater if it is given in short intense flashes separated by dark periods than if it is given continuously at lower intensity, convincingly confirmed Blackman's hypothesis, based on work with higher plants, that photosynthesis comprises chemical reactions which can proceed in the dark as well as photochemical reactions. The controversy about how many quanta are required for the reduction of a molecule of carbon dioxide to the carbohydrate level, which was initiated by Warburg and still rumbles on, has centred mainly on measurements made with *Chlorella*.

Further advances were made in the 1930s when the comparative biochemistry of photosynthesis was explored by van Niel. The seminal comparison was between higher plants and sulphur bacteria but an important contribution was made by Gaffron (1940) when he showed that after anaerobic adaptation the green alga *Scenedesmus* became capable of photoreduction of carbon dioxide using molecular hydrogen, a process similar to the photosynthesis of purple sulphur bacteria. A common pattern could be seen in these various forms of photosynthesis if it were assumed that the essential process is the splitting of the water molecule.

The important initial step towards separating photosynthesis from other processes in the cell was accomplished when Hill in 1939 showed that isolated chloroplasts could evolve oxygen in the light in the presence of a suitable hydrogen acceptor, but he got the chloroplasts from chickweed and algae generally have not proved a good source of active chloroplast preparations. However, the pioneer work of Bassham and Calvin (1957) on the path of carbon in photosynthesis, in which the status of phosphoglycerate as the first stable product and the cyclic regeneration of the carbon acceptor ribulose bisphosphate were demonstrated, was accomplished with *Chlorella* and *Scenedesmus*. Except for exposures to the ^{14}C-labelled carbon source of less than 2 s, for which leaves proved more convenient, these microalgae were ideal for tracer techniques. Many subsequent investigations have shown that their photosynthetic carbon metabolism is representative of that of all plants. The concept of the C_4 pathway, however, came from work on higher plants, algae being essentially C_3 plants. The associated problems of carbon dioxide uptake and concentration by algae, which are complex, have been the subject of recent discussion (Lucas and Berry, 1985).

The work of Bassham and Calvin gave striking confirmation to a point of view developed a few years earlier by Myers (1949). Myers had pointed out that the then accepted view that carbohydrate was the product of photosynthesis, in the sense that the carbon fixed went exclusively into this form before entering general metabolism, was based on experiments with non-growing material in which carbohydrate synthesis was the dominating process. He showed that photosynthesis is strongly influenced by nitrogen nutrition and that if experiments were done with actively growing *Chlorella*, in which there was no carbohydrate storage, then the gas exchanges became consistent with protein being the major product of photosynthesis. As soon as details of the carbon reduction cycle became known it was realized that photosynthetically fixed carbon could indeed be incorporated into amino acids and thence into protein without passing through an intermediate stage in carbohydrates. This was a great step forward in understanding since it showed that photosynthesis was not, as previously thought, a biochemically unique and segregated process, but one which intermeshes intimately with other metabolic processes. The integration of photosynthesis with other processes such as respiration, nitrate assimilation and nitrogen fixation and the varying balance of photosynthesis and other processes during the growth cycle are best seen in the algae (Fogg, 1972). They also show the plasticity of the photosynthetic processes and variations such as photoproduction of hydrogen, already mentioned, and photoassimilation of organic substrates are only found in these organisms and the photosynthetic bacteria.

Algae were also the major experimental material in investigations leading to understanding the role of accessory pigments and the recognition of the two photosystems. As long ago as 1884, Engelmann, again using motile bacteria as an indication of oxygen evolution, found that in the green, brown, red and blue–green algae there was correspondence between their absorption spectra and their photosynthetic action spectra. This suggested that light energy absorbed by the various different accessory pigments of the algal groups is used in photosynthesis and led to the formulation of the theory of chromatic adaptation, in which the pigmentation of an alga is supposed to be complementary to the light quality in the habitat in which it grows. With the blue–green alga *Chroococcus* Emerson and Lewis (1942) found that the quantum yield at 600 nm, where absorption by phycocyanin was about six times that by chlorophyll *a*, was about the same as between 660 and 680 nm where nearly all the light absorption was due to the chlorophyll. Comparison by Tanada (1951) of the quantum yield of the diatom *Navicula* as a function of wavelength, with the estimated distribution of absorbed light energy among the different pigments in the living cell showed that light energy absorbed by the carotenoid fucoxanthin is used in photosynthesis as well as that absorbed by the chlorophylls. Such work, combined with fluorescence studies, demonstrated that light energy absorbed by accessory pigments is transferred to chlorophyll *a*, the pigment directly involved in the photochemical act. Taken with the results of kinetic studies on *Chlorella* this led to the idea of a photosynthetic unit in which a single reaction centre is fed with energy derived from an assemblage of pigment molecules.

It was further observed that light of wavelengths longer than 680 nm was ineffective in photosynthesis although it was absorbed by the photosynthetic pigments. This result was obtained with monochromatic light but if it were supplemented with a low level of illumination of shorter wavelength then light of greater than 680 nm wavelength was utilized in photosynthesis. This enhancement effect was discovered by Emerson (Emerson *et al.*, 1957) with *Chlorella*. Another odd finding (Blinks, 1954) was that from the action spectrum of the red seaweed *Porphyra* it appeared that light absorbed by the accessory pigment phycoerythrin is more effective in photosynthesis than that absorbed by chlorophyll *a* itself. Such lines of enquiry, still largely dependent on algae as experimental material (Duysens and Sweers, 1963) led to the concept of two photochemical systems, using comparable amounts of energy and one of them dependent on the other if oxygen is to be evolved. This was embodied in the 'Z scheme' for the photosynthetic electron transport chain which is still generally accepted. In the physical separation of fractions responsible for the two photoreactions higher plant chloroplasts became the preferred experimental material rather than algae although the remarkable photosynthetic antenna system, the phycobilisome, found in blue–green and red algae, has attracted much attention (Scheer, 1982).

Workers on photosynthetic electron transport have now mostly got beyond mentioning actual organisms, but algae have played a notable part in their studies and a recent review entitled 'Photosynthetic electron transport in higher plants' (Haehnel, 1984) in fact refers substantially to results obtained with algae. In particular, mutants of *Chlamydomonas* and *Scenedesmus* have proved invaluable in the identification of the nature and function of the various carriers in the electron transport chain (Somerville, 1986). Such mutants can be grown conveniently in the dark on suitable organic substrates yet still produce chloroplasts – something that would be extremely difficult with similar mutants of higher plants – and are often able to carry out partial reactions if appropriate artificial donors or acceptors are provided to transport electrons in specific parts of the chain. By comparing mutants with regard to the components they lack and the partial reactions they can carry out, the sequence of transfers in the scheme has been worked out (Levine, 1974). In combination with electron microscopy such studies can be extended to relate particular chemical components to fine structure. Exciting recent work on the molecular architecture of the photosynthetic apparatus is, however, largely based on investigations with photosynthetic bacteria and higher plant chloroplasts.

Respiration and photorespiration

Because of problems in obtaining active cell-free extracts and fractions from the common laboratory algae, these organisms have played little part in the elucidation of biochemical pathways in dark respiration. It appears that these pathways are similar to those of higher plants (Lloyd, 1974) but most of our knowledge is of overall exchanges in respiration rather than of biochemical detail. Among such studies one may note those of Syrett (1962) on the interactions of respiration and assimilation of glucose and combined nitrogen.

Photorespiration, defined as light-dependent uptake of oxygen and release of carbon dioxide in photosynthetic tissues, occurs in algae in addition to dark respiration. Although it has now become of interest chiefly as a drain on higher plant productivity, it was first noticed in algae, as a factor in the oxygen inhibition of photosynthesis in *Chlorella* (Warburg, 1920). Studies on algal photorespiration have contributed substantially to our understanding of the mechanism of the process, which appears to be of the same general nature as it is in higher plants. The substrate for photorespiration, glycollic acid, was first detected as an early product of photosynthesis in *Chlorella, Scenedesmus* and barley (Benson and Calvin, 1950). Later, Tolbert and Zill (1956) found that a considerable proportion of the carbon fixed by *Chlorella* may be released from the cells in this form. The major source of glycollate appears to be ribulose bisphosphate, which is split and oxidized, to produce phosphoglycerate and phosphoglycollate, by ribulose bisphosphate carboxylase acting as an oxygenase under conditions of high oxygen partial pressure relative to that of carbon dioxide. This view arose from the work of Bowes and Berry (1972) with *Chlamydomonas*. In many algae the glycollate produced from the phosphoglycollate is released from the cell but in terrestrial plants it is metabolized via the glycollate pathway, yielding phosphoglycerate which is returned to the photosynthetic cycle, and carbon dioxide. There is a difference between unicellular algae and multicellular algae and angiosperms in that in the former it is glycollate dehydrogenase which introduces glycollate into the glycollate pathway whereas in the latter two groups it is glycollate oxidase (Tolbert, 1979). Several of the steps in the glycollate pathway have been established by experiments with algae.

Nitrogen fixation

Ability to use N_2 as a nitrogen source for growth is confined to prokaryotes and among the algae is therefore only found in the cyanobacteria, in which group, however, it is widespread. Good evidence that cyanobacteria possessed this property was available by 1889 but experimental investigation was slow, partly because of the mistaken ideas that these organisms were difficult to free from associated bacteria and grew slowly in culture. There was a great increase in interest around 1970 following the discovery that in some cyanobacteria nitrogen fixation is confined to special cells, the heterocysts, together with the advent of the sensitive acetylene reduction technique for assaying nitrogen fixation and the belated recognition by bacteriologists that blue–green algae are prokaryotic. A cell-free extract able to fix nitrogen had been obtained from the cyanobacterium *Mastigocladus* by Schneider *et al.* as early as 1960 but this did not lead immediately to further advances. On the basis of indirect evidence, Fay *et al.* (1968) argued that the heterocyst was the site of fixation in these organisms. They supposed that it afforded a means of protecting the oxygen-sensitive nitrogenase from photosynthetically produced oxygen, the heterocyst itself having only photosystem I and lacking the oxygenic photosystem II. It proved possible to obtain preparations of heterocysts free from vegetative cells by differential disruption in a French press (Fay and Walsby, 1966), sonication or lysozyme treatment (Fay and Lang, 1971). Such preparations are capable of fixing nitrogen if supplied with ATP as energy source and a hydrogen donor such as dithionite under dark anaerobic conditions (Stewart et al., 1969). The heterocyst is perhaps principally of interest in connection with nitrogen fixation as an elegant solution to the problem of carrying out an oxygen sensitive process under aerobic conditions. Some non-heterocystous cyanobacteria are capable of fixing nitrogen but have other means of achieving the same end. Much work has been done on heterocyst biochemistry and the mechanisms whereby energy, carbon skeletons and hydrogen donors for the fixation process and combined nitrogen for photosynthetic cells are supplied, elucidated (Haselkorn, 1978; Wolk, 1982), but this has contributed little to our understanding of the fixation process itself. However, cyanobacterial nitrogen fixation is of enormous potential importance in agriculture (Stewart et al., 1987).

Nitrogen assimilation

Eukaryotic microalgae have been extensively used as subjects for the study of the assimilation of combined nitrogen by green plants. Molisch in 1896 showed that the mineral requirements of these algae are much the same as those of higher plants. Nearly all of them can use either ammonia or nitrate as sources of nitrogen, although the former is usually used preferentially when both are supplied together (Syrett, 1981).

A major line of enquiry was started when Warburg and Negelein (1920) reported that nitrite was an intermediate in the reduction of nitrate by *Chlorella* and that the process was stimulated by light. Since nitrate is the major source of combined nitrogen for both terrestrial and aquatic plants its uptake accounts for a considerable proportion of the reducing power generated by photosynthesis – a matter of interest to both agronomists and aquatic biologists. The nitrate reductase of *Chlorella* is now perhaps the best characterized algal enzyme (Syrett, 1981). Guerrero et al. (1981) tabulate its properties along with those of similar enzymes from other algae, microorganisms and higher plants. It contains molybdenum and appears to function via the reduction of this by electrons from NAD(P)H. Substitution of molybdenum by tungsten produces an enzyme which is inactive in nitrate reduction but which, like the original molybdenum enzyme, possesses cytochrome *c* reductase activity. The tungsten enzyme has proved to be a useful experimental tool (Solomonson and Spehar, 1977). Nitrate reductase from *Anacystis* and other cyanobacteria cannot accept electrons directly from NAD(P)H but is dependent on reduced ferredoxin as its physiological electron donor (Guerrero et al., 1981). Both types of nitrate reductase operate in the dark if carbohydrate is available to generate electron donors but in green cells they appear closely coupled to photosynthesis and to use water as reductant in the light. Nitrate reductase in the eukaryotic algae is found in the soluble fraction of cell extracts and it is uncertain how it is associated physically with the chloroplast but in prokaryotes it is bound to chlorophyll-containing membranes. Nitrite reductase, which completes the reduction to ammonia, is an enzyme utilizing electrons from reduced ferredoxin. Although its localization in the cell is uncertain the observation that supply of nitrite (but not nitrate) to *Chlorella* or *Ankistrodesmus* causes a quenching of chlorophyll fluorescence, suggests a close coupling of nitrite reduction with the photochemical processes (Guerrero *et al.*, 1981; Syrett, 1981). The exact nature of the coupling between nitrate reduction and photosynthesis remains elusive. The regulation of nitrate assimilation – its induction, inhibition and the synthesis and disappearance of enzymes – has been a field of research in which algae have figured prominently as experimental subjects (Solomonson and Spehar, 1977; Guerrero *et al.*, 1981).

The uptake of ammonia by algae has been investigated by making use of its analogue, methylammonium (Wheeler, 1980). Studies of its assimilation have likewise depended greatly on algae, particularly nitrogen-deficient *Chlorella* (Syrett, 1981). Assimilation takes place at the expense of accumulated carbohydrate and its immediate product is glutamine in which the amide group is derived from the ammonia supplied. Glutamate is formed from ammonia and α-oxoglutarate by reduction, using NAD(P)H and catalysed by glutamic dehydrogenase (GDH). However, an alternative pathway, discovered first in bacteria, has as its first step synthesis of glutamine from glutamate and ammonia, catalysed by glutamine synthetase (GS) and requiring ATP, followed by transfer of the amide group to α-oxoglutarate to produce another molecule of glutamate, a reaction catalysed by glutamate synthase (GOGAT). Most algae contain the necessary enzymes and various lines of evidence suggest that the GS/GOGAT pathway is the predominant one in these organisms; one piece of evidence is the pattern of incorporation of ^{15}N supplied as ammonia, another that L-methionine-DL-sulphoxime (MSO or MSX), a powerful inhibitor of GS, and azaserine, an inhibitor of GOGAT, inhibit ammonia assimilation by various algae although they do not affect GDH. However, there are some algal strains which appear to depend entirely on the GDH pathway (Syrett, 1981).

Urea is an important source of nitrogen in aquatic habitats and is increasingly used as an agricultural fertilizer. Urease, which breaks down urea to ammonia and carbon dioxide, is present in some algae but many do not contain it and assimilate urea nitrogen by means of a urea carboxylase which converts urea into allophanate, which is then broken down to ammonia and carbon dioxide by allophanate hydrolase, the overall process being known as the urea amidolyase reaction. This achieves the same end result as the urease reaction but at the expense of a molecule of ATP for each molecule of urea (Syrett, 1981). The urea amidolyase system, which is active in cell-free extracts, is an inducible/repressible enzyme, which urease is not, but it is not evident what its biological advantage may be.

Interrelations of metabolic processes

The manner in which different metabolic processes are related and the pattern of metabolism regulated is best studied in organisms in which the complications of differentiation and translocation of metabolites are minimal. It was Beijerinck in 1904 who first tackled such problems when he showed that the accumulation of fats in diatoms results from a diversion, consequent on lack of supply of combined nitrogen, of photosynthetic product from protoplasmic synthesis. Changes in metabolism associated with variation in nitrogen supply were later studied in batch cultures of *Chlorella* by Pearsall and Loose (1937). They showed that the phase of exponential growth is dominated by protein synthesis. When the supply of combined nitrogen becomes exhausted cell division stops but assimilation of carbon continues, albeit at a reduced rate, and carbohydrate accumulates in the cells. They pointed out that this sequence is similar to that which takes place during the development of a green leaf. Myers (1949), considering the effect of nitrogen supply on the path of carbon in photosynthesis, had ideas along similar lines, as we have already seen. However, deficiency of nitrogen cannot be the only factor involved because in the same algal strain it some-times causes accumulation of fat and sometimes carbohydrate. Fogg (1959) explained this by reference to the work of Tamiya (Tamiya *et al.*, 1953) who had found, using synchronous cultures of *Chlorella*, that cells differed mark-edly in their composition and metabolic activities according to the stage reached in the division cycle. Thus abrupt deprivation by transfer to nitrogen-deficient medium might have a different effect from that of the gradual exhaustion of nitrogen accompanied by accumulation of staling products occurring in the normal course of population growth in batch culture. The division cycle is halted in different places and whether fat or carbohydrate accumulates is determined by the enzymic activities of the cells at those particular stages. Later investigations in which ^{14}C was used as a tracer to estimate carbon flux through various pools of intermediates at different stages in the division cycle of *Chlorella* (Kanazawa *et al.*, 1970) filled out this picture. Biochemical events in the cell cycle of several other microalgae have now been studied (Puiseux-Dao, 1981). A summary of the pattern of algal metabolism as revealed by synchronous culture work, and the related phenomenon of endogenous rhythms has been given by Fogg and Thake (1987).

The regulation of enzymic activity is a complex matter which has mostly been investigated in organisms other than algae. However, the cyanobac-teria have provided some interesting examples. It was suggested (Carr, 1973) that it is characteristic of these organisms to be unable to control enzyme synthesis by repression and derepression at the transcriptional level. This seems to be so with enzymes of the glycolytic, pentose phosphate, glyoxylate and interrupted tricarboxylic acid cycle pathways in *Anabaena*, in which levels of enzyme activities remained the same whether or not an organic carbon source was supplied. The fact that many cyanobacteria are obligately phototrophic leads one to suppose that a similar situation is general in the group. The release of substantial amounts of extracellular polypeptides was also seen as evidence of ill-controlled amino acid synthesis. However, *Anabaena* has both inductive and repressive control over formation of nitrate reductase (Carr, 1973). The formation of nitrogenase in heterocysts also provides an example of control at the transcriptional level in the cyanobacteria. In the DNA of vegetative cells of *Anabaena* the *nif*D gene, which encodes for a subunit of nitrogenase, is interrupted by a DNA element which prevents its expression. During the differentiation of the heterocyst this element is excised, restoring the *nif*D coding sequence and activating the whole transcription unit for nitrogenase. A gene responsible for the site specific recombinase which brings about the excision has been identified (Haselkorn *et al.*, 1987). Alteration of activity of existing enzymes has also been reported for cyanobacteria, for example end-product control of phosphofructokinase by ATP in *Anabaena* and of *N*-acetylglutamate-kinase by arginine in *Anacystis* (Carr, 1973). Thioredoxins, which appear to be involved in light-dependent modulation of key enzymes by protein dithiol-disulphide interconversions, have been found to be present in green algae and cyanobacteria (Holmgren, 1985). Light-induced changes in the redox state of thioredoxin in *Anabaena* are consistent with such a role in the regulation of metabolism in this organism (Darling *et al.*, 1986).

Ion uptake

In early studies on permeability and uptake of solutes by living cells, certain algae were favourite subjects for experimentation because they yielded enough vacuolar sap for direct analysis. As much as 5 ml of sap may be obtained from a single vesicle of the coenocytic marine algae belonging to the genera *Valonia* and *Halicystis* and the giant internodal cells of freshwater and brackish charophytes, *Chara* and *Nitella*, may yield about 0.1 ml each. Using them, Osterhout (1936) showed that the proportions of various ions in the sap are different from those in the bathing medium. Other workers have used these algae to study uptake of a wide variety of ions (see Davson and Danielli, 1943; Hope and Walker, 1975). Equally these algae are convenient for investigation with microelectrodes, having a cell wall with a thin layer of protoplasm inside it through which capillaries may be inserted into the vacuole. From these beginnings has developed a sophisticated line of investigation of ion uptake by plants (Hope and Walker, 1975). Reference to reviews (e.g. Glass, 1983; Raven, 1984) will show how greatly experiments with larger algae – and in recent years microalgae as well – have contributed to our knowledge of transport across protoplasmic membranes. A recently devised non-invasive technique for measuring action potentials in *Acetabularia* enables observations with a minimum of disturbance of the material and gives promise of further advances (Thavarungkul *et al.*, 1987).

A very practical use of the propensity of the larger algae to accumulate ions is in monitoring radionuclides and heavy metals in natural waters (Levine, 1984; Whitton, 1984; Say *et al.*, 1986). Since they are usually attached and collectable in quantity and because they readily accumulate a variety of substances from their environment they provide a good indication of pollution integrated over a period of time at the community level.

Differentiation and morphogenesis

Many algae because of special features provide opportunities for investigating general problems of differentiation and morphogenesis. Only a few can be mentioned here but reviews have been provided by Moss (1974) and Buggeln (1981). One of the first to be used was the fertilized egg of *Fucus*; Oltmanns reported in 1889 that the primary rhizoid originates on the side away from the source of incident light and thereafter this has been a favourite system for the investigation of polarity (Evans *et al.*, 1982). The classic experiments of Haemmerling, published in the 1930s, on *Acetabularia* initiated major advances in understanding the role of the nucleus in morphogenesis. *Acetabularia* has a 'cell' of distinctive form, with rhizoids, stalk and cap, up to 5 cm in height and with a single large nucleus which normally resides in a rhizoid. The nucleus can be removed and implanted into another nucleate or anucleate plant, even of a different *Acetabularia* species. In this way Haemmerling (1963) demonstrated that morphogenetic substances are liberated from the nucleus and persist in the cytoplasm. The nucleocytoplasmic interactions found in *Acetabularia* hold for other cell types and the body of knowledge which has developed from this work is at the core of molecular biology (see, for example, Schweiger *et al.*, 1984). However, we are still ignorant about some of the basic features of *Acetabularia's* life cycle; it is not well adapted for genetical experiments – having a life cycle of at least six months and a haploid stage that is difficult to manipulate – and no mutants have been isolated (Lewin, 1976). Among the microalgae the desmids have elaborate and characteristic shapes and are particularly suitable for morphogenetic investigation, being comparatively large and easy to handle and grow in culture. The genus *Micrasterias* has thin, flat cells in which details of development can be viewed clearly and the large nucleus is easy to manipulate. The cell is divided into two semi-cells which on cell division separate, each regenerating a new semi-cell. The differentiation of wings on the new semi-cell depends on the existence of corresponding old ones, indicating that besides nuclear control there is some plasmatic structural continuity within the cell (Brook, 1981). Cytoplasmic tubules seem to be implicated in this but, since morphogenesis continues after their disorganization by treatment with colchicine or vinblastine, their significance must be indirect. Kiermayer and Meindl (1984) introduced special fixation procedures for electron microscopy of *Micrasterias* enabling studies of morphogenesis at the ultrastructural level. Observed interactions between different products of the Golgi apparatus and the plasma membrane support the hypothesis that the latter functions as a template for cytomorphogenesis. Another ultrastructural aspect of morphogenesis is exemplified by the many microalgae which produce external scales or other structures, often of elaborate form, composed of silica, calcium carbonate or organic material. An example given by Schnepf (1984) is that of the freshwater colonial flagellate *Synura*. Siliceous scales on the surface of the cell are produced internally in special vesicles which act as templates for their morphogenesis. The vesicles themselves are formed by complicated interaction of the chloroplast envelope with microtubules and microfilaments.

A different morphogenetic situation is presented by the heterocystous cyanobacteria. In *Anabaena* spp. heterocysts are formed midway between two pre-existing heterocysts as a filament growing on N_2 elongates. The hypothesis that this pattern arises because heterocysts inhibit nearby vegetative cells from turning into heterocysts covers the facts to a first approximation. Combined nitrogen, particularly if in the form of ammonium ion, inhibits heterocyst formation (Fogg, 1949). Nitrogen fixed in the heterocyst is exported to adjacent cells, decreasing progressively in concentration with distance from the heterocyst, until it falls below the level inhibitory to heterocyst differentiation. There is thus a diffusion based mechanism. It seems that the inhibitory substance is glutamine, the product of nitrogen fixation, or a derivative of it, rather than ammonia. Investigated more closely, this beguiling simplicity begins to disappear. Some undifferentiated cells are more disposed than others to turn into heterocysts and, in spite of the formation of active nitrogenase lagging behind the morphological differentiation of the heterocyst, a proheterocyst in an early stage of differentiation is able to inhibit adjacent cells from differentiating. The process of differentiation in the heterocyst itself is complicated, involving profound ultrastructural and biochemical reorganization beyond the morphological changes seen under the microscope (see Wolk, 1982; Bothe *et al.*, 1984; Haselkorn *et al.*, 1987).

Flagella

Because they are easily grown and manipulated in culture, *Chlamydomonas* spp. are favoured material for investigation of flagella. Both ultrastructurally and chemically the flagella of *Chlamydomonas* are complex yet they can be regenerated within 70–90 min (Trainor and Cain, 1986). Mutants in which there are lesions in particular ultrastructural components have provided valuable information about the control of flagellar movement, regeneration and regression (Lewin, 1976). Following the removal of flagella there is a rapid induction of synthesis of the tubulin subunits which provide the main structural elements in the flagellar axonemal microtubules and the coordinated expression of tubulin genes and post-translational modification of the proteins involved in this process have been investigated (Brunke *et al.*, 1984). Flagella are involved in the mating of *Chlamydomonas*; when gametes of opposite mating types are brought together there is rapid adhesion via their flagella, the points of adhesion shift to the tips of the flagella, cell walls are shed, then sexual fusion begins. Many of the ultrastructural details of these processes have been described and a beginning has been made in the characterization of the biochemical and genetical mechanisms involved. This work seems of importance for the general understanding of specific cell recognition mechanisms and adhesion-induced events (Snell, 1985).

Bioassay

In bioassays algae are used as tools rather than as objects of research but nevertheless mention must be made of this role since it has contributed to many areas of science. Nearly always, the concentration of some factor, growth-promoting or growth-inhibiting, has been measured in terms of final population in batch culture. One of the first uses of this sort was of the photosynthetic flagellates *Euglena* and *Ochromonas* for assay of cobalamin, vitamin B_{12}. Whereas *E. gracilis* has the *Lactobacillus* type of specificity towards vitamin B_{12} analogues, *O. malhamensis* has a similar specificity to that of mammals (Provasoli and Carlucci, 1974) and for this reason was adopted as the official assay organism in Britain. It has to be borne in mind that many algae, even if they do not have a requirement for the vitamin, release a vitamin B_{12} binding factor which inhibits its action (Provasoli and Carlucci, 1974). Algae have as well been used in the bioassay of thiamin and biotin (Gold, 1973; Swift, 1984). Algal bioassays have also been much used for estimating the capacity of natural waters to sustain unwanted algal growth. A favourite for this is the green alga *Selenastrum capricornutum*. Procedures have been extensively investigated, standardized and officially adopted in the United States under the name of the 'algal assay procedure bottle test' (Environmental Protection Agency, 1971) and a substantial bibliography is available (Environmental Protection Agency, 1979). Extension and refinement of this approach has led to much discussion about its scope and limitations (Shubert, 1984). For successful routine application the method of estimating biomass is critical. Butterwick *et al.* (1982) compared eight different methods, all of which gave acceptable precision. Estimation *in vivo* of optical properties such as fluorescence were most rapid, chemical methods required relatively large samples and were slow but enabled batches of samples to be processed together, while visual cell counts although slow were unsurpassed for the low limit of detection obtained and economy of material. Estimation of relative growth rate rather than final yield sometimes gives more useful information but it is time consuming and comparisons of bioassays made in the two different ways suggest that there is normally little difference between them (Fogg and Thake, 1987). Short-term bioassays in which assimilation of ^{14}C from bicarbonate is determined after a few hours yield results which are sometimes, but not always, similar to those obtained from biomass determinations (Li and Goldman, 1981; Fogg and Thake, 1987).

Investigations of the physiology and biochemistry of algae

Besides being used in the study of general biological problems the algae are sufficiently different from other kinds of organisms and sufficiently varied to warrant investigation for their own sakes (Stewart, 1974). This not only enriches comparative physiology and biochemistry but leads to better understanding of the role of these organisms in the environment and to much of biotechnological value. However, in doing such research one ought not to select the species most amenable to the techniques available, otherwise knowledge will be biased; special difficulties presented by some algae must be recognized. A compendium of physiological and biochemical methods adapted for the algae is available (Hellebust and Craigie, 1978) but a few points should be mentioned briefly.

It has often been suggested that the inability of algae to assimilate certain organic substrates should be ascribed to impermeability of their cell membranes. Most often, one suspects, the reason is more deep seated but it is a general observation that *Chlorella* spp., which have relatively thick and tough cell walls, are often unstainable by many of the usual cytological reagents and intact cells are unaffected by a variety of different enzymes. The resistant carotenoid polymer, sporopollenin, has been found in the outer wall layer of some strains (Atkinson *et al.*, 1972) and this would be likely to confer such impermeability. *Chlorella* has also proved a particularly difficult subject for the preparation of cell-free extracts retaining enzymic activity and of undamaged organelle fractions. Millbank (1957) tried a variety of different techniques but was unsuccessful in obtaining homogenates of *Chlorella* capable of oxidizing tricarboxylic acid cycle intermediates. Mitochondrial preparations can be obtained from these algae (Grant, 1978), although with more difficulty than with most organisms. Intact chloroplasts do not seem to have been obtained from *Chlorella* but they can be prepared from *Euglena* and *Chlamydomonas*, and, particularly easily, from many coenocytic green algae. Brown algae present a special difficulty with cell-free extracts in that they contain copious amounts of polyphenols, stored in vesicles known as physodes, which on homogenization of the cell bind irreversibly to proteins and so inactivate enzymes (Ragan and Glombitza, 1986). In fact, few studies on the enzymes of brown algae have been published.

As already indicated, the culture of algae has been selective and many of the most frequent strains in collections may be fairly described as weeds rather than representative of natural algal populations. This may be satisfactory from some points of view because robust, rapidly growing forms are often the most useful in biotechnology, but it does mean that the vast variety of metabolism exhibited in the algae is not being exploited. Differences between conditions in culture and those in the natural environment have been discussed by Fogg and Thake (1987). One that is most often overlooked is that of turbulence patterns; whether a culture is gently stirred or violently agitated may be critical for growth of some species and it is difficult to reproduce the small-scale circulation patterns of the surface waters of lakes or the sea even in large artificial enclosures. The behaviour of an alga may be different according to the type of culture in which it is grown. Uptake rates for nitrate by *Oscillatoria*, for example, differ considerably according to whether measurements are made with material in batch culture or steady-state chemostats (see Fogg and Thake, 1987). There is also increasing evidence that under natural conditions some algae may have rather specific relationships with bacteria, their behaviour being different according to whether these bacteria are present or not. An example is *Botryococcus*, a planktonic alga of potential economic importance, the growth and hydrocarbon production of which is strongly influenced by associated bacteria (Chirac *et al.*, 1985). The often elaborate requirements for culture of the larger seaweeds, many of which are sources of valuable chemicals (Chapman, 1973), have inhibited experimental work on them and farming in the sea is usually the only practical way of rearing them for economic purposes (Tseng, 1981).

Conclusions

Certain algae have become standard experimental organisms for research in many fields of physiology and biochemistry and a vast body of information about their behaviour and composition is available. Many are of a size which permits them to be handled by biotechnological methods similar to those used for bacteria and fungi. The idea that algae are slow growing, which has often been advanced against their use on the commercial scale in the past, has little foundation – many species have doubling times of hours rather than days. There is no doubt that within the group there is enormous variety of metabolic pattern and products and that many forms have great potential for use in biotechnology.

References

Atkinson A W Jr, Gunning B E S, John P C L 1972 Sporopollenin in the cell of *Chlorella* and other algae: Ultrastructure, chemistry, and incorporation of ^{14}C-acetate, studied in synchronous cultures. *Planta* **107**: 1–32

Bassham J A, Calvin M 1975 *The Path of Carbon in Photosynthesis*, Prentice-Hall, Englewood Cliffs, N J

Benson A A, Calvin M 1950 The path of carbon in photosynthesis. VII. Respiration and photosynthesis. *J. Exp. Bot.* **1**: 63–8

Bisalputra T 1974 Plastids. In Stewart W D P (ed) *Algal Physiology and Biochemistry*, Blackwell, Oxford, pp 124–60

Blinks L R 1954 The role of accessory pigments in photosynthesis. In Fry B A, Peel J L (eds) *Autotrophic Micro-organisms*, Cambridge University Press, pp 224–46

Bothe H, Nelles H, Kentemich T, Papen H, Neuer G 1984 Recent aspects of heterocyst biochemistry and differentiation. In Wiessner W, Robinson D G, Starr R C (eds) *Compartments in Algal Cells and their Interaction*, Springer-Verlag, Berlin, pp 218–32

Bowes G, Berry J A 1972 The effect of oxygen on photosynthesis and glycolate excretion in *Chlamydomonas reinhardtii*. *Carnegie Inst. Yr. Bk.* **71**: 148–58

Brook A J 1981 *The Biology of the Desmids*, Blackwell, Oxford

Brunke K, Anthony J, Sternberg E, Weeks D 1984 Regulation of tubulin gene expression in *Chlamydomonas reinhardtii* following flagellar excision. In Wiessner W, Robinson D G, Starr R C (eds) *Compartments in Algal Cells and their Interaction*, Springer-Verlag, Berlin, pp 88–95

Buggeln R G 1981 Morphogenesis and growth regulators. In Lobban C S, Wynne M J (eds) *The Biology of Seaweeds*, Blackwell, Oxford, pp 627–60

Butterwick C, Heaney S I, Talling J F 1982 A comparison of eight methods for estimating the biomass and growth of planktonic algae. *Br. Phycol. J.* **17**: 69–79

Carr N G 1973 Metabolic control and autotrophic physiology. In Carr N G, Whitton B A (eds) *The Biology of Blue-Green Algae*, Blackwell, Oxford, pp 39–65

Castelfranco P A, Beale S I 1983 Chlorophyll biosynthesis: Recent advances and areas of current interest. *A. Rev. Pl. Physiol.* **34**: 241–78

Chapman A R O 1973 Methods for macroscopic algae. In Stein J R (ed) *Handbook of Phycological Methods: Culture Methods and Growth Measurements*, Cambridge University Press, pp 87–104

Chirac G, Casadevall E, Largeau C, Metzger P 1985 Bacterial influence upon growth and hydrocarbon production of the green alga *Botryococcus braunii. J. Phycol.* **21**: 380–7

Darling A J, Rowell P, Stewart W D P 1986 Light-induced transitions in the redox state of thioredoxin in the N_2-fixing cyanobacterium *Anabaena cylindrica. Biochim. Biophys. Acta* **850**: 116–20

Davson H, Danielli J F 1943 *The Permeability of Natural Membranes*, Cambridge University Press

Duysens L N M, Sweers H E 1963 Mechanisms of two photochemical reactions in algae as studied by means of fluorescence. In *Studies on Microalgae and Photosynthetic Bacteria*, University of Tokyo Press, pp 353–72

Emerson R, Chalmers R, Cederstrand C 1957 Some factors influencing the long-wave limit of photosynthesis. *Proc. Nat. Acad. Sci. USA* **43**: 133–43

Emerson R, Lewis C M 1942 The photosynthetic efficiency of phycocyanin in *Chroococcus* and the problem of carotenoid participation in photosynthesis. *J. Gen. Physiol.* **25**: 579–95

Environmental Protection Agency 1971 *Algal assay procedure bottle test*, National Technical Information Service, Springfield, Virginia

Environmental Protection Agency 1979 *Bibliography of literature pertaining to the genus* Selenastrum, National Technical Information Service, Springfield, Virginia

Evans L V 1974 Cytoplasmic organelles. In Stewart W D P (ed) *Algal Physiology and Biochemistry*, Blackwell, Oxford, pp 86–123

Evans L V, Callow J A, Callow M E 1982 The biology and biochemistry of reproduction and early development in *Fucus*. In Round F E, Chapman D J (eds) *Progress in Phycological Research 1*, Elsevier , Amsterdam, pp 67–110

Fay P, Lang N J 1971 The heterocysts of blue–green algae I. Ultrastructural integrity after isolation. *Proc. R. Soc. Lond.* B **178**: 185–92

Fay P, Stewart W D P, Walsby A E, Fogg G E 1968 Is the heterocyst the site of nitrogen fixation in blue–green algae? *Nature, Lond.* **220**: 810–2

Fay P, Walsby A E 1966 Metabolic activities of isolated heterocysts of the blue–green alga *Anabaena cylindrica. Nature, Lond.* **209**: 94–5

Fogg G E 1949 Growth and heterocyst production in *Anabaena cylindrica* Lemm. II. In relation to carbon and nitrogen metabolism. *Ann. Bot.* **13**: 241–59

Fogg G E 1959 Nitrogen nutrition and metabolic patterns in algae. *Symp. Soc. Exp. Biol.* **13**: 106–25

Fogg G E 1972 *Photosynthesis*, 2nd edn, English Universities Press, London

Fogg G E 1986 Picoplankton. *Proc. R. Soc. Lond.* B **228**: 1–30

Fogg G E, Thake B A 1987 *Algal Cultures and Phytoplankton Ecology*, University of Wisconsin Press

Gaffron H 1940 Carbon dioxide reduction with molecular hydrogen in green algae. *Amer. J. Bot.* **27**: 273–83

Gallagher J C 1982 Physiological variation and electrophoretic banding patterns of genetically different seasonal populations of *Skeletonema costatum* (Bacillariophyceae). *J. Phycol.* **18**: 148–62

Glass A D M 1983 Regulation of ion transport. *A. Rev. Pl. Physiol.* **34**: 311–26

Gold K 1973 Bioassay: thiamine. In Stein J R (ed) *Handbook of Phycological Methods: Culture Methods and Growth Measurements*, Cambridge University Press, pp 395–403

Goldschmidt-Clermont M, Dron M, Erickson J M, Rochaix J -D, Schneider M, Spreitzer R, Vallet J -M 1984 Structure and expression of chloroplast and nuclear genes in *Chlamydomonas reinhardtii*. In Wiessner W, Robinson D G, Starr R C (eds) *Compartments in Algal Cells and their Interaction*, Springer-Verlag, Berlin, pp 23–7

Grant N G 1978 Mitochondria from *Chlorella*. In Hellebust J A. Craigie J S (eds) *Handbook of Phycological Methods: Physiological and Biochemical Methods*, Cambridge University Press, pp 25–30

Guerrero M G, Vega J M, Losada M 1981 The assimilatory nitrate-reducing system and its regulation. *A. Rev. Pl. Physiol.* **32**: 169–204

Haehnel W 1984 Photosynthetic electron transport in higher plants. *A. Rev. Pl. Physiol.* **35**: 659–93

Haemmerling 1963 Nucleo-cytoplasmic interactions in *Acetabularia* and other cells. *A. Rev. Pl. Physiol.* **14**: 65–92

Haselkorn R 1978 Heterocysts. *A. Rev. Pl. Physiol.* **29**: 319–44

Haselkorn R, Golden J W, Lammers P J, Mulligan M E 1987 Rearrangement of *nif* genes during cyanobacterial heterocyst differentiation. *Phil. Trans. R. Soc.* B **317**: 173–81

Hellebust J A, Craigie J S 1978 (eds) *Handbook of Phycological Methods: Physiological and Biochemical Methods*, Cambridge University Press

Holmgren A 1985 Thioredoxin. *A. Rev. Biochem.* **54**: 237–71

Hope A B, Walker N A 1975 *The Physiology of Giant Algal Cells*, Cambridge University Press

Kanazawa T, Kanazawa K, Kirk M R, Bassham J A 1970 Regulation of photosynthetic carbon metabolism in synchronously growing *Chlorella pyrenoidosa*. *Pl. Cell Physiol., Tokyo* **11**: 149–60

Kiermayer O, Meindl U 1984 Interaction of the Golgi apparatus and the plasmalemma in the cytomorphogenesis of *Micrasterias*. In Wiessner W, Robinson D G, Starr R C (eds) *Compartments in Algal Cells and their Interaction*, Springer-Verlag, Berlin, pp 175–82

Levine H G 1984 The use of seaweeds for monitoring coastal waters. In Shubert L E (ed) *Algae as Ecological Indicators*, Academic Press, London pp 189–212

Levine R P 1974 Mutant studies on photosynthetic electron transport. In Stewart W D P (ed) *Algal Physiology and Biochemistry*, Cambridge University Press, pp 424–33

Lewin R A 1949 Genetics of *Chlamydomonas*–paving the way. *Biol. Bull.* **97**: 243–4

Lewin R A 1976 (ed) *The Genetics of Algae*, Blackwell, Oxford

Li W K W, Goldman J C 1981 Problems in estimating growth rates of marine phytoplankton from short-term ^{14}C assays *Microbiol. Ecol* **7**: 113–21

Lien S 1978 Hill reaction and photophosphorylation with chloroplast preparations from *Chlamydomonas reinhardi*. In Hellebust J A, Craigie J S (eds) *Handbook of Phycological Method: Physiological and Biochemical Methods*, Cambridge University Press, pp. 305–16

Lloyd D 1974 Dark respiration. In Stewart W D P (ed) *Algal Physiology and Biochemistry*, Blackwell, Oxford, pp 505–29

Lucas W J, Berry J A 1985 (eds) *Inorganic Carbon Uptake by Aquatic Photosynthetic Organisms*, American Society of Plant Physiologists

McCandless E L 1981 Polysaccharides of seaweeds. In Lobban C S, Wynne M J (eds) *The Biology of Seaweeds*, Blackwell, Oxford, pp 559–88

Mierle G 1985 Kinetics of phosphate transport by *Synechococcus leopoliensis* (Cyanophyta): evidence for diffusion limitation of phosphate uptake. *J. Phycol.* **21**: 177–81

Millbank J W 1957 Studies on the preparation of a respiratory cell-free extract of the alga *Chlorella*. *J. Exp. Bot.* **8**: 96–104

Morris G J, Coulson G, Clark A 1979 The cryopreservation of *Chlamydomonas*. *Cryobiology* **16**: 401–10

Morris I Glover H E 1974 Questions on the mechanism of temperature adaptation in marine phytoplankton. *Mar. Biol.* **24**: 147–54

Moss B 1974 Morphogenesis. In Stewart W D P (ed) *Algal Physiology and Biochemistry*, Cambridge University Press, pp. 788–813

Myers J 1949 The pattern of photosynthesis in *Chlorella*. In Franck J, Loomis W E (eds) *Photosynthesis in Plants*, Iowa State College Press, pp 349–64

Myers J, Clark L B 1944 Culture conditions and the development of the photosynthetic mechanism. II. An apparatus for the continuous culture of *Chlorella*. *J. Gen. Physiol.* **28**: 103–12

Osterhout W J V 1936 The absorption of electrolytes in large plant cells. *Bot. Rev.* **2**: 283–315

Pearsall W H, Loose L 1937 Growth of *Chlorella vulgaris* in pure culture. *Proc. R. Soc. Lond.* B **121**: 451–501

Pringsheim E G 1946 *Pure Cultures of Algae*, Cambridge University Press

Provasoli L, Carlucci A F 1974 Vitamins and growth regulators. In Stewart W D P (ed) *Algal Physiology and Biochemistry*, Blackwell, Oxford, pp. 741–81

Puiseux-Dao S 1981 Cell-cycle events in unicellular algae. *Can. Bull. Fish. Aquat. Sci.* **210**: 130–49

Ragan M A, Glombitza K -W 1986 Phlorotannins, brown algal polyphenols. In Round F E, Chapman D J (eds) *Progress in Phycological Research 4* Biopress, Bristol, pp 129–41

Raven J A 1984 *Energetics and Transport in Aquatic Plants*, Alan R Liss, New York

Raven J A 1986 Physiological consequences of extremely small size for autotrophic organisms in the sea. *Can. Bull. Fish. Aquat. Sci.* **214**: 1–70

Say P J, Burrows I G, Whitton B A 1986 *Enteromorpha as a Monitor of Heavy Metals in Estuarine and Coastal Intertidal Waters*, Northern Environmental Consultants Ltd, Consett, Co. Durham

Scheer H 1982 Phycobiliproteins: molecular aspects of a photosynthetic antenna system. In Fong F K (ed) *Light Reaction Path in Photosynthesis*, Springer-Verlag, Berlin, p 7

Schneider K C, Bradbeer C, Singh R N, Wang L C, Wilson P W, Burris R H 1960 Nitrogen fixation by cell-free preparations from micro-organisms. *Proc. Nat. Acad. Sci. USA* **46**: 726–33

Schnepf E 1984 The cytological viewpoint of functional compartmentation. In Wiessner W, Robinson D G, Starr R C (eds) *Compartments in Algal Cells and their Interaction*, Springer-Verlag, Berlin, pp 1–10

Schweiger H G, Bannwarth H, Berger S, De Groot E, Neuhaus G, Neuhaus-Url G 1984 Interactions between compartments in *Acetabularia* during gene expression. In Wiessner W, Robinson D G, Starr R C (eds) *Compartments in Algal Cells and their Interaction*, Springer-Verlag, Berlin, pp 28–35

Shubert L E 1984 *Algae as Ecological Indicators*, Academic Press, London

Snell W J 1985 Cell–cell interactions in *Chlamydomonas*. *A. Rev. Pl. Physiol.* **36**: 287–315

Solomonson L P, Spehar A M 1977 Model for the regulation of nitrate assimilation. *Nature, (Lond.)* **265**: 373–5

Somerville C R 1986 Analysis of photosynthesis with mutants of higher plants and algae. *A. Rev. Pl. Physiol.* **37**: 467–507

Spanswick R M 1981 Electrogenic ion pumps. *A. Rev. Pl. Physiol* **32**: 267–89

Stein J R 1980 *Handbook of Phycological Methods. Culture Methods and Growth Measurements*, Cambridge University Press

Stewart W D P 1974 (ed) *Algal Physiology and Biochemistry*, Blackwell, Oxford

Stewart W D P, Haystead A, Pearson H W 1969 Nitrogenase activity in heterocysts of blue–green algae. *Nature, Lond.* **224**: 226–8

Stewart W D P, Rowell P, Kerby N W, Reed R H, Machray G C 1987 N$_2$-fixing cyanobacteria and their potential applications. *Phil. Trans. R. Soc. Lond.* B **317**: 245–58

Swift D G 1984 Algal assays for vitamins. In Shubert L E (ed) *Algae as Ecological Indicators*, Academic Press, London, pp 281–316

Syrett P J 1962 Nitrogen assimilation. In Lewin R A (ed) *Physiology and Biochemistry of Algae*, Academic Press, New York, pp 171–88

Syrett P J 1981 Nitrogen metabolism of microalgae. *Can. Bull. Fish. Aquat. Sci.* **210**: 182–210

Tamiya H, Iwamura T, Shibata K, Hase E, Nihei T 1953 Correlation between photosynthesis and light-independent metabolism in the growth of *Chlorella*. *Biochim. Biophys. Acta* **12**: 23–40

Tanada T 1951 The photosynthetic efficiency of carotenoid pigments in *Navicula minima*. *Amer. J. Bot.* **38**: 276–83

Tandeau de Marsac N, Houmard J 1987 Advances in cyanobacterial molecular genetics. In Fay P, Van Baalen C (eds) *The Cyanobacteria*, Elsevier, Amsterdam, pp 251–302

Thavarungkul P, Lertsithichai S, Sherlock R A 1987 Spontaneous action potential initiation and propagation in regenerating cell segments of *Acetabularia mediterranea*. *J. Exp. Bot.* **38**: 1541–56

Tolbert N E 1979 Glycolate metabolism by higher plants and algae. In Gibbs M, Latzko E (eds) *Photosynthesis II. Photosynthetic Carbon Metabolism and Related Processes*, Springer-Verlag, Berlin, pp 338–52

Tolbert N E, Zill L P 1956 Excretion of glycolic acid by algae during photosynthesis. *J. Biol. Chem.* **222**: 895–906

Trainor F R, Cain J R 1986 Famous algal genera. I. *Chlamydomonas*. In Round F E, Chapman D J (eds) *Progress in Phycological Research 4*, Biopress, Bristol, p 81

Tseng C K 1981 Commercial cultivation. In Lobban C S, Wynne M J (eds) *The Biology of Seaweeds*, Blackwell, Oxford, pp 680–725

Warburg O 1919 Über die Geschwindigkeit der photochemischen Kohlensäurezersetzung in lebenden Zellen. I. *Biochem. Z.* **100**: 230–70

Warburg O 1920 Über die Geschwindigkeit der photochemischen Kohlensäurezersetzung in lebenden Zellen. II. *Biochem. Z.* **103**: 188–217

Warburg O, Negelein E 1920 Über die Reduktion der Salpetersäure in grünen Zellen. *Biochem. Z.* **110**: 66–115

Wheeler P A 1980 Use of methylammonium as an ammonium analogue in nitrogen transport and assimilation studies with *Cyclotella cryptica* (Bacillariophyceae). *J. Phycol.* **16**: 328–34

Whitton B A 1984 Algae as monitors of heavy metals. In Shubert L E (ed) *Algae as Ecological Indicators*, Academic Press, London, pp 257–80

Wiessner W, Robinson D G, Starr R C 1984 (eds) *Compartments in Algal Cells and their Interaction*, Springer-Verlag, Berlin

Wolk C P 1982 Heterocysts. In Carr N G, Whitton B A (eds) The *Biology of Cyanobacteria*, Blackwell, Oxford, pp 359–86

3

The transport, assimilation and production of nitrogenous compounds by cyanobacteria and microalgae

N. W. Kerby, P. Rowell and W. D. P. Stewart

Introduction

Cyanobacteria are photosynthetic prokaryotes and more closely resemble bacteria than microalgae. Due to certain similarities in their potential applications, nutritional requirements, methods of mass culture, modes of metabolism and in keeping with tradition, we have considered cyanobacteria and microalgae together.

The use of microorganisms to produce a variety of chemicals using inexpensive carbon compounds is of economic interest. Photosynthesis is a unique means of primary production of organic compounds using air, light and water, and cyanobacteria and microalgae provide an efficient means of converting solar energy into biomass. The large-scale culture of photosynthetic microorganisms is theoretically cheaper than that of heterotrophic organisms since they do not require organic substrates. However, all organisms require a nitrogen source for growth. This requirement can be met in most cases by supplying ammonium salts, nitrates or urea. The use of these nitrogen sources in the mass culture of microalgae has recently been discussed (see Richmond, 1986; Borowitzka and Borowitzka, 1988). Although ammonia is used preferentially it can be toxic at high concentrations. Certain cyanobacteria have an added advantage due to their ability to fix atmospheric nitrogen making their mass cultivation potentially cheaper. Nitrogen-fixing cyanobacteria are also particularly useful for the production of nitrogenous compounds and for use as biofertilizers.

In this review we discuss the transport, assimilation and, in certain specialized cases, the liberation of nitrogenous compounds by cyanobacteria and microalgae. Equal emphasis is given to applied and fundamental studies since a thorough knowledge of molecular aspects of nitrogen transport and metabolism is essential for further effective utilization of these organisms. The genetic manipulation of microalgae and cyanobacteria to produce nitrogenous compounds of economic potential also relies on a thorough understanding of the processes of transport and metabolism of nitrogenous compounds.

The uptake of nitrogenous compounds

The uptake of nitrogenous compounds by microalgae and cyanobacteria is a series of metabolic processes which includes membrane transport, reduction of dinitrogen, nitrate and nitrite, and ammonia assimilation. Most studies on the uptake of nitrogenous compounds by microalgae and cyanobacteria have considered the summation of these processes and the mechanism of transport into the cells has received relatively little attention.

Transport systems serve to move many different solutes into or out of cells to maintain the optimum *in vivo* concentration of nutrients. Solutes can enter cells by one or more transport systems that can be distinguished by substrate specificity and uptake kinetics. Plasma membranes consist of lipids modified by proteins with most of the phospholipid being arranged as a bilayer with polar heads between two aqueous phases. In the fluid-mosaic model of membrane structure (Singer and Nicholson, 1972), membrane proteins may be either peripheral or integral and some of these may span the membrane and allow protein-catalysed transport. For an ion to be transported across a membrane a pathway and a driving force are required. The driving force may be a concentration gradient, electrical potential, metabolic energy or a combination of these (see Nicholls, 1982). Transport processes fall into three major categories: passive diffusion, facilitated diffusion and active transport. Certain solutes (e.g. CO_2, NH_3, O_2, N_2, H_2O and possibly urea) can cross lipid bilayers by diffusion (see Raven, 1980). Such non-mediated transport is the non-specific transport of a solute down an electrochemical gradient and does not show saturation kinetics. The hydrophobic nature of membranes creates a barrier to the movement of many important nutrients (e.g. glucose, amino acids, NO_3^-, NH_4^+, HPO_4^{2-}, SO_4^{2-}, K^+ and Mg^{2+}). This impermeability also extends to protons which are vital for energy transduction (see Nicholls, 1982). For such solutes the net flux is enhanced by the presence of specific proteins (porters) in the membrane. Transport mechanisms include: *uniport* which is transport involving a single ion; *symport* or *co-transport* which is the obligatory coupling of the transport of two or more ions in parallel; *antiport* which is the transport of one ion coupled to that of another ion in the opposite direction. Transport may be electroneutral, with no net charge transfer across the membrane, or electrogenic. *Active transport* may be *primary*, requiring a direct input of energy or *secondary*, being driven by an ion gradient established by a primary system.

In this chapter we discuss only ammonium, nitrate and amino acid transport. The transport and assimilation of other nitrogen sources, including urea, are discussed by Syrett (1981) and for more detailed accounts of bioenergetics and microalgal nutrient transport see Nicholls (1982) and Raven (1980), respectively.

Ammonia uptake

Ammonia is taken up by microalgae and cyanobacteria in preference to other nitrogen sources (Morris, 1974; Ohmori et al., 1977; Guerrero et al., 1981). The transport of ammonia is of particular interest since ammonia is a powerful repressor of nitrate uptake and assimilation in microalgae and cyanobacteria and of nitrogenase activity and synthesis in N_2-fixing cyanobacteria. The terms ammonia and methylamine, which is often used as an analogue of ammonia, are used in this paper unless the charged (NH_4^+ or $CH_3NH_3^+$) or uncharged (NH_3 or CH_3NH_2) species are meant specifically.

NH_3 is a small, uncharged molecule which can pass freely through lipid bilayers by passive diffusion. A mechanism for the accumulation of ammonia at external pH values higher than that of the cytoplasm (pH 6.5–7.5, Raven, 1980; Reed et al., 1980; Peschek et al., 1985), such as are found during bloom conditions, could involve simple diffusion and ion trapping through protonation, given a membrane permeable to NH_3 but not NH_4^+. At external pH values near neutral, as found in most naturally occurring water bodies, a specific energy-dependent uptake mechanism is essential. Additionally, an NH_4^+ transport system may also be required for the retention of ammonia generated internally by nitrogen fixation, nitrate reduction and photorespiration.

Microalgae and cyanobacteria are capable of growing at very low external ammonia concentrations and uptake is by a very high affinity system (apparent K_m ~1 μM (Eppley et al., 1969; Eppley and Renger, 1974; Healey, 1977)). However, despite the fact that ammonia is one of the most important nitrogen sources for microalgae and cyanobacteria, relatively little is known about the mechanism and regulation of uptake. This is due partly to the lack of a suitable analogue for use in tracer studies and also to the difficulties in separating transport and assimilation.

Methylammonium ($CH_3NH_3^+$) has been used both as a substrate for NH_4^+ transport (see Kleiner, 1981, 1985; Kerby et al., 1987a) and in the measurement of intracellular pH (Kashket and Wilson, 1973; Barnes, 1980) in microorganisms. Methylamine has been used in the study of ammonia transport in Chlorella pyrenoidosa (Pelley and Bannister, 1979), Chlorella vulgaris (Schlee and Komor, 1986), plankton communities as a field assay (Vincent, 1979), Cyclotella cryptica (Wheeler, 1980; Wheeler and Hellebust, 1981), Phaeodactylum tricornutum (Wright and Syrett, 1983; Cresswell and Syrett, 1984) the cyanobacterium Anacystis nidulans (Boussiba et al., 1984a,b; Ritchie and Gibson, 1987) and various N_2-fixing cyanobacteria (Kashyap and Johar, 1984; Rai et al., 1984; Turpin et al., 1984; Singh et al., 1985, 1986; Kerby et al., 1986a).

Methylamine uptake by C. cryptica showed saturation kinetics and evidence was presented for the presence of two transport systems, one with high affinity (apparent K_m 6 μM) and low capacity and the other with low affinity (apparent K_m 39 μM) and high capacity (Wheeler, 1980). However, no evidence was obtained for the presence of two transport systems in P. tricornutum which showed saturable kinetics of methylamine transport with an apparent K_m of 35 μM (Wright and Syrett, 1983). Uptake of methylamine in both C. cryptica and P. tricornutum was dependent on Na^+ and was stimulated by nitrogen starvation. Methylamine transport was inhibited by ammonia and by high external K^+ concentrations. Uptake was linear over a period of 1 h in *C. cryptica* and *P. tricornutum*, in contrast to the kinetics of uptake in cyanobacteria which are usually biphasic (Boussiba *et al.*, 1984b; Rai *et al.*, 1984; Kerby *et al.*, 1986a). The initial rapid phase is attributed, at pH 7.0, to uptake via a $CH_3NH_3^+$ (NH_4^+) transport system and, at pH 9.0, probably mainly to passive diffusion and trapping by protonation (see Walker *et al.*, 1979). Addition of NH_4Cl inhibited uptake and caused a rapid efflux of ^{14}C-labelled methylamine from cyanobacteria (see Kerby *et al.*, 1987a). The uptake of ^{14}C-labelled methylamine by *A. cylindrica* was different from that of other nitrogen-fixing cyanobacteria, there being a rapid initial uptake at pH 7.0 and pH 9.0 with no or little evidence of a second slower phase. Possible reasons for the absence or reduction of this second phase will be discussed below.

Similar biphasic patterns of $^{14}CH_3NH_3^+$ uptake have been observed in *Anacystis nidulans* R-2 (Boussiba *et al.*, 1984b), *Anabaena azollae* (Rai *et al.*, 1984), *Anabaena cycadae* (Singh *et al.*, 1985), *Anabaena flos-aquae* (Turpin *et al.*, 1984) and *Anabaena doliolum* (Singh *et al.*, 1986), although the kinetics of uptake are interpreted in various ways. The view of some authors is that the initial rapid phase represents transport into the cell and the second slower phase represents metabolism of the analogue, whilst others have suggested the presence of two different transport systems.

Ammonia concentration gradients of between 10 and 3000 between the cells and the external medium have been reported for cyanobacteria (Boussiba *et al.*, 1984b; Kashyap and Johar, 1984). Such variations can be attributed to problems associated with the determination of intracellular pools which are subject to rapid turnover and enlargement due to the lability of metabolites such as glutamine. The reported apparent K_m values for uptake of $CH_3NH_3^+$ are in the region of 7 μM and those for uptake of NH_4^+ are in the range 1–66 μM for cyanobacteria.

Uncouplers such as CCCP (carbonylcyanide *m*-chlorophenylhydrazone) and FCCP (*p*-trifluoromethoxycarbonyl-cyanidephenylhydrazone), $TPMP^+$ (triphenylmethylphosphonium) which at high concentrations collapses $\Delta\psi$, and ATPase inhibitors such as DCCD (dicyclohexylcarbodiimide) have been shown to inhibit methylamine and ammonia transport (Boussiba *et al.*, 1984a,b; Rai *et al.*, 1984; Wright and Syrett, 1983; Schlee and Komor, 1986). This indicates that the transport of NH_4^+ and $CH_3NH_3^+$ ions is an active, electrogenic process which may be similar to the better characterized system of *Escherichia coli* where the mechanism of transport appears to be a $CH_3NH_3^+$ (NH_4^+)/K^+ antiport driven by the electrochemical K^+ gradient (Jayakumar *et al.*, 1985). Uptake of $^{14}CH_3NH_3^+$ and NH_4^+ occurs in the dark (Balch, 1985) but not under dark anaerobic conditions (Wright and Syrett, 1983; Boussiba *et al.*, 1984a). The sustained uptake of $^{14}CH_3NH_3^+$ (second slower phase) was inhibited following incubation of *A. variabilis* aerobically in the dark (Kerby *et al.*, 1986a) without reduction in the initial phase of uptake and this may be due to the assimilation of methylamine being rate-limiting.

The biphasic pattern of $^{14}CH_3NH_3^+$ uptake in many cyanobacteria has been interpreted as transport into cells (the first rapid phase) followed by assimilation of methylamine by GS (the second slower phase) (Boussiba *et al.*, 1984b; Rai *et al.*, 1984; Boussiba and Gibson, 1985; Kerby *et al.*, 1986a). γ-Methylglutamine is generally considered to be synthesized via GS in microorganisms (Kleiner, 1981; Kleiner and Fitzke, 1981; Barnes *et al.*, 1983) and accumulates, but is not metabolized further, in cyanobacteria. Evidence for this comes from the inhibition of GS with L-methionine-D, L-sulphoximine (MSX) and more recently, from the use of mutant strains of *A. variabilis* resistant to lethal concentrations of methylamine and ethylenediamine (EDA, 1,2-diaminoethane). Following incubation of cells with MSX to inhibit GS, the first phase of transport is unaffected, the second slower phase is absent and no γ-methylglutamine is formed. Strains resistant to EDA, which have reduced GS activity (see page 70) showed a corresponding reduction in the second phase but not the first phase of transport (Reglinski *et al.*, 1989). A mutant strain resistant to methylamine (strain 4m3) had both a reduced first phase and second phase of ^{14}C-labelled methylamine uptake and the rate of ammonia uptake was lower than in the parent strain at pH 7.0 but not at pH 9.0. This strain apparently has a deficient active $CH_3NH_3^+/NH_4^+$ transport system but normal uptake of methylamine and ammonia at pH 9.0, where passive diffusion and ion trapping may be largely responsible for their intracellular accumulation (Reglinski *et al.*, 1989). Strain 4m3 did not liberate ammonia extracellularly in contrast with the findings of an NH_4^+-transport deficient mutant of *Klebsiella pneumoniae* (Castorph and Kleiner, 1984) but in agreement with the findings for an NH_4^+-transport deficient mutant of *Chlamydomonas reinhardtii* (Franco *et al.*, 1987). This suggests that an active NH_4^+ transport system is not essential for the retention of internally generated ammonia. No evidence for the metabolism of methylamine has been found in *Phaeodactylum tricornutum* (Wright and Syrett, 1983; Cresswell and Syrett, 1984) but γ-methylglutamine is formed in *C. reinhardtii* (Franco *et al.*, 1984). The absence of γ-methylglutamine formation in certain species may be due to differences in the affinity of GS from different organisms for methylamine. Sustained rates of ammonia and methylamine uptake are dependent on metabolism. Since methylamine is metabolized to γ-methylglutamine by GS in certain species of cyanobacteria and microalgae, but not others, care must be taken in the interpretation of the kinetics of uptake and it is important to distinguish clearly between effects on the uptake of a compound and on its subsequent metabolism. The reliability of kinetic constants for transport of NH_4^+ and $CH_3NH_3^+$ is questionable since entry into cells is so rapid and is followed by metabolism. Kinetic constants for uptake may represent the affinity of the porter and/or the affinity of GS for the substrate.

Transport of nitrate

Nitrate is probably the commonest source of nitrogen for utilization by microalgae and cyanobacteria in the environment and only the ionized form is encountered at physiological or ecological pH values. Uptake of nitrate leads to an increase in the pH of the medium whereas uptake of ammonia leads to a decrease (Goldman and Brewer, 1980). On entering the cell, nitrate is rapidly metabolized making measurement of transport difficult. Various techniques have been employed to measure nitrate transport although no convenient radiotracer is available. Some have attempted to minimize the effects of metabolism, e.g. by growth on nitrite to prevent activation or induction of nitrate reductase (Eppley and Coatsworth, 1968), use of membrane fractions to separate transport from metabolism (Falkowski, 1975a,b) and use of metabolic inhibitors of nitrate reductase which minimize metabolism (Serra et al., 1978a,b). Studies on transport in the presence of inhibitors to reduce assimilation must be treated with caution especially if the internal concentrations of solutes such as nitrate are regulated at fairly low levels (<1 mM internal nitrate concentration in many microalgae, Raven, 1980) because inhibition of assimilation may then inhibit net influx (Pistorius et al., 1978).

As environmental nitrogen concentrations are often low and uptake displays saturation kinetics, it appears that an active influx at the plasmalemma is ubiquitous in those algae assimilating nitrate (for reviews see Raven, 1980; Syrett, 1981; Wheeler, 1983) and that this influx requires ATP (Ahmed and Morris, 1967, 1968; Falkowski, 1975a,b; Falkowski and Stone, 1975).

Most experiments designed to study nitrate transport have been based on measuring either nitrate depletion from the growth medium including the use of ion-specific electrodes (Tischner and Lorenzen, 1979a; Fuggi et al., 1981; Schlee et al., 1985) or $^{15}NO_3^-$ uptake, both methods measuring uptake rather than transport. Therefore, there is a need for short-term tracer experiments and $^{36}ClO_3^-$ has been used as an analogue of NO_3^- in this context. Ideally, analogues which are used as tracers to study transport should not be metabolized. Transport of $^{36}ClO_3^-$ by the diatoms Skeletonema costatum and Nitzschia closterium (Balch, 1987) appeared to be active, showed saturation kinetics and was inhibited by nitrate, providing evidence that both molecules compete for the same transport site. The affinity of cells was considerably higher for NO_3^- than for ClO_3^- in these two diatoms.

Field observations have shown that $^{15}NO_3^-$ uptake by phytoplankton is stimulated by light and saturates at high light intensities (MacIsaac and Dugdale, 1972; Bates, 1976). In contrast, laboratory experiments have shown that light has little effect on the initial accumulation of nitrate (Eppley and Rogers, 1970; Cresswell and Syrett, 1981) and chlorate (Balch, 1987). It appears that the transport of nitrate and chlorate are unaffected by light but the assimilation of nitrate is light dependent. The utilization of nitrate is dependent not only on light for the supply of reductant and ATP and for enzyme activation but also on the availability of CO_2 (Romero et al., 1985, 1987).

Nitrate uptake by microalgae is repressed in the presence of ammonia in both laboratory (Eppley and Coatsworth, 1968; Conway, 1977; Pistorius et al., 1978; Cresswell and Syrett, 1981) and field populations (Dugdale and MacIsaac, 1971; McCarthy and Eppley, 1972) and ammonia is preferentially assimilated (see Syrett, 1981). An active nitrate reductase and the nitrate transport system are not formed in the presence of ammonia and ammonia addition can lead to the rapid inhibition of nitrate utilization. The inhibition of nitrate uptake by ammonia is too rapid to be explained by the inacti-vation of nitrate reductase (Syrett, 1981) and a product of ammonia assimilation may be the inhibitor of nitrate uptake. Support for this comes from studies using MSX which irreversibly inhibits GS, there being no inhibition by ammonia of nitrate uptake in cells pretreated with MSX (Rigano *et al.*, 1979; Flores *et al.*, 1980; Cullimore and Sims, 1981a). Others have proposed that nitrate uptake may be regulated by ammonia *per se* (Florencio and Vega, 1982; Bagchi *et al.*, 1985). The effects of adding ammonia to cells metabolizing nitrate are complex and the interpretation of results is often difficult due to the regulatory effect of one nutrient on the transport and assimilation of another.

Despite numerous studies on the uptake of nitrate, very little is actually known about the transport mechanism and regulation of the transport process. In the acidophilic alga *Cyanidium caldarium* (Fuggi *et al.*, 1984) evidence has been obtained for high ($K_m < 1$ μM) and low (K_m 0.45 mM) affinity transport systems. In both *C. caldarium* (Fuggi, 1985) and *Chlorella vulgaris* (Schlee *et al.*, 1985) transport appears to be a secondary active process involving co-transport of protons.

Uptake of amino acids

Amino acids may act as sources of nitrogen and certain algae may also utilize them as sources of carbon (Neilson and Lewin, 1974). The inability of some microalgae to utilize certain amino acids may reflect a deficiency in the transport of these amino acids. Recent reviews have dealt with plant amino acid transport (Reinhold and Kaplan, 1984) and the molecular biology of bacterial amino acid transport (Antonucci and Oxender, 1986).

Algae can use a variety of amino acids as the sole nitrogen source (Neilson and Lewin, 1974; Neilson and Larsson, 1980). Dark heterotrophic growth with amino acids acting as sources of nitrogen has occasionally been demonstrated (Kempner and Miller, 1965; Lewin and Hellebust, 1976; Rigano *et al.*, 1977; Neilson and Larsson, 1980). Growth rates of microalgae and cyanobacteria obtained using amino acids as sources of nitrogen are often comparable to those obtained with ammonia or nitrate but vary greatly (Kapp *et al.*, 1975; Neilson and Larsson, 1980). Unfortunately, the inability to utilize a particular organic nitrogenous compound does not distinguish between an ineffective transport system and deficiencies in metabolism.

The ability to utilize amino acids as sources of both carbon and nitrogen has been studied using a mutant strain of the cyanobacterium *Nostoc muscorum* which was heterocystous but non-nitrogen fixing (Vaishampayan, 1982). In the presence of DCMU this strain required both a carbon and a nitrogen source. Certain amino acids were toxic, some were utilized as sources of nitrogen and/or carbon whereas others were not utilized. Further evidence that amino acids can serve as nitrogen sources has been obtained by studying growth rates and the effect of amino acids on nitrogenase activity in the cyanobacterium *Anabaena cylindrica* (Rawson, 1985). Results indicated that certain amino acids may partially satisfy the nitrogen requirement whereas others appeared to be toxic. Such studies provide little information on the transport of amino acids and few generalizations can be made on their utilization by cyanobacteria, except for the apparent toxicity of glutamate.

Amino acid transport has been extensively studied in *Chlorella vulgaris*. Hexoses and hexose analogues stimulate a hexose transport system and also two amino acid transport systems in autotrophically grown *C. vulgaris* (Cho *et al.*, 1981). Following glucose addition the rates of uptake of glycine, alanine, proline and serine (the proline system), which compete with each other for entry into cells, increased more than 100-fold while the rates of arginine and lysine transport (the arginine system) increased by a factor of 25–50. Glucose does not serve solely as a metabolic substrate since 6-deoxyglucose, a non-metabolizable glucose analogue, also caused induction of these amino acid transport systems. Induction was prevented by treatment with cycloheximide (Cho *et al.*, 1981). A similar induction of amino acid transport systems can be observed when *Chlorella* is starved of nitrogen (Sauer *et al.*, 1983) in contrast to the lack of such an effect on nitrate transport by *Chlorella* (Schlee *et al.*, 1985). During nitrogen starvation the internal sucrose concentration increases and certain free-pool amino acid concentrations decrease. These changes may be responsible for the induction of the two amino acid transport systems described above via changes in the C:N ratio. The basic amino acid transport system was very specific for arginine and lysine with apparent K_m values of 2 μM and 7 μM, respectively (Cho and Komor, 1984). Maximal accumulation ratios of 600–1200-fold for arginine have been observed with an external concentration of 1 mM. These are far beyond the electrochemical equilibration potential (Komor and Tanner, 1976). This accumulation may possibly be explained by the binding of arginine to intracellular components or sequestration in organelles (Cho and Komor, 1984). Uptake of positively charged arginine is accompanied by an immediate efflux of protons which decreases as the efflux of K^+ increases (Cho and Komor, 1984). The ratio of cation efflux (H^+ and K^+) to arginine uptake was found to be close to one in these studies although the ratio of H^+ efflux to K^+ efflux changed during the time course of arginine uptake. No evidence was found for an arginine/H^+ co-transport (Cho and Komor, 1984).

A third amino acid transport system, which is induced in the presence of glucose and relatively high concentrations of nitrogen (10 mM nitrate or ammonia), has been described for *C. vulgaris* (Sauer, 1984; Cho and Komor, 1985). This system has a rather broad specificity and transports at least ten neutral and acidic amino acids, three of which are also transported by the previously described proline system (Cho *et al.*, 1981). In addition to this general amino acid transport system other systems for transporting basic amino acids (arginine, lysine and histidine), acidic amino acids (aspartate and glutamate) and also systems for asparagine, threonine, methionine and glutamine have been reported (Cho and Komor, 1985). These results indicate that uptake of amino acids by *Chlorella* is as complex as in other free-living microorganisms such as bacteria and yeasts. As yet the intracellular signals that regulate amino acid transport in *Chlorella* and the role of glucose in induction are not known. Recent progress has been made in the preliminary characterization of amino acid porters in *Chlorella*. Mutant strains of *Chlorella*, deficient in the arginine system and the proline system, have been obtained following chemical mutagenesis and selection against amino acid analogues (Sauer and Tanner, 1985). Such strains had lost the ability not only to transport arginine and proline but also the other amino acids transported by these systems. The loss of transport activities in mutant strains correlated with the loss of certain radiolabelled protein bands from plasmalemma enriched fractions, following SDS-polyacrylamide gel electrophoresis, which were assumed to be responsible for the different transport systems (Sauer, 1986).

The energy-dependent accumulation of ^{14}C-labelled aminoisobutyrate, which is not metabolized, and ^{14}C-leucine by *Anacystis nidulans* showed saturation kinetics (Lee-Kaden and Simonis, 1982). Two uptake systems, one with high and the other with low affinity (apparent K_m values of 6 μM and 56 μM, repectively), were responsible for the transport of aminoisobutyrate whereas only one was demonstrated for the transport of leucine (apparent K_m 125 μM). Competition experiments revealed that *A. nidulans* has amino acid transport systems similar to those described in bacteria which are the DAG system for the transport of D-alanine, aminoisobutyrate and glycine and the LIV system for the transport of leucine, isoleucine and valine. Photosystem I activity was required for uptake of aminoisobutyrate which appeared to be an electrogenic process dependent on $\Delta\psi$. Transport was proposed to be by a H^+/aminoisobutyric acid co-transport system (Lee-Kaden and Simonis, 1982).

High and low affinity glutamine (apparent K_m values of 13.8 μM and 1.1 mM, repectively) and glutamate (apparent K_m values of 100 μM and 1.4 mM, repectively) transport systems have been demonstrated in the N_2-fixing cyanobacterium *Anabaena variabilis* (Chapman and Meeks, 1983). The analogue MSX appears to be transported by the same high affinity system as glutamate and glutamine. On the basis of competition experiments other amino acids transported by this system were alanine, methionine, histidine, aspartate, asparagine, citrulline and γ-methylhydrazide and this system may be similar to the general amino acid transport system of *Chlorella* (Sauer, 1984; Cho and Komor, 1985). Mutants resistant to MSX had reduced rates of MSX, glutamine and glutamate transport (Chapman and Meeks, 1983).

The presence of a high and low affinity glutamine transport system in the unicellular cyanobacterium *Synechocystis* PCC 6803 has recently been reported (Labarre *et al.*, 1987). Unlike the situation described for *Chlorella*, amino acid uptake by *Synechocystis* 6803 was not stimulated by glucose and only two major amino acid permeases have been demonstrated. These permeases have a relatively broad but non-overlapping substrate specificity. Mutant strains of *Synechocystis* 6803 resistant to toxic levels of 4-azaleucine and histidine were isolated. Strains resistant to histidine were unable to transport arginine, histidine and lysine whereas strains resistant to azaleucine were able to transport these basic amino acids but had lost the ability to transport all other amino acids except glutamate which may therefore be transported via a third permease (Labarre *et al.*, 1987).

Nitrogen assimilation

The assimilation of nitrogenous compounds (via ammonia) by microalgae and cyanobacteria proceeds mainly via the glutamine synthetase–glutamate synthase (GS–GOGAT) cycle (see Miflin and Lea, 1976; Stewart, 1980; Syrett, 1981). Evidence for this comes mainly from isotope tracer experiments, the demonstration of appropriate enzyme activities and inhibitor studies.

The pathways of ammonia assimilation

In early studies on the enzymes of primary ammonia assimilation in cyanobacteria (Haystead *et al.*, 1973; Batt and Brown, 1974) it was shown that there was an appreciable activity of GS, a low activity of alanine dehydrogenase (ADH) and barely detectable activities of glutamate dehydrogenase (GDH) in cell-free extracts of *Anabaena cylindrica*. Unequivocal evidence for the presence of GOGAT (NADPH-dependent) was not obtained in these studies but a ferredoxin-dependent GOGAT was subsequently demonstrated (Lea and Miflin, 1975).

Evidence that GS and GOGAT are involved in ammonia assimilation in cyanobacteria was obtained by the use of inhibitors such as MSX and azaserine (Stewart and Rowell, 1975; Rowell *et al.*, 1977) and by examining, in conjunction with the use of enzyme inhibitors, the kinetics of incorporation of $^{13}N_2$ (Thomas *et al.*, 1975; Wolk *et al.*, 1976) and ^{13}N-ammonia (Meeks *et al.*, 1977) by *Anabaena cylindrica*. Subsequent isotope tracer experiments have confirmed the role of GS and GOGAT in the pathway of primary ammonia assimilation in cyanobacteria, although a fraction of the ^{13}N-ammonia assimilated was apparently via GDH or ADH in certain species (Meeks *et al.*, 1978). A cyanobacterial ADH has been characterized (Rowell and Stewart, 1976), but the consensus is that this enzyme is not of major importance in primary ammonia assimilation and it is therefore not discussed further.

Isolated heterocysts have been shown to form radiolabelled glutamine from ^{13}N-ammonia, $^{13}N_2$ or ^{14}C-glutamate and, on the basis of this and enzymological data, it was concluded that heterocysts were the sites of ammonia (derived from nitrogen fixation) assimilation via GS, that glutamine was exported to vegetative cells where it was metabolized via GOGAT and that part of the glutamate thus formed was returned to the heterocysts as substrate for GS (Thomas *et al.*, 1977). However, this scheme has been questioned since substantial GOGAT activity has been demonstrated in heterocysts which may therefore, potentially at least, be capable of glutamate synthesis from dinitrogen and 2-oxoglutarate (Häger *et al.*, 1983; Bothe *et al.*, 1984a). The level of GOGAT in heterocysts remains uncertain with reports of substantial (Gupta and Carr, 1981) and very low (Thomas *et al.*, 1977; Rai *et al.*, 1982) activities in isolated heterocysts. The relatively slow transfer of radioactivity from glutamine to glutamate by *A. cylindrica* when $^{13}N_2$ is the nitrogen source compared with $^{13}NH_4^+$ or $^{13}NO_3^-$ is consistent with the transfer of glutamine from heterocysts to vegetative cells, which is unnecessary when exogenous ammonia or nitrate is available and can be assimilated entirely in vegetative cells. Further studies on the distribution of GOGAT (Gupta and Carr, 1981; Rai *et al.*, 1982; Häger *et al.*, 1983) demonstrated substantially higher activities in vegetative cells than in heterocysts. It is not known, however, whether heterocysts synthesize the glutamate required for their own metabolism (Häger *et al.*, 1983).

A quantitative investigation of the pathway of ammonia assimilation in *Chlamydomonas reinhardtii* (Cullimore and Sims, 1981b) indicated that ammonia is assimilated exclusively via the GS-GOGAT cycle and that GDH has a catabolic role.

The enzymes GS and GOGAT have been demonstrated in virtually all strains so far tested. GDH is widely distributed in microalgal strains (see Syrett, 1981; Falkowski, 1983), but not in cyanobacteria, and this enzyme apparently plays only a minor role in primary ammonia assimilation. It has also been reported that GS, carbamylphosphate synthase and glutaminase may provide a route of primary ammonia assimilation in *Anabaena* strain 1F, which apparently lacks GOGAT (Chen *et al.*, 1987).

Nitrogenase (EC 1.18.2.1)

The ability to fix dinitrogen which is widespread among cyanobacteria and has been considered in several recent reviews (Stewart *et al.*, 1979; Stewart, 1980; Bothe *et al.*, 1984b; Gallon and Chaplin, 1987) is not discussed in detail here. Aerobic nitrogen fixation is invariably a property of heterocystous cyanobacteria and most of the non-heterocystous cyanobacteria which are known to fix dinitrogen do so only under anaerobic or microaerobic conditions or have a temporal separation of photosynthetic and nitrogen fixing activities.

Nitrogenase of cyanobacteria is a FeMo-containing enzyme, similar to that of other diazotrophs, which catalyses the reaction:

$$N_2 + 8H^+ + 8e^- + 16MgATP \rightarrow 2NH_3 + H_2 + 16MgADP + 16Pi$$

The enzyme, which is extremely oxygen sensitive, is protected from irreversible inactivation by oxygen by means of spatial or temporal separation of photosynthetic oxygen evolution and nitrogen fixation.

Nitrogenase is composed of two redox proteins, both of which are essential for substrate reduction. The larger FeMo protein is an FeMoScontaining tetramer (M_r 220 000–240 000) consisting of two pairs of non-identical subunits. The smaller Fe protein is an FeS dimer of two identical subunits (each of M_r 30 000–32 000). For further information on the structure of nitrogenase, the biochemistry and physiology of nitrogen fixation and the organization of the genes for nitrogen fixation see Bothe *et al.* (1984b), Burgess (1985), Stewart and Rowell (1986); Gallon and Chaplin (1987) and Haselkorn *et al.* (1987).

Fogg (1949) demonstrated that ammonia represses the differentiation of heterocysts, the site of nitrogenase in certain filamentous cyanobacteria, and inhibits nitrogen fixation by *Anabaena* species. These findings have been confirmed by numerous studies on a variety of cyanobacteria (see for reviews Stewart, 1980; Carr and Whitton, 1982). However, the regulatory mechanisms controlling nitrogenase biosynthesis and activity in heterocystous cyanobacteria are still largely unknown. Since ammonia or its metabolites repress the formation of heterocysts, it has been difficult to establish whether the inhibition of nitrogen fixation is due to an inhibition of heterocyst differentiation or to a repression of nitrogenase biosynthesis. Furthermore, ammonia may inhibit nitrogenase activity *in vivo* by interrupting substrate supply (Ohmori and Hattori, 1974, 1978; Murry and Benemann, 1979).

Nitrate, but not ammonia, has been reported to repress the synthesis of heterocysts and nitrogenase in *Anabaena* CA (Bottomley *et al.*, 1979). Meeks *et al.* (1983), using several freshwater cyanobacteria, confirmed the variable repressive effect of nitrate and found that only in *Anabaena* PCC 7120 was heterocyst differentiation consistently repressed. In a detailed comparison of nitrate and dinitrogen assimilation in *Anabaena* PCC 7120 and *A. cylindrica* it was found that the effects of nitrate on heterocyst and nitrogenase synthesis and of ammonia on nitrate assimilation were stronger in *A. cylindrica* (Meeks *et al.*, 1983). The different effects of nitrate on nitrogen fixation did not correlate with the rates of nitrate transport, ammonia assimilation or growth.

Heterocyst differentiation becomes insensitive to repression by ammonia at an intermediate stage of development (Bradley and Carr, 1977; Murry and Benemann, 1979). Nitrogenase is only inhibited by ammonia in the early stages of heterocyst development in *A. cylindrica*. Addition of ammonia does not inhibit synthesis of the FeMo protein, after a point when heterocyst differentiation becomes irreversible (Murry *et al.*, 1983). Addition of ammonia to log-phase cultures resulted in a decrease in specific activity of nitrogenase and a loss of FeMo protein at a rate equivalent to growth. Ammonia also has a strong inhibitory effect on light-limited linear phase cultures of both *A. cylindrica* (Murry *et al.*, 1983) and an *Anabaena* sp. (Yoch and Gotto, 1982), probably due to an ammonia-effected inhibition of substrate supply (Ohmori and Hattori, 1978).

GS is involved in the regulation of heterocyst differentiation and of nitrogenase synthesis and activity in cyanobacteria. Evidence for this comes from studies using MSX. Inhibition of nitrogenase activity and heterocyst production by ammonia and ammonia analogues can be prevented in cells pretreated with MSX (Stewart and Rowell, 1975; Ownby, 1977; Kerby *et al.*, 1987a).

In addition to this long-term effect of ammonia on nitrogenase a rapid and reversible inhibition of nitrogenase, due to covalent modification by ammonia, termed 'ammonia switch-off' has been studied in a variety of photosynthetic bacteria (Jouanneau *et al.*, 1983; Kanemoto and Ludden, 1984; Hartman *et al.*, 1986). A similarly rapid inactivation by ammonia of nitrogenase in the cyanobacterium *A. variabilis* has been reported to occur at pH 10.0 but not at pH 7.2 (Reich *et al.*, 1986) and may be due to an uncoupling effect.

Nitrate reductase (EC 1.6.6.2)

Nitrate is reduced to ammonia via nitrite by the enzymes nitrate reductase and nitrite reductase (for reviews see Guerrero *et al.*, 1981; Syrett, 1981; Flores *et al.*, 1983a).

Nitrate reductase of eukaryotic algae (e.g. *Chlorella vulgaris*) (Solomonson *et al.*, 1975; Solomonson, 1979; Solomonson and Barber, 1987) catalyses the reaction:

$$NO_3^- + NAD(P)H + H^+ \rightarrow NO_2^- + NAD(P)^+ + H_2O$$

The enzyme from *Chlorella* has a M_r of 360 000 and is composed of four identical subunits (Solomonson and Barber, 1987). It is probably a cytoplasmic enzyme and contains molybdenum, haem and FAD. In contrast, nitrate reductase of cyanobacteria (Hattori and Myers, 1967; Manzano *et al.*, 1976, 1978; Losada and Guerrero, 1979; Ida and Mikami, 1983) is a particulate enzyme in cell-free extracts possibly associated with thylakoid membranes. It is smaller than the algal enzyme (a single polypeptide of M_r 75 000), does not contain haem or FAD and uses reduced ferredoxin rather than a reduced pyridine nucleotide as the electron donor. The apparent K_m value of nitrate reductase for nitrate is in the region of 110 μM. Nitrate reductase is absent from heterocysts of *Anabaena* PCC 7120 (Kumar *et al.*, 1985; Rai and Bergman, 1986).

Nitrite reductase (EC 1.7.7.1)

Nitrite reductase of algae (Eppley and Rogers, 1970; Grant 1970; Zumft, 1972) and cyanobacteria (Hattori and Myers, 1966; Hattori and Uesugi, 1968; Peschek, 1979; Mendez *et al.*, 1981) contains sirohaem and an iron–sulphur centre, uses reduced ferredoxin as the electron donor, is partially membrane associated and catalyses the reaction:

$$NO_2^- + 8H^+ + 6Fd(reduced) \rightarrow NH_4^+ + 6Fd(oxidized) + 2H_2O$$

In algae, nitrite reductase is a single polypeptide of M_r 63 000 (Zumft, 1972) and may be located in the chloroplast (Kessler and Zumft, 1973). In cyanobacteria it is also a single polypeptide of M_r 68 000 (Hattori and Uesugi, 1968) or 50 000 (Flores *et al.*, 1983a) and occurs in both heterocysts and vegetative cells (Rai and Bergman, 1986).

Regulation of the reduction of nitrate to ammonia

Active nitrate reductase is absent from ammonia-grown cells and is synthesized on transfer to a nitrogen-free medium (Morris and Syrett, 1965). Repression of nitrate reductase synthesis probably involves a product of ammonia assimilation rather than ammonia itself (see Guerrero et al., 1981; Syrett, 1981, 1987; Flores et al., 1983a). In addition, nitrate per se may function in the induction of synthesis of nitrate reductase (see Florencio and Vega, 1983a). In general, the cellular level of nitrite reductase is higher than that of nitrate reductase indicating that the latter may be rate limiting. Synthesis of active nitrate reductase on removal of ammonia does not require de novo synthesis of the whole enzyme (see Syrett, 1981). There are several mechanisms of inactivation of nitrate reductase which are either reversible or irreversible. Irreversible inactivation, involving proteolytic degradation, occurs on addition of ammonia or transfer of Chlamydomonas reinhardtii to darkness (Thacker and Syrett, 1972). A rapid inactivation of Chlorella nitrate reductase on adding ammonia to nitrate-grown cells (Losada et al., 1970) is reversed on removing ammonia in vivo and on treatment with the oxidizing agent ferricyanide in vitro. Inactivation of the enzyme requires its reduction, is prevented by nitrate and is greatest at high O_2 and low CO_2 concentrations suggesting an involvement of photorespiration (Pistorius et al., 1976; Solomonson and Spehar, 1977). In synchronous cultures of Chlorella spp. an activation of nitrate reductase following transfer to the light, which differs from the above oxidative activation, has been observed (Hodler et al., 1972; Tischner and Hütterman, 1978). The reversible inactivation of nitrate reductase in Cyanidium caldarium is prevented by MSX (Rigano et al., 1979). The light-dependent activation of nitrite reductase in Synechococcus leopoliensis (Tischner and Schmidt, 1984) may be mediated by thioredoxin.

Ohmori and Hattori (1970) reported that nitrate and nitrite reductases of Anabaena cylindrica are induced by nitrate and nitrite, respectively. Synthesis of cyanobacterial nitrate reductase is repressed by ammonia (Stevens and Van Baalen, 1974; Manzano et al., 1976) and, in addition, ammonia causes a rapid, reversible short-term inhibition of nitrate utilization by Anacystis nidulans (Flores et al., 1980, 1983a). This short-term effect of ammonia is blocked by MSX or azaserine indicating that regulation requires a product of ammonia assimilation and the target for the regulatory effect may be the nitrate uptake system (Flores et al., 1983a). In A. nidulans nitrate uptake is also dependent on a supply of CO_2 (Flores et al., 1983b; Romero et al., 1987). The availability of carbon skeletons may limit the amount of nitrate which can be assimilated which in turn may regulate the influx of nitrate. The long-term effects of ammonia on nitrate uptake involve the repression of enzyme synthesis by ammonia, and nitrate is not required as an inducer in the non-heterocystous cyanobacteria so far tested (Herrero et al., 1981, 1985). In nitrogen fixing, heterocystous cyanobacteria (Ohmori and Hattori, 1970; Herrero et al., 1981, 1985; Flores et al., 1983a) nitrate appears to be required as an inducer and again, MSX treatment abolishes the ammonia inhibition of nitrate reductase. In Anabaena PCC 7119

(Mendez *et al.*, 1981) and *A. nidulans* (Flores *et al.*, 1983a) nitrite reductase synthesis is also repressed by ammonia and the repression is prevented by MSX treatment.

Glutamate dehydrogenase (EC 1.4.1.2–4)

Glutamate dehydrogenase, which catalyses the reaction:

2-oxoglutarate + NH_3 + NAD(P)H + H^+ → glutamate + $NAD(P)^+$ + H_2O

has been purified and characterized from *Synechocystis* PCC 6803 (Florencio *et al.*, 1987) and *Phormidium laminosum* (Martinez-Bilbao *et al.*, 1987). The M_r of the enzyme from *Synechocystis* was 208 000 and it was composed of four identical subunits while the M_r of the *Phormidium* enzyme was 280 000. The enzyme from both sources was NADPH-specific. Activities in cell-free extracts of *Synechocystis* have been reported to be similar to those of GOGAT and it has been suggested that GDH may function in ammonia assimilation at high ammonia concentrations. However, synthesis of the enzyme in *Synechocystis* was not induced by growth on ammonia.

The NADPH-dependent enzyme of *Chlorella* is, in contrast, induced by ammonia (Talley *et al.*, 1972; Shatilov and Kretovich, 1977; Gronostajski *et al.*, 1978) and the NAD(P)H-enzyme is synthesized constitutively. The constitutive GDH has been reported to be either tetrameric with a M_r of 180 000 (4 × 45 000) (Meredith *et al.*, 1978) or hexameric with a M_r of 294 000 (6 × 49 000) and stable (Shatilov and Kretovich, 1977), whereas the inducible, labile, allosteric enzyme (Shatilov and Kretovich, 1977) has been reported to have a M_r of about 290 000–410 000 (5–7 × 58 000) (Gronostajski *et al.*, 1978). In contrast, the activity of the NADPH-dependent enzyme of *Euglena gracilis* is high in glutamate grown cells and negligible in ammonia grown cells (Parker *et al.*, 1985). Reported apparent K_m values for ammonia for different species vary from 4–68 mM and are substantially higher than the K_m of GS (see below).

Glutamine synthetase (EC 6.3.1.2)

Glutamine synthetase (GS) catalyses the reaction:

$$\text{L-glutamate} + \text{ATP} + NH_3 \xrightarrow{Me^{2+}} \text{L-glutamine} + \text{ADP} + \text{Pi}$$

in vivo. *In vitro* studies frequently utilize the γ-glutamyl transferase reaction which is not physiologically significant but is sensitive and convenient.

GS has been purified from several cyanobacteria (Stacey *et al.*, 1977, 1979; Sawhney and Nicholas, 1978a; Emond *et al.*, 1979; Sampaio *et al.*, 1979; Tuli and Thomas, 1980; Orr and Haselkorn, 1981; Orr *et al.*, 1981; Ip *et al.*, 1983; Florencio and Ramos, 1985). The structure of the enzyme (M_r 600 000) is similar to that of other prokaryotes, consisting of 12 identical subunits arranged in two superimposed hexagonal rings (Sampaio *et al.*, 1979; Orr *et al.*, 1981). The available evidence suggests that cyanobacteria possess only a single GS, the product of the *gln*A gene, which is transcribed from different promoters when *Anabaena* PCC 7120 is grown on dinitrogen or ammonia (Tumer *et al.*, 1983), ensuring expression of *gln*A in the presence or absence of ammonia.

Algal GS enzymes are structurally similar to those of other eukaryotes (Cullimore and Sims, 1981b; Florencio and Vega, 1983b; Beudeker and Tabita, 1985): the M_r being 380 000 and having 8 identical subunits. Recent studies have demonstrated the presence of two isoforms of GS of similar structure in *Chlamydomonas reinhardtii* (Florencio and Vega, 1983b) and *Chlorella sorokiniana* (Beudeker and Tabita, 1985). The *Chlorella* enzymes also have similar immunological and catalytic properties and it has been suggested that one of them (GS_{11}) is involved in the photorespiratory reassimilation of ammonia whereas the other may be involved in ammonia assimilation in the dark (Beudeker and Tabita, 1985). GS_{11} of *C. reinhardtii* had a higher apparent K_m for ammonia (244 μM) than GS_1 (83 μM), was present at higher intracellular levels in cells grown in the light and was absent from cells incubated in the dark when GS_1 increased (Florencio and Vega, 1983b). Reported apparent K_m values of microalgal and cyanobacterial GS for ammonia vary from less than 20 μM to 1.6 mM depending on the species and the conditions of assay. It has been suggested that a rapid, reversible deactivation of *C. reinhardtii* GS, on applying ammonia and darkness to nitrate grown cells, is important in controlling the rates of nitrate and ammonia assimilation (see Cullimore, 1981; Cullimore and Sims, 1981a). GS is also important for the reassimilation of ammonia produced internally in a photorespiratory N-cycle (Cullimore and Sims, 1980; Beudeker and Tabita, 1984, 1985). It was demonstrated that N-recycling in *C. reinhardtii*, under photorespiratory stress can be accounted for by protein turnover and is quantitatively less important than primary ammonia assimilation (Cullimore and Sims, 1980). Hipkin *et al.* (1982) suggested that in nitrogen-deficient *C. reinhardtii*, ammonia generated in a proteolytic ammonia cycle provides a supply of N for the synthesis of the necessary amino acids and proteins in the absence of an exogenous nitrogen source.

GS is a major cellular protein and constitutes about 2% of the soluble protein in cell-free extracts of cyanobacteria (Stacey *et al.*, 1977). The enzyme is subject to regulation at the levels of synthesis and activity. The specific activity is about two-fold higher in nitrogen-fixing or nitrate grown cells of cyanobacteria than in ammonia grown cells (Dharmawardene *et al.*, 1973; Rowell *et al.*, 1977, 1979; Stacey *et al.*, 1977; Emond *et al.*, 1979; Tuli and Thomas, 1980; Orr and Haselkorn, 1982; Florencio and Ramos, 1985) as a result of repression of synthesis. The enzyme is present in both heterocysts and vegetative cells (Dharmawardene *et al.*, 1973; Thomas *et al.*, 1977; Rowell *et al.*, 1985a; Stewart *et al.*, 1985). Similar variations in activity, in response to nitrogen availability, have been reported for algal GS (Rigano *et al.*, 1979, 1987; Tischner and Lorenzen, 1979b; Tischner and Hütterman, 1980; Cullimore, 1981).

The light-dependent activation of GS reported for some cyanobacteria and algae (Rowell *et al.*, 1979; Tischner and Hütterman, 1980; Cullimore,

1981; Tischner and Schmidt, 1982; Florencio and Vega, 1983b) may be mediated by the thioredoxin system (Schmidt, 1981; Ip *et al.*, 1984; Papen and Bothe, 1984; Tischner and Schmidt, 1984). The enzyme may be regulated by feedback inhibition by several end products of nitrogen metabolism with the amino acids alanine, glycine and serine and the nucleotide AMP usually being the most effective inhibitors (Dharmawardene *et al.*, 1973; Rowell *et al.*, 1977, Sawhney and Nicholas, 1978b; Tuli and Thomas, 1980; Stewart *et al.*, 1982; Florencio and Vega, 1983b; Beudeker and Tabita, 1985; Florencio and Ramos, 1985) and with inhibition being greater at higher concentrations of the substrate ammonia (Sawhney and Nicholas, 1978b; Stewart *et al.*, 1982). Glyoxylate protects GS of *Anabaena* L-31 against inhibition by these amino acids (Tuli and Thomas, 1980) which may be of significance in relation to the functioning of GS in the photorespiratory nitrogen cycle.

Divalent cation availability may be an important factor in regulating GS activity, with activity being dependent on Mg^{2+} (Sawhney and Nicholas, 1978a,b; Stacey *et al.*, 1979; Tuli and Thomas, 1980; Ip *et al.*, 1983) and Co^{2+} having a marked stimulatory effect on the Mg^{2+} dependent biosynthetic activity (Stacey *et al.*, 1979; Ip *et al.*, 1983). Light-dependent changes in the free Mg^{2+} concentrations in cyanobacterial cells have been reported and since changes in the Mg^{2+} concentrations within the physiological concentration range (Bornefeld and Weis, 1981) markedly affect GS activity, it is likely that Mg^{2+} availability is important in regulating the activity of the enzyme.

Glutamate synthase

C. reinhardtii has two distinct glutamate synthases (GOGAT), one specific for NADH and the other for ferredoxin (Cullimore and Sims, 1981c; Vega *et al.*, 1987). The M_r of the NADH-dependent enzyme (EC 1.4.1.14) is reported to be 240 000 (Cullimore and Sims, 1981c) or 368 000 (Márquez *et al.*, 1984), it is oxygen sensitive and can be partially protected by thiols. It is very susceptible to photoinhibition (Gotor *et al.*, 1987). The ferredoxin-dependent enzyme (EC 1.4.7.1) is reported to have a single subunit (M_r 165 000) (Cullimore and Sims, 1981c; Galván *et al.*, 1984) and to contain FAD, FMN, and an iron–sulphur cluster (Márquez *et al.*, 1986). Reported apparent K_m values for glutamine are in the range 190–900 μM for the ferredoxin–GOGAT and 900 μM for the NADH–GOGAT.

In contrast to GS, GOGAT is present at similar levels in ammonia and nitrate grown *Synechocystis* sp. (Florencio *et al.*, 1987). The level of NADH–GOGAT and, to a lesser extent, ferredoxin–GOGAT in *Chlamydomonas reinhardtii* increases on nitrogen starvation (Vega *et al.*, 1987). The level of NADH–GOGAT is high in the dark under CO_2-limited conditions and low under conditions where the photorespiration rate is high whereas the level of ferredoxin–GOGAT is high under CO_2-limited conditions and under optimal photosynthetic conditions. It has therefore been suggested that NADH–GOGAT is involved in ammonia assimilation in the dark and in the reassimilation of ammonia released in protein turnover, and that the ferredoxin–GOGAT is involved in the reassimilation of ammonia released in photorespiration and primary ammonia assimilation in the light (Vega *et al.*, 1987). A light-dependent ferredoxin–GOGAT activity is associated with photosynthetic membranes isolated from *C. reinhardtii* (Márquez *et al.*, 1987) indicating that it is located in the chloroplast. GOGAT of *Chlorella sorokiniana* is activated in the light (Tischner and Schmidt, 1982), activation possibly being mediated via thioredoxin.

The photoproduction of nitrogenous compounds

Cyanobacteria and microalgae have potential applications for the photoproduction of nitrogenous compounds which relies on the supply of an appropriate nitrogen source (N_2, ammonia, nitrate and urea).

Cyanobacteria as biofertilizers

Nitrogen-fixing cyanobacteria can use sunlight as the sole energy source for the fixation of carbon and nitrogen and, therefore, have potential as biofertilizers. The agronomic potential of cyanobacteria, either free-living or in symbiotic association with the water fern *Azolla*, has long been recognized (De, 1939; Singh, 1961; Stewart *et al.*, 1979; Roger and Kulasooriya, 1980; Achtnich *et al.*, 1986; Venkataraman, 1986) and the natural fertility of tropical paddy fields has been attributed to nitrogen-fixing cyanobacteria. Indeed, the cyanobacteria–*Azolla* association can fix nitrogen at rates in excess of the rhizobium–legume symbiosis (Lumpkin and Plucknett, 1982). The basic significance of Singh's (1961) ecological observations only became apparent when it was recognized that heterocysts were the sites of nitrogen fixation in heterocystous strains (Stewart *et al.*, 1969; Peterson and Wolk, 1978; Stewart, 1980) and that various non-heterocystous forms fixed nitrogen anaerobically (Stewart and Lex, 1970; Rippka and Waterbury, 1977). Many of the dominant forms of cyanobacteria in paddy soils are nitrogen-fixing species, which explains how rice has been grown for centuries in paddy soils in the absence of chemical fertilizers.

To improve crop productivity there has been a tendency to breed and select high-yielding varieties which require a large amount of nitrogen to express their full potential. However, the production and distribution of nitrogen fertilizer is expensive and in developing countries the availability of nitrogen fertilizer is a major factor limiting crop production. There has therefore been renewed interest in the role of biological nitrogen fixation as a source of nitrogen fertilizer, to reduce the dependency of agriculture on fossil fuel supply.

Inoculants of fresh (100 kg ha^{-1}) or air-dried (10 kg ha^{-1}) cyanobacteria are applied to rice fields shortly after transplanting rice seedlings to act as a biofertilizer. In India cyanobacterial inoculants are grown in ponds, dried and distributed onto rice fields where they grow and fix nitrogen (Venkataraman, 1981). Cyanobacterial (and microalgal) biomass increases during rice cultivation, often resulting in dense algal blooms. The length of time required for a cyanobacterial bloom to develop is variable and dependent on the nature of the inoculant and environmental and climatic conditions such as fertilizer application (N and P), soil chemistry, ploughing, temperature, irradiance and water availability. Since cyanobacterial nitrogen only becomes available to rice plants through mineralization, the timing of degradation of the cyanobacterial bloom is important if rice plants are to obtain the maximum benefit of the cyanobacteria as a biofertilizer. Ideally, this would be during tillering to meet the nitrogen demand for rapid rice growth. Up to 40% of the cyanobacterial nitrogen can be utilized by rice within 60 days as demonstrated by ^{15}N-tracer experiments (Mian and Stewart, 1985). At present, grain and nitrogen yields of rice treated with cyanobacteria are comparable to those obtained following application of 30 kg ha^{-1} nitrogen fertilizer (Venkataraman, 1979; Singh and Singh, 1986, 1987). Cyanobacteria are also applied together with combined nitrogen, resulting in a response equal to an additional application of 20–30 kg ha^{-1} nitrogen fertilizer (Singh and Singh, 1986). However, cyanobacteria and microalgae will compete with rice for the available nitrogen, and nitrogenase activity may be depressed by the presence of combined nitrogen. Few attempts have been made to select specific strains for use in different localities or to use mutant strains with derepressed nitrogenase.

Interest is also being shown in the mass culture of nitrogen-fixing cyanobacteria for use as a slow release biofertilizer either in combination with or instead of chemical fertilizers for agricultural applications (Tucker, 1985; Karuna-Karan, 1987).

This section will consider some of our recent work on the potential applications of cyanobacteria with particular reference to the selection of mutant strains which liberate nitrogenous compounds and/or are derepressed with respect to nitrogen fixation.

Induced ammonia liberation

Nitrogen-fixing cyanobacteria convert dinitrogen to ammonia which is assimilated by the primary ammonia assimilating enzyme GS. Such nitrogen-fixing cyanobacteria normally use the nitrogen which they fix for further growth. Ammonia liberation by nitrogen-fixing cyanobacteria is not a normal physiological function, other than in some symbiotic associations (Stewart and Rowell, 1977; Stewart et al., 1983; Rowell et al., 1985b) but can be achieved either by the addition of an enzyme inhibitor, or by the selection of ammonia liberating strains. Stewart and Rowell (1975) demonstrated that over 90% of the nitrogen fixed was released extracellularly as ammonia when GS was inhibited by the glutamate analogue MSX. Additionally, MSX alleviates the inhibitory effects of exogenous ammonia and ammonia analogues on nitrogenase activity and synthesis (Stewart and Rowell, 1975; Ownby, 1977; Kerby et al., 1987a). Prolonged exposure to MSX inhibits growth and results in nitrogen starvation. Other inhibitors of

GS include 5-hydroxylysine (Ladha *et al.*, 1978) and phosphinothricin (2-amino-4-(methylphosphinyl)-butanoic acid) (Lea *et al.*, 1984) which have been shown to induce ammonia liberation by cyanobacteria. Using MSX, Musgrave *et al.* (1982) demonstrated the technical feasibility of continuous ammonia photoproduction by the nitrogen-fixing cyanobacterium *Anabaena* ATCC 27893 immobilized in Ca-alginate gel beads, and sustained ammonia liberation has been achieved using a variety of continuous flow reactors of different configurations including: packed beds, fluidized beds, parallel plates and air-lift reactors (Musgrave *et al.*, 1982, 1983a,b; Kerby *et al.*, 1983). Air-lift reactors proved the most suitable for the laboratory scale optimization of ammonia photoproduction (Musgrave, 1985) due to their homogeneous nature, low shear forces and efficient gas exchange. MSX was applied either continuously or intermittently to bioreactors and specific rates of production varied from 4 to 40 μmol of ammonia mg chla^{-1} h^{-1}, depending on the conditions employed, with durations of over 800 h (Kerby *et al.*, 1983, 1986b; Musgrave *et al.*, 1983b; Musgrave, 1985). These rates compare favourably with those obtained for the photoproduction of ammonia from nitrate by free-living *Anacystis nidulans* (Ramos *et al.*, 1982a,b), free-living nitrogen-fixing *Anabaena* ATCC 33047 (Ramos *et al.*, 1984) and free-living *Anabaena azollae* (Zimmerman and Boussiba, 1987). Important factors which determine the duration of ammonia production and the productivity of the immobilized system include: (1) biomass loading of the Ca-alginate beads; (2) MSX concentration, the optimum dose being that which allows the biomass concentration to remain constant; (3) pulses of MSX, which rather than a continuous dose prolong the duration; (4) the rate of MSX supply; (5) nitrogen limiting conditions, which increase hetero-cyst frequency and induce high nitrogenase activity; (6) dilution rate, which affects volumetric productivity and duration; (7) pH; (8) temperature; and (9) irradiance (Musgrave, 1985). Similar conclusions have been reached using free-living cultures of *Anabaena* ATCC 33047 (Ramos and Madueno, 1986; Ramos *et al.*, 1987).

Other groups have recently been employing MSX to promote ammonia production by immobilized cyanobacteria (Hall *et al.*, 1985; Brouers and Hall, 1986; Jeanfils and Loudeche, 1986; Vincenzini *et al.*, 1986) with similar results to those obtained by this laboratory. Immobilization techniques have included entrapment in Ca-alginate gels, growth into polyurethane and polyvinyl foams and enclosure of cells in dialysis tubing.

Ammonia produced following MSX treatment of cyanobacteria and microalgae is probably derived from newly fixed nitrogen (Stewart and Rowell, 1975; Spiller *et al.*, 1986) or from the assimilation of nitrate, but may also result from an inability to re-assimilate ammonia produced by the photorespiratory nitrogen cycle, protein turnover and catabolism of exogen-ous amino acids (Cullimore and Sims, 1980; Flores *et al.*, 1982; Hipkin *et al.*, 1982; Larsson *et al.*, 1982; Bergman, 1984).

In the absence of mutant strains of cyanobacteria which liberate ammonia, treatment of cells with GS inhibitors (usually MSX) seems to be the preferred procedure for inducing the excretion of ammonia. However, the use of MSX has three main disadvantages. First, it is difficult to sustain prolonged ammonia production in its presence in bioreactors; second, since the analogue is toxic it would have to be separated from the extracellularly produced ammonia prior to the use of the latter as a biofertilizer; third, it is a very expensive chemical relative to ammonia. The alternative approaches are the selection of mutant strains with GS activity reduced to a level resulting in ammonia release but with sufficient GS activity to sustain metabolism, or the isolation of naturally occurring strains which normally liberate ammonia.

Uninduced ammonia liberation

Cyanobacteria couple the production and assimilation of ammonia to cell growth and, therefore, ammonia is not normally detected in the growth medium unless assimilation is inhibited by compounds such as MSX or the strain is deficient in ammonia assimilation. Ammonia liberation by a paddy-field strain of *Anabaena* in the absence of metabolic inhibitors has been reported (Subramanian and Shanmugasundaram, 1986) although the rates of ammonia production were very low ($0.2 \ \mu$mol mg chla^{-1} h^{-1}) when compared to rates obtained with MSX. Ammonia liberation can also occur, in the absence of MSX, following the immobilization of *Anabaena azollae* in polyurethane and polyvinyl foams (Shi *et al.*, 1987). Again the rates of ammonia production were very low (up to $0.1 \ \mu$mol mg chla^{-1} h^{-1}). Interestingly, this method of immobilization stimulated heterocyst frequency (up to 16%) to a level similar to that found in the *Anabaena–Azolla* symbiosis although nitrogenase activities, as measured by acetylene reduction, were very low ($3 \ \mu$mol mg chla^{-1} h^{-1}). Addition of MSX stimulated ammonia release from immobilized cells grown in the absence of combined nitrogen. Following 35 days of immobilization and MSX treatment the reported nitrogenase activities and rates of ammonia production were low.

An alternative approach that we and others have adopted is the selection of mutant strains of nitrogen-fixing cyanobacteria which have a reduced ammonia assimilating capacity resulting in ammonia liberation. Such strains have been selected for resistance to ethylenediamine (EDA) following chemical mutagenesis (Polukhina *et al.*, 1982; Kerby *et al.*, 1986b). EDA is toxic to cyanobacteria and is metabolized via GS to a glutamine analogue (aminoethylglutamine) which accumulates (Kerby *et al.*, 1985). Additionally, EDA causes a rapid inhibition of nitrogenase at pH 9.0 similar to that caused by ammonia but, unlike ammonia, it does not inhibit nitrogenase at pH 7.0 (Kerby *et al.*, 1987a). EDA is transported by passive diffusion in response to a pH gradient by *A. variabilis*. The transport of dimethylamine by *Phaeodactylum tricornutum* (Wright and Syrett, 1983) and ethylamine and dimethylamine by *Cyclotella cryptica* (Wheeler and Hellebust, 1981) are also by passive diffusion and not by the $CH_3NH_3^+/NH_4^+$ transport system demonstrated in these species (see above). Thus, when EDA-resistant mutant strains are obtained at high pH values they are not unwanted transport mutants but generally prove to be strains with reduced GS activity. This is in contrast to strains which have been selected for resistance to methyl-amine which may be deficient in the $CH_3NH_3^+/NH_4^+$ transport system (Reglinski *et al.*, 1989).

Mutant strains ED81 and ED92, which are resistant to normally lethal concentrations of EDA, have reduced growth rates when grown under nitrogen-fixing conditions or in the presence of ammonia but show enhanced growth rates in the presence of glutamine (Sakhurieva *et al.*, 1982; Kerby *et al.*, 1986b). Nitrogenase activity was higher in mutant strains ED81 and ED92 than in the parent strain and was derepressed with respect to ammonia. The GS activity of ED92 was much reduced as measured by both the biosynthetic and transferase activities. However, the transferase activity of ED81 was comparable to the parent strain although the biosynthetic activity was much reduced. The GS protein of strain ED92 was reduced to the same extent as the GS activity, that is, approximately 25% of that of the parent strain, whereas ED81 had a similar content to the parent strain. SDS-polyacrylamide gel electrophoresis and immunoblotting confirmed the reduction in GS protein in strain ED92 and showed that there was no marked alteration in the size of the GS subunit in either mutant strain as compared to the wild type. Northern blot analysis of GS mRNA showed that there was a similar reduction in the amount of GS mRNA in mutant strain ED92 as compared to the parent strain and ED81 (Hien *et al.*, 1988). A similar reduction in GS mRNA has been demonstrated in the symbiotic cyanobacterium *Anabaena azollae* and relates to the liberation of ammonia by this cyanobiont when associated with the water fern *Azolla* (Nierzwicki-Bauer and Haselkorn, 1986). This infers that ED92 may be a regulatory mutant with less GS mRNA and consequently less GS protein whereas ED81 may be a GS structural mutant. Photoproduction of ammonia occurred at rates comparable to those obtained using MSX and was sustained for periods in excess of 600 h.

Another approach for the selection of mutants capable of liberating ammonia is selection for resistance to MSX (Singh *et al.*, 1983; Spiller *et al.*, 1986). MSX-resistant strains which liberate ammonia, like those resistant to EDA, have derepressed nitrogenase (Spiller *et al.*, 1986). An MSX-resistant strain of *A. variabilis* (strain SA1) was also found to have low levels of GS biosynthetic activity and the ratio of biosynthetic activity to transferase activity was reduced as compared to the parent strain, as was the case for ED81. The GS activity of strain SA1 was not inhibited by MSX and the kinetic properties of the parent and mutant strains were similar. Strain SA1 failed to grow in a medium containing glutamate or aspartate as the sole nitrogen source and the uptake of these amino acids was much reduced. These results infer that resistance to MSX in this strain affected not only GS activity but also the transport of glutamate and aspartate which apparently share the same transport system as MSX (Chapman and Meeks, 1983). Therefore, resistance to MSX can result in transport mutants, strains with an MSX-resistant GS and/or strains deficient in GS biosynthetic activity. It has been our experience that selecting strains resistant to MSX does not usually result in strains capable of photoproducing ammonia and this may be due to the selection of transport mutants and/or strains whose GS activity is unaffected by MSX. Strain SA1 enhanced the growth of rice plants grown in nitrogen-free medium, and both the dry weight and nitrogen content of rice plants were increased by this strain but not the parent strain (Latorre et al., 1986).

Additionally, it is important to develop strains with elevated levels of nitrogen fixation including those with derepressed nitrogenase which continue to fix dinitrogen in the presence of combined nitrogen, as do certain strains resistant to EDA and MSX. We have selected strains of A. variabilis resistant to methylamine which show impaired NH_4^+ transport, and nitrogenase activity in these strains was less sensitive to exogenous ammonia than that of the parent strain (Reglinski et al., 1989). The development of nitrogen-fixing strains for use in particular ecosystems is discussed by Stewart et al., (1979, 1987).

Photoproduction of amino acids

Cyanobacteria liberate small quantities of amino acids, polypeptides and proteins into their medium (Watanabe, 1951; Fogg, 1952, 1966, 1971; Venkataraman and Saxena, 1963; Fogg et al., 1973). The liberation of nitrogenous compounds by cyanobacteria occurs irrespective of whether they are grown in the presence of dinitrogen or combined nitrogen. The amount of extracellular nitrogen produced is dependent on the stage of growth and is highest during the lag and stationary phases (Fogg, 1952; Stewart, 1963; Pattnaik, 1966). Many of these studies have identified classes of compounds liberated but their precise identity is unknown. Fogg (1952) reported the occurrence of traces of free amino acids in culture filtrates of Anabaena cylindrica but concluded that these were not of quantitative significance. The extracellular nitrogen released was found to consist mainly of polypeptides with a high content of threonine and serine (Walsby, 1965). More recently, the production of extracellular amino acids by Anabaena siamensis, isolated from rice paddies, has been quantified and individual amino acids identified (Antarikanonda, 1984). The most abundant amino acids liberated were phenylalanine, threonine, glutamate and glycine, irrespective of the nitrogen source on which the cells were grown. The total amount of amino acids liberated was very small (17 μmol 100 g protein $^{-1}$) at late logarithmic phase of growth.

Microalgae also liberate amino acids into their growth media (Hellebust, 1974). The release of amino acids has been monitored in the diatom Chaetoceros debile during different growth phases (Poulet and Martin-Jézéquel, 1983) and the major amino acids liberated were aspartate, histidine, serine, threonine, alanine, phenylalanine together with ammonia, leucine and ornithine. The amounts and relative composition of amino acids varied during different growth phases. The total amino acids liberated varied from 0.01 to 1.0 μM and reached a maximum during the transition from exponential growth phase to stationary phase. Productivities at this stage were

(Jensen and Hall, 1982). We have isolated two forms of DAHP synthase by ion-exchange chromatography from *A. variabilis* (Niven *et al.*, 1988a). One form was sensitive to inhibition by tyrosine and the other was sensitive to inhibition by phenylalanine. Only the tyrosine-sensitive form was detected in mutant strains FT-2 and FT-6 and both forms were detected in mutant strain FT-7 but the phenylalanine form was deregulated with respect to phenylalanine inhibition. Hall and Jensen (1981) have shown that mutant strains of *Anabaena* ATCC 29151 resistant to 4-fluorophenylalanine had deregulated DAHP synthase and mutant strains of *Anacystis nidulans* resistant to 4-fluorophenylalanine and 2-amino-3-phenylbutanoic acid had DAHP synthase activity which was deregulated with respect to tyrosine (Phares and Chapman, 1975).

The uptake and incorporation of ^{14}C-labelled tryptophan by *A. variabilis* and the alanine liberating strain FT-7 were compared (our unpublished results). Virtually all the tryptophan taken up by the parent strain was rapidly incorporated into protein, indicating a small free pool with rapid turnover. The rates of uptake in FT-7 and the amount of ^{14}C incorporated into protein were reduced and it was shown that tryptophan was metabolized to glutamate, alanine and three unidentified non-protein amino acids. These findings may explain, in part at least, the liberation of a broad range of amino acids, including alanine, by this and other 6-FT resistant strains.

Growth of mutant strain FT-9, which liberated predominantly alanine, in the presence of the detergent MYRJ 45 resulted in the liberation of tryptophan (Niven *et al.*, 1988b). This detergent probably enhances tryptophan liberation by causing changes in the composition of the plasmalemma which may result in the uncoupling of specific amino acid transport systems as has been shown for glutamate production by *Corynebacterium glutamicum* (Clément *et al.*, 1984; Clément and Lanéelle, 1986). This selective release of tryptophan overcomes the feedback inhibition of anthranilate synthase which in FT-resistant strains is still subject to inhibition by tryptophan (see Niven *et al.*, 1988a).

Low molecular weight metabolites, including amino acids may be released from free-living and immobilized cyanobacteria following osmotic shock (Reed *et al.*, 1986). Transfer of *Synechocystis* PCC 6714 and *Synechococcus* PCC 6311 from a medium of high salt to one of low salt resulted in a transitory loss of plasmalemma integrity and a loss of organic compounds normally retained intracellularly. This method could be repeated and after three cycles there was no evidence of long-term damage to the cells. Since this method does not involve harvesting or death of cells it is a useful means of recovering metabolites from viable microorganisms. The production of proline by *Chlorella* sp. 580 which accumulates proline as a compatible solute in response to osmotic stress has been studied (Leavitt, 1986). *Nannochloris bacillaris*, which also produces proline in response to osmotic shock, and mutant strains resistant to azetidine-2-carboxylic acid produce higher levels of proline than the parent strain when subjected to a range of salinities (Vanlerberghe and Brown, 1987).

Conclusions

Cyanobacteria and microalgae have long been recognized as sources of food, chemicals and biofertilizers but very few commercial ventures have exploited them. This may be due to one of several factors, the most important being economic methods of mass cultivation, identification of suitable high value products and selection of suitable strains.

The effective isolation or selection of strains for commercial application is dependent on a knowledge of transport and metabolism. Characterization of such strains can increase our basic knowledge thus permitting further strain improvement. Advances are being made in both fundamental and applied aspects by use of mutant strains of cyanobacteria and microalgae. Recent advances in molecular genetics, particularly of cyanobacteria (see Porter, 1987; Stewart *et al.*, 1987), can be expected to have a major impact in this field.

Acknowledgements

This work was supported by the Agricultural and Food Research Council and we would like to thank Gordon Niven, Hilary Powell and Gail Alexander for their helpful discussions and comments.

References

Achtnich W, Moawad A M, Johal C S 1986 *Azolla*, a biofertilizer for rice. *Int. J. Trop. Agric.* **4**: 188–211

Ahmed J, Morris I 1967 Inhibition of nitrate and nitrite reduction by 2,4-dinitrophenol in *Ankistrodesmus*. *Arch. Mikrobiol.* **56**: 219–24

Ahmed J, Morris I 1968 The effects of 2,4-dinitrophenol and other uncoupling agents on the assimilation of nitrate and nitrite by *Chlorella*. *Biochim. Biophys. Acta* **162**: 32–8

Antarikanonda P 1984 Production of extracellular free amino acids by cyanobacterium *Anabaena siamensis* Antarikanonda. *Curr. Microbiol.* **11**: 191–6

Antonucci T K, Oxender D L 1986 The molecular biology of amino-acid transport in bacteria. In Rose A H, Tempest D W (eds) *Advances in Microbial Physiology*, vol 28, Academic Press, London pp 154–80

Bagchi S N, Rai U N, Rai A N, Singh H N 1985 Nitrate metabolism in the cyanobacterium *Anabaena cycadae*: Regulation of nitrate uptake and reductase by ammonia. *Physiol. Plant.* **63**: 322–6

Balch W M 1985 Lack of an effect of light on methylamine uptake by phytoplankton. *Limnol. Oceanogr.* **30**: 665–74

Balch W M 1987 Studies of nitrate transport by marine phytoplankton using $^{36}ClO_3^-$ as a transport analogue. I. Physiological findings. *J. Phycol.* **23**: 107–18

Barnes E M 1980 Proton-coupled calcium transport by intact cells of *Azotobacter vinelandii*. *J. Bact.* **143**: 1086–9

Barnes E M, Zimniak P, Jayakumar A 1983 Role of glutamine synthetase in the uptake and metabolism of methylammonium by *Azotobacter vinelandii*. *J. Bact.* **156**: 752–7

Bates S S 1976 Effect of light and ammonium on nitrate uptake by two species of estuarine phytoplankton. *Limnol. Oceanogr.* **21**: 212–8

Batt T, Brown D H 1974 The influence of inorganic nitrogen supply on amination and related reactions in the blue–green alga *Anabaena cylindrica* Lemm. *Planta* **116**: 27–37

Bergman B 1984 Photorespiratory ammonium release by the cyanobacterium *Anabaena cylindrica* in the presence of methionine sulfoximine. *Arch. Microbiol.* **137**: 21–5

Beudeker R F, Tabita F R 1984 Glycolate metabolism is under nitrogen control in *Chlorella*. *Pl. Physiol.* **75**: 516–20

Beudeker R F, Tabita F R 1985 Characterization of glutamine synthetase isoforms from *Chlorella*. *Pl. Physiol.* **77**: 791–4

Bloom F R, Kretschmer P J 1983 Effects of genetic engineering of microorganisms on the future production of amino acids from a variety of sources. In Wise D L (ed) *Organic Chemicals from Biomass*, Benjamin/Cummings Publishing Company, pp 145–71

Bornefeld T, Weis U 1981 Adenylate energy charge and phosphorylation potential in the blue–green bacterium *Anacystis nidulans*. *Biochem. Physiol. Pf.* **176**: 71–82

Borowitzka M A, Borowitzka L J (eds) 1988 *Micro-algal Biotechnology*, Cambridge University Press, Cambridge

Bothe H, Nelles H, Hager K-P, Papen H, Neuer P 1984b Physiology and biochemistry of N₂-fixation by cyanobacteria. In Veeger C, Newton W E (eds) *Advances in Nitrogen Fixation Research*, Martinus Nijhoff/Junk Publishers, The Hague pp 199–210

Bothe H, Nelles H, Kentemich T, Papen H, Neuer G 1984a Recent aspects of heterocyst biochemistry and differentiation. In Wiessner W, Robinson D, Starr R C (eds) *Compartments in Algal Cells and their Interaction*, Springer-Verlag, Berlin pp 218–32

Bottomley P J, Grillo J F, Van Baalen C, Tabita F R 1979 Synthesis of nitrogenase and heterocysts by *Anabaena* CA in the presence of high levels of ammonia. *J. Bact.* **140**: 938–43

Boussiba S, Gibson J 1985 The role of glutamine synthetase activity in ammonium transport in *Anacystis nidulans* R-2. *FEBS Lett.* **180**: 13–16

Boussiba S, Resch C M, Gibson J 1984a Ammonia uptake and retention in some cyanobacteria. *Arch. Microbiol.* **138**: 287–92

Boussiba S, Dilling W, Gibson J 1984b Methylammonium transport in *Anacystis nidulans* R-2. *J. Bact.* **160**: 204–10

Bradley S, Carr N G 1977 Heterocyst development in *Anabaena cylindrica*: the necessity of light as an initial trigger and sequential stages of commitment. *J. Gen. Microbiol.* **101**: 291–7

Brouers M, Hall D O 1986 Ammonium and hydrogen production by immobilized cyanobacteria. *J. Biotechnol.* **3**: 307–21

Burgess B K 1985 Nitrogenase mechanism – an overview. In Evans H J, Bottomley P J, Newton W E (eds) *Nitrogen Fixation Research Progress*, Martinus Nijhoff Publishers, Dordrecht, pp 543–50

Carr N G, Whitton B A (eds) 1982 *The Biology of Cyanobacteria*, Blackwell Scientific Publishers, Oxford

Castorph H, Kleiner D 1984 Some properties of a *Klebsiella pneumoniae* ammonium transport negative mutant (Amt⁻). *Arch. Microbiol.* **139**: 245–7

Chapman J S, Meeks J C 1983 Glutamine and glutamate transport by *Anabaena variabilis*. *J. Bact.* **156**: 122–9

Chen C, Van Baalen C, Tabita F R 1987 DL-7-azatryptophan and citrulline metabolism in the cyanobacterium *Anabaena* sp. strain 1F. *J. Bact.* **169**: 1114–19

Cho B-H, Komor E 1984 Mechanism of arginine transport in *Chlorella*. *Planta* **162**: 23–9

Cho B-H, Komor E, 1985 The amino acid transport systems of the autotrophically grown green alga *Chlorella*. *Biochim. Biophys. Acta* **821**: 384–92

Cho B-H, Sauer N, Komor E, Tanner W 1981 Glucose induces two amino acid transport systems in *Chlorella. Proc. Nat. Acad. Sci. USA* **78**: 3591–4

Clément Y, Escoffier B, Trombe M C, Lanéelle G 1984 Is glutamate excreted by its uptake system in *Corynebacterium glutamicum*? A working hypothesis. *J. Gen. Microbiol.* **132**: 925–9

Clément Y, Lanéelle G 1986 Glutamate excretion mechanism in *Corynebacterium glutamicum*: triggering by biotin starvation or by surfactant addition. *J. Gen. Microbiol.* **132**: 925–29

Conway H L 1977 Interactions of inorganic nitrogen in the uptake and assimilation by marine phytoplankton. *Mar. Biol.* **39**: 221–32

Cresswell R C, Syrett P J 1981 Uptake of nitrate by the diatom *Phaeodactylum tricornutum. J. Exp. Bot.* **32**: 19–25

Cresswell R C, Syrett P J 1984 Effects of methylammonium and L-methionine-DL-sulfoximine on the growth and nitrogen metabolism of *Phaeodactylum tricornutum. Arch. Microbiol.* **139**: 67–71

Cullimore J V 1981 Glutamine synthetase of *Chlamydomonas*: rapid reversible deactivation. *Planta* **152**: 587–91

Cullimore J V, Sims A P 1980 An association between photorespiration and protein catabolism: studies with *Chlamydomonas. Planta* **150**: 587–92

Cullimore J V, Sims A P 1981a Glutamine synthetase of *Chlamydomonas*: its role in the control of nitrate assimilation. *Planta* **153**: 18–24

Cullimore J V, Sims A P 1981b Pathway of ammonia assimilation in illuminated and darkened *Chlamydomonas reinhardii. Phytochemistry* **20**: 933–40

Cullimore J V, Sims A P 1981c Occurrence of two forms of glutamate synthase in *Chlamydomonas reinhardii. Phytochemistry* **20**: 597–600

De P K 1939 The role of blue–green algae in nitrogen fixation in rice fields. *Proc. R. Soc. Lond. Series* B **127**: 121–39

Dharmawardene M W N, Haystead A, Stewart W D P 1973 Glutamine synthetase of the nitrogen-fixing alga *Anabaena cylindrica. Arch. Mikrobiol.* **90**: 281–95

Dugdale R C, MacIsaac J J 1971 A computation model for the uptake of nitrate in the Peru upwelling region. *Invest. Pescq.* **35**: 299–308

Emond D, Rondeau N, Cedergren R J 1979 Distinctive properties of glutamine synthetase from the cyanobacterium *Anacystis nidulans. Can. J. Biochem.* **57**: 843–51

Eppley R W, Coatsworth J L 1968 Uptake of nitrate and nitrite by *Ditylum brightwellii* – Kinetics and mechanisms. *J. Phycol.* **4**: 151–6

Eppley R W, Renger E H 1974 Nitrogen assimilation of an oceanic diatom in nitrogen-limited continuous culture. *J. Phycol.* **10**: 15–23

Eppley R W, Rogers J N 1970 Inorganic nitrogen assimilation of *Ditylum brightwellii*, a marine plankton diatom. *J. Phycol.* **6**: 344–51.

Eppley R W, Rogers J N, McCarthy J J 1969 Half-saturation constants for uptake of nitrate and ammonium by marine phytoplankton. *Limnol. Oceanogr.* **14**: 912–20

Falkowski P G 1975a Nitrate uptake in marine phytoplankton: comparison of half saturation constants from seven species. *Limnol. Oceanogr.* **20**: 412–7

Falkowski P G 1975b Nitrate uptake in marine phytoplankton: (nitrate: chloride)-activated adenosine triphosphatase from *Skeletonema costatum* (Bacillariophyceae). *J. Phycol.* **11**: 323–6

Falkowski P G 1983 Enzymology of nitrogen assimilation. In Carpenter E J, Capone D G (eds) *Nitrogen in the Marine Environment*, Academic Press, New York, pp 839–68

Falkowski P G, Stone D P 1975 Nitrate uptake in marine phytoplankton: energy sources and the interaction with carbon fixation. *Mar. Biol.* **32**: 77–84

Florencio F J, Marqués S, Candau P 1987 Identification and characterization of a glutamate dehydrogenase in the unicellular cyanobacterium *Synechocystis* PCC 6803. *FEBS Lett.* **223**: 37–41

Florencio F J, Ramos J L 1985 Purification and characterization of glutamine synthetase from the unicellular cyanobacterium *Anacystis nidulans*. *Biochim. Biophys. Acta* **838**: 39–48

Florencio F J, Vega J M 1982 Regulation of the assimilation of nitrate in *Chlamydomonas reinhardii*. *Phytochemistry* **21**: 1195–200

Florencio F J, Vega J M 1983a Regulation of the synthesis of the NAD(P)H-nitrate reductase complex in *Chlamydomonas reinhardii*. *Z. Pflanzenphysiol.* **111**: 223–32

Florencio F J, Vega J M 1983b Separation, purification and characterization of two isoforms of glutamine synthetase from *Chlamydomonas reinhardtii*. *Z. Naturfor.* **38c**: 531–8

Flores E, Guerrero M G, Losada, M 1980 Short-term ammonium inhibition of nitrate utilization by *Anacystis nidulans* and other cyanobacteria. *Arch. Microbiol.* **128**: 137–44

Flores E, Herrero A, Guerrero M G 1982 Production of ammonium dependent on basic L-amino acids by *Anacystis nidulans*. *Arch. Microbiol.* **131**: 91–4

Flores E, Ramos J L, Herrero A, Guerrero M G 1983a Nitrate assimilation by cyanobacteria. In Papageorgiou G C, Packer L, (eds) *Photosynthetic Prokaryotes. Cell Differentiation and Function*, Elsevier Biomedical, New York, pp 363–88

Flores E, Romero J, Guerrero M G, Losada M 1983b Regulatory interaction of photosynthetic nitrate utilization and carbon dioxide fixation in the cyanobacterium *Anacystis nidulans*. *Biochim. Biophys. Acta* **725**: 529–32

Fogg G E 1949 Growth and heterocyst production in *Anabaena cylindrica* Lemm. II. In relation to carbon and nitrogen metabolism. *Ann. Bot.* **13**: 241–59

Fogg G E 1952 The production of extracellular nitrogenous substances by a blue–green alga. *Proc. R. Soc. Lond. Series B* **139**: 372–97

Fogg G E 1966 The extracellular products of algae. *Oceanogr. Mar. Biol. A. Rev.* **4**: 195–212

Fogg G E 1971 Extracellular products of algae in freshwater. *Arch. Hydrobiol.* **5**: 1–25

Fogg G E, Stewart W D P, Fay P, Walsby A E 1973 *The Blue–Green Algae*, Academic Press, London

Franco A R, Cardénas J, Fernández E 1984 Ammonium (methylammonium) is the co-repressor of nitrate reductase in *Chlamydomonas reinhardtii*. *FEBS Lett.* **176**: 453–6

Franco A R, Cardénas J, Fernández E 1987 A mutant of *Chlamydomonas reinhardtii* altered in the transport of ammonium and methylammonium. *Mol. Gen. Genet.* **206**: 414–8

Fuggi A 1985 Mechanism of proton-linked nitrate uptake in *Cyanidium caldarium*, an acidophilic non-vacuolated alga. *Biochim. Biophys. Acta* **815**: 392–8

Fuggi A, Di Martino Rigano V, Vona V, Rigano C 1981 Nitrate and ammonium assimilation in algal cell-suspensions and related pH variations in the external medium, monitored by electrodes. *Pl. Sci. Lett.* **23**: 129–38

Fuggi A, Vona V, Di Martino Rigano V, Di Martino C, Martello A, Rigano C 1984 Evidence for two transport systems for nitrate in the acidophilic thermophilic alga *Cyanidium caldarium*. *Arch. Microbiol.* **137**: 281–5

Gallon J R, Chaplin A E 1987 *An Introduction to Nitrogen Fixation*, Cassell Educational Ltd, London

Galván F, Márquez A J, Vega J M 1984 Purification and molecular properties of ferredoxin-glutamate synthase from *Chlamydomonas reinhardii*. *Planta* **162**: 180–7

Goldman J C, Brewer P G 1980 Effect of nitrogen source and growth rate on phytoplankton-mediated changes in alkalinity. *Limnol. Oceanogr.* **25**: 352–7

Gotor C, Márquez A J, Vega J M 1987 Studies on the *in vitro* O_2-dependent inactivation of NADH-glutamate synthase from *Chlamydomonas reinhardii* stimulated by flavins. *Photochem. Photobiol.* **46**: 353–8

Grant B R 1970 Nitrite reductase in *Dunaliella tertiolecta*: isolation and properties. *Pl. Cell Physiol.* **11**: 55–64

Gronostajski R M, Yeung A T, Schmidt R R 1978 Purification and properties of the inducible nicotinamide adenine nucleotide phosphate-specific glutamate dehydrogenase from *Chlorella sorokiniana*. *J. Bact.* **134**: 621–8

Guerrero M G, Vega J M, Losada M 1981 The assimilatory nitrate reducing system and its regulation. *A. Rev. Pl. Physiol.* **32**: 169–204

Gupta M, Carr N G 1981 Detection of glutamate synthase in heterocysts of *Anabaena* sp. strain 7120. *J. Bact.* **148**: 980–2

Hager K-P, Danneberg G, Bothe H 1983 The glutamate synthase in heterocysts of *Nostoc muscorum*. *FEMS Microbiol. Lett.* **17**: 179–83

Hall D O, Affolter D A, Brouers M, Shi D J, Wang L W, Rao K K 1985 Photobiological production of fuels and chemicals by immobilized algae. *A. Proc. Phytochem. Soc. Eur.* **26**: 161–85

Hall G, Flick M B, Jensen R A 1980 Approach to recognition of regulatory mutants of cyanobacteria. *J. Bact.* **143**: 981–8

Hall G C, Jensen R A 1981 Regulatory isozymes of 3-deoxy-D-arabinoheptulosonate 7-phosphate synthase in the cyanobacterium *Anabaena* sp. strain ATCC 29151. *J. Bact.* **148**: 361–4

Hartman A, Fu H, Burris R H 1986 Regulation of nitrogenase activity by ammonium chloride in *Azospirillum* spp. *J. Bact.* **165**: 864–70

Haselkorn R, Golden J W, Lammers P J, Mulligan M E 1987 Rearrangement of *nif* genes during cyanobacterial heterocyst differentiation. *Phil. Trans. R. Soc. Lond. Series B* **317**: 173–81

Hattori A, Myers J 1966 Reduction of nitrate and nitrite by subcellular preparations of *Anabaena cylindrica*. 1. Reduction of nitrate to ammonia. *Pl. Physiol.* **41**: 1031–6

Hattori A, Myers J 1967 Reduction of nitrate and nitrite by subcellular preparations of *Anabaena cylindrica* 2. Reduction of nitrate to nitrite. *Pl. Cell Physiol.* **8**: 327–37

Hattori A, Uesugi I 1968 Purification and properties of nitrite reductase from the blue–green alga *Anabaena cylindrica*. *Pl. Cell Physiol.* **9**: 689–99

Haystead A, Dharmawardene M W N, Stewart W D P 1973 Ammonia assimilation in a nitrogen-fixing blue–green alga. *Pl. Sci. Lett.* **1**: 439–45

Healey F P 1977 Ammonium and urea uptake by some freshwater algae. *Can. J. Bot.* **55**: 61–9

Hellebust J A 1974 Extracellular products. In Stewart W D P (ed) *Algal Physiology and Biochemistry*, Blackwell Scientific Publishers, Oxford, pp 838–63

Herrero A, Flores E, Guerrero M G 1981 Regulation of nitrate reductase levels in the cyanobacteria *Anacystis nidulans*, *Anabaena* sp. strain 7119, and *Nostoc* sp. strain 6719. *J. Bact.* **145**: 175–80

Herrero A, Flores E, Guerrero M G 1985 Regulation of nitrate reductase cellular levels in the cyanobacteria *Anabaena variabilis* and *Synechocystis* sp. *FEMS Microbiol. Lett.* **26**: 21–5

Hien N T, Kerby N W, Machray G C, Rowell P, Stewart W D P 1988 Expression of glutamine synthetase in mutant strains of the cyanobacterium *Anabaena variabilis* which liberate ammonia. *FEMS Microbiol. Lett.* **56**: 337–42

Hipkin C R, Everest S A, Rees T A V, Syrett P J 1982 Ammonium generation by nitrogen-starved cultures of *Chlamydomonas reinhardii*. *Planta* **154**: 587–92

Hodler M, Morgenthaler J-J, Eichenberger W, Grob E C 1972 The influence of light on the activity of nitrate reductase in synchronous cultures of *Chlorella pyrenoidosa*. *FEBS Lett.* **28**: 19–21

Ida S, Mikami B 1983 Purification and characterization of the assimilatory nitrate reductase from the cyanobacterium *Plectonema boryanum. Pl. Cell Physiol.* **24**: 649–58

Ip S-M, Rowell P, Stewart W D P 1983 The role of specific cations in regulation of cyanobacterial glutamine synthetase. *Biochem. Biophys. Res. Commun.* **114**: 206–13

Ip S-M, Rowell P, Aitken A, Stewart W D P 1984 Purification and characterization of thioredoxin from the N_2-fixing cyanobacterium *Anabaena cylindrica. Eur. J. Biochem.* **141**: 497–504

Jayakumar A, Epstein W, Barnes E M 1985 Characterization of ammonium (methylammonium)/potassium antiport in *Escherichia coli. J. Biol. Chem.* **260**: 7528–32

Jeanfils J, Loudeche R 1986 Photoproduction of ammonia by immobilized heterocystic cyanobacteria. Effect of nitrite and anaerobiosis. *Biotechn. Lett.* **8**: 265–70

Jensen R A, Hall G C 1982 Endo-oriented control of pyramidally arranged metabolic branchpoints. *Trends Biochem. Sci.* **7**: 177–85

Jouanneau Y, Meyer C M, Vignais P M 1983 Regulation of nitrogenase activity through iron protein interconversion into an active and inactive form in *Rhodopseudomonas capsulata. Biochim. Biophys. Acta* **749**: 318–28

Kanemoto R H, Ludden P W 1984 Effect of ammonia, darkness and phenazine methosulfate on whole-cell nitrogenase and Fe protein modification in *Rhodospirillum rubrum. J. Bact.* **158**: 713–20

Kapp R, Stevens S E, Fox J L 1975 A survey of available nitrogen sources for the growth of the blue–green alga, *Agmenellum quadruplicatum. Arch. Microbiol.* **104**: 135–8

Karuna-Karan A 1987 Product formulations from commercial scale culture of microalgae. In *The World Biotech Report* 1987, Vol 1, Part 4, Online Publications Ltd, London, pp 37–44

Kashket E R, Wilson T H 1973 Proton-coupled accumulation of galactoside in *Streptococcus lactis* 7962. *Proc. Nat. Acad. Sci. USA* **70**: 2866–9

Kashyap A K, Johar G 1984 Genetic control of ammonium transport in the nitrogen fixing cyanobacterium *Nostoc muscorum. Mol. Gen. Genet.* **197**: 509–12

Kempner E S, Miller J H 1965 The molecular biology of *Euglena gracilis.* III. General carbon metabolism. *Biochemistry* **4**: 2735–9

Kerby N W, Musgrave S C, Codd G A, Rowell P, Stewart W D P 1983 Photoproduction of ammonia by immobilized cyanobacteria. In *'Biotech '83' Proceedings of the international conference on the commercial applications and implications of biotechnology*, Online publications Limited, Northwood, UK, pp 1029–36

Kerby N W, Musgrave S C, Rowell P, Shestakov S V, Stewart W D P 1986b Photoproduction of ammonium by immobilised mutant strains of *Anabaena variabilis. Appl. Microbiol. Biotechnol.* **24**: 42–6

Kerby N W, Niven G W, Rowell P, Stewart W D P 1987b Photoproduction of amino acids by mutant strains of N_2-fixing cyanobacteria. *Appl. Microbiol. Biotechnol.* **25**: 547–52

Kerby N W, Niven G W, Rowell P, Stewart W D P 1988 Ammonia and amino acid production by cyanobacteria. In Stadler T, Mollion J, Verdus M-C, Karamanos Y, Morvan H, Christiaen D (eds) *Algal Biotechnology*, Elsevier Applied Science Publishers, Barking pp 277–86

Kerby N W, Rowell P, Stewart W D P 1985 Ethylenediamine uptake and metabolism in the cyanobacterium *Anabaena variabilis. Arch. Microbiol.* **141**: 244–8

Kerby N W, Rowell P, Stewart W D P 1986a The uptake and metabolism of methylamine by N_2-fixing cyanobacteria. *Arch. Microbiol.* **143**: 353–8

Kerby N W, Rowell P, Stewart W D P 1987a Cyanobacterial transport, ammonium assimilation, and nitrogenase regulation. *N.Z. J. Mar. Fresh Wat. Res.* **21**: 447–55

Kessler E, Zumft W G 1973 Effect of nitrite and nitrate on chlorophyll fluorescence in green algae. *Planta* **111**: 41–6

Kleiner D 1981 The transport of NH_3 and NH_4^+ across biological membranes. *Biochim. Biophys. Acta* **639**: 41–52

Kleiner D 1985 Bacterial ammonium transport. *FEMS Microbiol. Rev.* **32**: 87–100

Kleiner D, Fitzke E 1981 Some properties of a new electrogenic transport system: the ammonium (methylammonium) carrier from *Clostridium pasteurianum*. *Biochim. Biophys. Acta* **641**: 138–47

Komor E, Tanner W 1976 The determination of the membrane potential of *Chlorella vulgaris*. *Eur. J. Biochem.* **70**: 197–204

Kumar A P, Rai A N, Singh H N 1985 Nitrate reductase activity in isolated heterocysts of the cyanobacterium *Nostoc muscorum*. *FEBS Lett.* **179**: 125–8

Labarre J, Thuriaux P, Chauvat F 1987 Genetic analysis of amino acid transport in the facultatively heterotrophic cyanobacterium *Synechocystis* sp. strain 6803. *J. Bact.* **169**: 4668–73

Ladha J K, Rowell P, Stewart W D P 1978 Effects of 5-hydroxylysine on acetylene reduction and NH_4^+ assimilation in the cyanobacterium *Anabaena cylindrica*. *Biochem. Biophys. Res. Commun.* **83**: 688–96

Larsson M, Larsson C-M, Ullrich W R 1982 Regulation by amino acids of photorespiratory ammonia and glycollate release from *Ankistrodesmus* in the presence of methionine sulphoximine. *Pl. Physiol.* **70**: 1637–40

Latorre C, Lee J H, Spiller H, Shanmugam K T 1986 Ammonium ion-excreting cyanobacterial mutant as a source of nitrogen for the growth of rice: a feasibility study. *Biotechnol. Lett.* **8**: 507–12.

Lea P J, Joy K W, Ramos J L, Guerrero M G 1984 The action of 2-amino-4-(methylphosphinyl)-butanoic acid (phosphinothricin) and its 2-oxo-derivative on the metabolism of cyanobacteria and higher plants. *Phytochemistry* **23**: 1–6

Lea P J, Miflin B J 1975 Glutamate synthase in blue–green algae. *Biochem. Soc. Trans.* **3**: 381–84

Leavitt R I 1986 Osmotic regulation in *Chlorella* sp. 580 as a mechanism for the production of L-proline. *Nova Hedwigia* **83**: 139–41

Lee-Kaden J, Simonis W 1982 Amino acid uptake and energy coupling dependent on photosynthesis in *Anacystis nidulans*. *J. Bact.* **151**: 229–36

Lewin J, Hellebust J A 1976 Heterotrophic nutrition of the marine pennate diatom *Nitzchia angularis* var. affinis. *Mar. Biol.* **36**: 313–20

Losada M, Guerrero M G 1979 The photosynthetic reduction of nitrate and its regulation. In Barber J (ed) *Photosynthesis in Relation to Model Systems*, Elsevier Biomedical Press, North Holland, pp 366–408

Losada M, Paneque A, Aparacio P J, Vega J M, Cardénas J, Herrera J 1970 Inactivation and repression by ammonium of the nitrate reducing system in *Chlorella*. *Biochem. Biophys. Res. Commun.* **38**: 1009–15

Lumpkin T A, Plucknett D L 1982 *Azolla as a Green Manure. Use and Management in Crop Production*, Westview Press, Bowker Publishing Company, Epping, Essex

MacIsaac J J, Dugdale R C 1972 Interactions of light and inorganic nitrogen in controlling nitrogen uptake in the sea. *Deep-Sea Res.* **19**: 209–32

Manzano C, Candau P, Gomez-Moreno G, Relimpio A M, Losada M 1976 Ferredoxin-dependent photosynthetic reduction of nitrate and nitrite by particles of *Anacystis nidulans*. *Mol. Cell. Biochem.* **10**: 161–9

Manzano C, Candau P, Guerrero M G 1978 Affinity chromatography of *Anacystis nidulans* ferredoxin–nitrate reductase and NADP reductase on ferredoxin–sepharose. *Anal. Biochem.* **90**: 408–12

Márquez A J, Gálvan F, Vega J M 1984 Purification and characterization of NADH–glutamate synthase from *Chlamydomonas reinhardii*. *Pl. Sci. Lett.* **34**: 305–14

Márquez A J, Gotor C, Romero L C, Galván F, Vega J M 1986 Ferredoxin–glutamate synthase from *Chlamydomonas reinhardii*. Prosthetic groups and preliminary study of mechanism. *Int. J. Biochem.* **18**: 531–5

Márquez A J, Serra M A, Vega J M 1987 Characterization of a light-dependent glutamate synthase activity in *Chlamydomonas reinhardtii. Photosyn. Res.* **12**: 73–81

Martinez-Bilbao M, Urkijo I, Serra J 1987 Purification and properties of the NADPH–glutamate dehydrogenase from the filamentous cyanobacterium *Phormidium laminosum.* Conference abstract, *Second International Symposium on Nitrate Assimilation. Genetic and Molecular Aspects*, St Andrews, Scotland, p B20

McCarthy J J, Eppley R W 1972 A comparison of chemical, isotopic and enzymatic methods for measuring nitrogen assimilation of marine phytoplankton. *Limnol. Oceanogr.* **17**: 371–82

Meeks J C, Wolk C P, Thomas J, Lockau W, Shaffer P W, Austin S M, Chien W-S, Galonsky A 1977 The pathways of assimilation of $^{13}NH_4^+$ by the cyanobacterium *Anabaena cylindrica. J. Biol. Chem.* **252**: 7894–900

Meeks J C, Wolk C P, Lockau W, Schilling N, Shaffer P W, Chien W-S 1978 Pathways of assimilation of $[^{13}N]N_2$ and $^{13}NH_4^+$ by cyanobacteria with and without heterocysts. *J. Bact.* **134**: 125–30

Meeks J C, Wycoff K L, Chapman J S, Enderlin C S 1983 Regulation and expression of nitrate and dinitrogen assimilation by *Anabaena* species. *Appl. Env. Microbiol.* **45**: 1351–9

Mendez J M, Herrero A, Vega J M 1981 Characterization and catalytic properties of nitrite reductase from *Anabaena* sp. 7119. *Z. Pflanzenphysiol.* **4**: 305–15

Meredith M J, Gronostajski R M, Schmidt R R 1978 Physical and kinetic properties of nicotinamide adenine dinucleotide-specific glutamate dehydrogenase purified from *Chlorella sorokiniana. Pl. Physiol.* **61**: 967–74

Mian M H, Stewart W D P 1985 Fate of nitrogen applied as *Azolla* and blue–green algae (cyanobacteria) in waterlogged rice soils – a ^{15}N tracer study. *Pl. Soil* **83**: 363–70

Miflin B J, Lea P J 1976 The pathway of nitrogen assimilation in plants. *Phytochemistry* **15**: 1–13

Morris I 1974 Nitrogen assimilation and protein synthesis. In Stewart W D P (ed) *Algal Physiology and Biochemistry*, Blackwell Scientific Publishers, Oxford, pp 583–609

Morris I, Syrett P J 1965 The effect of nitrogen starvation on the activity of nitrate reductase and other enzymes in *Chlorella. J. Gen. Microbiol.* **38**: 21–8

Murry M A, Benemann J R 1979 Nitrogenase regulation in *Anabaena cylindrica. Pl. Cell Physiol.* **20**: 1391–401

Murry M A, Jensen D B, Benemann J R 1983 Role of ammonia in the regulation of nitrogenase synthesis and activity in *Anabaena cylindrica. Biochim. Biophys. Acta* **756**: 13–9

Musgrave S C 1985 Technological applications of cyanobacteria. Ph.D dissertation (Thesis), University of Dundee

Musgrave S C, Kerby N W, Codd G A, Stewart W D P 1982 Sustained ammonia production by immobilized filaments of the nitrogen-fixing cyanobacterium *Anabaena* 27893. *Biotechnol. Lett.* **4**: 647–52

Musgrave S C, Kerby N W, Codd G A, Stewart W D P 1983a Structural features of calcium alginate entrapped cyanobacteria modified for ammonia production. *Appl. Microbiol. Biotechnol.* **17**: 133–6

Musgrave S C, Kerby N W, Codd G A, Rowell P, Stewart W D P 1983b Reactor types for the utilization of immobilized photosynthetic microorganisms. *Process Biochemistry* (Supplement), Wheatland Journals, London, pp 184–90

Neilson A H, Larsson T 1980 The utilization of organic nitrogen for the growth of algae: physiological aspects. *Physiol. Plant.* **48**: 542–53

Neilson A H, Lewin R A 1974 The uptake and utilisation of organic carbon by algae: an essay in comparative biochemistry. *Phycologia* **13**: 227–64

Nicholls DG 1982 *Bioenergetics: An Introduction to the Chemiosmotic Theory*, Academic Press, London

Nierzwicki-Bauer S A, Haselkorn R 1986 Differences in mRNA levels in *Anabaena* living freely or in symbiotic association with *Azolla*. *EMBO J.* **5**: 29–35

Niven G W, Kerby N W, Rowell P, Stewart W D P 1988a The regulation of aromatic amino acid biosynthesis in amino acid liberating mutant strains of *Anabaena variabilis*. *Arch. Microbiol.* **150**: 272–7

Niven G W, Kerby N W, Rowell P, Foster C A, Stewart W D P 1988b The effect of detergents on amino acid liberation by the N_2-fixing cyanobacterium *Anabaena variabilis* and 6-fluorotryptophan-resistant mutant strains. *J. Gen. Microbiol.* **134**: 689–95

Ohmori M, Hattori A 1970 Induction of nitrate and nitrite reductases in *Anabaena cylindrica*. *Pl. Cell Physiol.* **11**: 873–8

Ohmori M, Hattori A 1974 Effect of ammonia on nitrogen fixation by the blue–green alga *Anabaena cylindrica*. *Pl. Cell Physiol.* **15**: 131–42

Ohmori M, Hattori A 1978 Transient change in the ATP pool of *Anabaena cylindrica* associated with ammonia assimilation. *Arch. Microbiol.* **117**: 17–20

Ohmori M, Ohmori K, Strotmann H 1977 Inhibition of nitrate uptake by ammonia in a blue–green alga *Anabaena cylindrica*. *Arch. Microbiol.* **114**: 225–9

Orr J, Keefer L M, Keim P, Nguyen T D, Wellems T, Heinrikson L, Haselkorn R 1981 Purification, physical characterization and NH_2-terminal sequence of glutamine synthetase from the cyanobacterium *Anabaena* 7120. *J. Biol. Chem.* **256**: 13091–8

Orr J, Haselkorn R 1981 Kinetic and inhibition studies of glutamine synthetase from the cyanobacterium *Anabaena* 7120. *J. Biol. Chem.* **256**: 13099–104

Orr J, Haselkorn R 1982 Regulation of glutamine synthetase activity and synthesis in free-living and symbiotic *Anabaena* spp. *J. Bact.* **152**: 626–35

Ownby J D 1977 Effects of amino acids on methionine-sulfoximine-induced heterocyst formation. *Planta* **136**: 277–9

Papen H, Bothe H 1984 The activation of glutamine synthetase from *Anabaena cylindrica* by thioredoxin. *FEMS Microbiol. Lett.* **23**: 41–6

Parker J E, Javed Q, Merret M J 1985 Glutamate dehydrogenase (NADP-dependent) mRNA in relation to enzyme synthesis in *Euglena gracilis*. Evidence for post-transcriptional control. *Eur. J. Biochem.* **153**: 573–8

Pattnaik H 1966 Studies on nitrogen fixation by *Westiellopsis prolifica* Janet. *Ann. Bot.* **30**: 231–8

Pelley J L, Bannister T T 1979 Methylamine uptake in the green alga *Chlorella pyrenoidosa*. *J. Phycol.* **15**: 110–2

Peschek G A 1979 Nitrate and nitrite reductase and hydrogenase in *Anacystis nidulans* grown in Fe- and Mo-deficient media. *FEMS Microbiol. Lett.* **6**: 371–4

Peschek G A, Czerny T, Schmetterer G, Nitschmann W H 1985 Transmembrane proton electrochemical gradients in dark aerobic and anaerobic cells of the cyanobacterium (blue–green alga) *Anacystis nidulans*. *Pl. Physiol.* **79**: 278–84

Peterson R B, Wolk C P 1978 High recovery of nitrogenase activity of [55]Fe-labeled nitrogenase in heterocysts isolated from *Anabaena variabilis*. *Proc. Nat. Acad. Sci. USA* **75**: 6271–5

Phares W, Chapman L F 1975 *Anacystis nidulans* mutants resistant to aromatic amino acid analogues. *J. Bact.* **122**: 943–8

Pistorius E K, Funkhouser E A, Voss H 1978 Effect of ammonium and ferricyanide on nitrate utilisation by *Chlorella vulgaris*. *Planta* **141**: 279–82

Pistorius E K, Gewitz H-S, Voss H, Vennesland B 1976 Reversible inactivation of nitrate reductase in *Chlorella vulgaris* in vivo. *Planta* **128**: 73–80

Polukhina L E, Sakhurieva G N, Shestakov S V 1982 Ethylenediamine-resistant *Anabaena variabilis* mutants with derepressed nitrogen-fixing system. *Microbiology* **51**: 90–5

Porter R D 1987 Transformation in cyanobacteria. *CRC Crit. Rev. Microbiol.* **13**: 11–132

Poulet S A, Martin-Jézéquel V 1983 Relationships between dissolved free amino acids, chemical composition and growth of the marine diatom *Chaetoceros debile*. *Mar. Biol.* **77**: 93–100

Rai A N, Bergman B 1986 Modification of NO_3^- metabolism in heterocysts of the N_2-fixing cyanobacterium *Anabaena* 7120 (ATCC 27893). *FEMS Microbiol. Lett.* **36**: 133–7

Rai A N, Rowell P, Stewart W D P 1982 Glutamate synthase activity of heterocysts and vegetative cells of the cyanobacterium *Anabaena variabilis* Kutz. *J. Gen. Microbiol.* **128**: 2203–5

Rai A N, Rowell P, Stewart W D P 1984 Evidence for an ammonium transport system in free-living and symbiotic cyanobacteria *Arch. Microbiol.* **137**: 241–6

Ramos J L, Guerrero M G, Losada M 1982a Photoproduction of ammonia from nitrate by *Anacystis nidulans* cells. *Biochim. Biophys. Acta* **679**: 323–30

Ramos J L, Guerrero M G, Losada M 1982b Sustained photoproduction of ammonia from nitrate by *Anacystis nidulans*. *Appl. Env. Microbiol.* **44**: 1020–5

Ramos J L, Guerrero M G, Losada M 1984 Sustained photoproduction of ammonia from dinitrogen and water by the nitrogen-fixing cyanobacteria *Anabaena* sp. strain ATCC 33047. *Appl. Env. Microbiol.* **48**: 114–18

Ramos J L, Guerrero M G, Losada M 1987 Factors affecting the photoproduction of ammonia from dinitrogen and water by the cyanobacterium *Anabaena* sp. strain ATCC 33047. *Biotechnol. Bioeng.* **24**: 566–71

Ramos J L, Madueno F 1986 Induction of increase in the heterocyst frequency of *Anabaena* sp. strain ATCC 33047. Effect on ammonium photoproduction. *FEMS Microbiol. Lett.* **36**: 73–6

Raven J A 1980 Nutrient transport in microalgae. In Rose A H, Morris J G (eds) *Advances in Microbial Physiology*, vol 21, Academic Press, London, pp 47–226

Rawson D M 1985 The effect of exogenous amino acids on growth and nitrogenase activity in the cyanobacterium *Anabaena cylindrica* PCC 7122. *J. Gen. Microbiol.* **131**: 2549–54

Reed R H, Rowell P, Stewart W D P 1980 Components of the proton electrochemical potential gradient in *Anabaena variabilis*. *Biochem. Soc. Trans.* **8**: 707–8

Reed R H, Warr S R C, Kerby N W, Stewart W D P 1986 Osmotic shock-induced release of low molecular weight metabolites from free-living and immobilised cyanobacteria. *Enzyme Microbial Technol.* **8**: 101–4

Reglinski A, Rowell P, Kerby N W, Stewart W D P 1989 Characterisation of the methylammonium/ammonium transport system in mutant strains of *Anabaena variabilis* resistant to ammonium analogues. *J. Gen. Microbiol.* **135**: (in press)

Reich S, Almon H, Böger P 1986 Short-term effect of ammonia on nitrogenase activity of *Anabaena variabilis* (ATCC 29413). *FEMS Microbiol. Lett.* **34**: 53–6

Reinhold L, Kaplan A 1984 Membrane transport of sugars and amino acids. *A. Rev. Pl. Physiol.* **35**: 45–83

Riccardi G, Sora S, Ciferri O 1981a Production of amino acids by analog-resistant mutants of the cyanobacterium *Spirulina platensis*. *J. Bact.* **147**: 1002–7

Riccardi G, Sanangelantoni D, Carbonera D, Savi A, Ciferri O 1981b Characterization of mutants of *Spirulina platensis* resistant to amino acid analogues. *FEMS Microbiol. Lett.* **12**: 333–6

Richmond A (ed) 1986 *CRC Handbook of Microalgal Mass Culture*, CRC Press Inc., Boca Raton, Florida

Rigano C, Di Martino Rigano V, Fuggi A 1987 Nitrogen metabolism in thermophilic algae. In Ullrich W R, Apparicio P J, Syrett P J, Castillo F (eds) *Inorganic Nitrogen Metabolism*, Springer-Verlag, Berlin, pp 210–16

Rigano C, Di Martino Rigano V, Fuggi A, Vona V 1977 Heterotrophic growth patterns in the unicellular alga *Cyanidium caldarium*. *Arch. Microbiol.* **113**: 191–7

Rigano C, Di Martino Rigano V, Vona V, Fuggi A 1979 Glutamine synthetase activity, ammonia assimilation and control of nitrate reduction in the unicellular red alga Cyanidium caldarium. Arch. Microbiol. **121**: 117–20

Rippka R, Waterbury J B 1977 The synthesis of nitrogenase by non-heterocystous cyanobacteria. FEMS Microbiol. Lett. **2**: 83–6

Ritchie R J, Gibson J 1987 Permeability of ammonia, methylamine and ethylamine in the cyanobacterium, Synechococcus R-2 (Anacystis nidulans) PCC 7942. J. Mem. Biol. **95**: 131–42

Roger P A, Kulasooriya S A 1980 Blue–Green Algae and Rice, IRRI, Los Banos, Laguna, Philippines

Romero J M, Coronil T, Lara C, Guerrero M G 1987 Modulation of nitrate uptake in Anacystis nidulans by the balance between ammonium assimilation and CO_2 fixation. Arch. Biochem. Biophys. **256**: 578–84

Romero J M, Lara C, Guerrero M G 1985 Dependence of nitrate utilization upon active CO_2 fixation in Anacystis nidulans: a regulatory aspect of the interaction between photosynthetic carbon and nitrogen metabolism. Arch. Biochem. Biophys. **237**: 396–401

Rowell P, Enticott S, Stewart W D P 1977 Glutamine synthetase and nitrogenase activity in the blue–green alga Anabaena cylindrica. New Phytol. **79**: 41–54

Rowell P, Kerby N W, Darling A J, Stewart W D P 1985a Physiological and biochemical aspects of N_2-fixing cyanobacteria. In Evans H J, Bottomley P J, Newton W E (eds) Nitrogen Fixation Research Progress, Martinus Nijhoff Publishers, Dordrecht, pp 387–93

Rowell P, Rai A N, Stewart W D P 1985b Studies on the nitrogen metabolism of the lichens Peltigera aphthosa and Peltigera canina. In Brown D H (ed) Lichen Physiology and Cell Biology, Plenum Press, New York, pp 145–60

Rowell P, Sampaio M J A M, Ladha J K, Stewart W D P 1979 Alteration of cyanobacterial glutamine synthetase activity in vivo in response to light and NH_4^+. Arch. Microbiol. **120**: 195–200

Rowell P, Stewart W D P 1976 Alanine dehydrogenase of the N_2-fixing blue–green alga Anabaena cylindrica. Arch. Microbiol. **107**: 115–24

Sakhurieva G N, Polukhina L E, Shestakov S V 1982 Glutamine synthetase in Anabaena variabilis mutants with derepressed nitrogenase. Microbiology **51**: 308–12

Sampaio M J A M, Rowell P, Stewart W D P 1979 Purification and some properties of glutamine synthetase from the nitrogen-fixing cyanobacteria Anabaena cylindrica and a Nostoc sp. J. Gen. Microbiol. **111**: 181–91

Sauer N 1984 A general amino-acid permease is inducible in Chlorella vulgaris. Planta **161**: 425–31

Sauer N 1986 Hexose-transport-deficient mutants of Chlorella vulgaris. Lack of transport activity correlates with absence of inducible proteins. Planta **168**: 139–44

Sauer N, Komor E, Tanner W 1983 Regulation and characterization of two inducible amino-acid transport systems in Chlorella vulgaris. Planta **159**: 404–10

Sauer N, Tanner W 1985 Selection and characterization of Chlorella mutants deficient in amino acid transport. Further evidence for three independent systems. Pl. Physiol. **79**: 760–4

Sawhney S K, Nicholas D J D 1978a Some properties of glutamine synthetase from Anabaena cylindrica. Planta **139**: 289–99

Sawhney S K, Nicholas D J D 1978b Effects of amino acids, adenine nucleotides and inorganic pyrophosphate on glutamine synthetase from Anabaena cylindrica. Biochim. Biophys. Acta **527**: 485–96

Schlee J, Cho B-H, Komor E 1985 Regulation of nitrate uptake by glucose in Chlorella. Pl. Sci. **39**: 25–30

Schlee J, Komor E 1986 Ammonium uptake by Chlorella. Planta **168**: 232–8

Schmidt A 1981 A thioredoxin activated glutamine synthetase in *Chlorella*. *Z. Naturfor.* **36c**: 396–9

Serra J L, Llama M J, Cardenas E 1978a Nitrate utilisation by the diatom *Skeletonema costatum*. I. Kinetics of nitrate uptake. *Pl. Physiol.* **62**: 987–90

Serra J L, Llama M J, Cardenas E 1978b Nitrate utilisation by the diatom *Skeletonema costatum*. II. Regulation of nitrate uptake *Pl. Physiol.* **62**: 991–4

Shatilov V R, Kretovich W L 1977 Glutamate dehydrogenases from *Chlorella*: forms, regulation and properties. *Mol. Cell. Biochem.* **15**: 210–12

Shi D-J, Brouers M, Hall D O, Robbins R J 1987 The effects of immobilization on the biochemical, physiological and morphological features of *Anabaena azollae*. *Planta* **172**: 298–308

Singer S J, Nicholson G L 1972 The fluid mosaic model of the structure of cell membranes. *Science* **175**: 720–31

Singh A L, Singh P K 1986 Comparative effects of *Azolla* and blue–green algae in combination with chemical N fertilizer on rice crop. *Proc. Ind. Acad. Sci. (Plant Sciences)* **96**: 147–52

Singh A L, Singh P K 1987 Comparative study on *Azolla* and blue–green algae dual culture with rice. *Israel J. Bot.* **36**: 53–61

Singh D T, Modi D R, Singh H N 1986 Evidence for glutamine synthetase and methylammonium (ammonium) transport system as two distinct primary targets of methionine sulfoximine inhibitory action in the cyanobacterium *Anabaena doliolum*. *FEMS Microbiol. Lett.* **37**: 95–8

Singh D T, Rai A N, Singh H N 1985 Methylammonium (ammonium) uptake in a glutamine auxotroph of the cyanobacterium *Anabaena cycadae*. *FEBS Lett.* **186**: 51–3

Singh H N, Singh R K, Sharma R 1983 An L-methionine-D,L-sulphoximine-resistant mutant of the cyanobacterium *Nostoc muscorum* showing inhibitor-resistant γ-glutamyl-transferase, defective glutamine synthetase and producing extracellular ammonia during N_2 fixation. *FEBS Lett.* **154**: 10–13

Singh R N 1961 *The Role of Blue–Green Algae in Nitrogen Economy of Indian Agriculture*, Indian Council of Agricultural Research, New Delhi, India

Soeder C J, Bolze A 1981 Sulphate deficiency stimulates release of dissolved organic matter in synchronous cultures of *Scenedesmus obliquus*. *Physiol. Plant.* **52**: 233–8

Solomonson L P 1979 Structure of *Chlorella* nitrate reductase. In Hewitt E J, Cutting C V (eds) *Nitrogen Assimilation of Plants*, Academic Press, New York and London, pp 199–205

Solomonson L P, Barber M J 1987 Structure–function relationships of assimilatory nitrate reductase. In Ullrich W R, Apparicio P J, Syrett P J, Castillo F (eds) *Inorganic Nitrogen Metabolism*, Springer-Verlag, Berlin, pp 71–5

Solomonson L P, Lorimer G H, Hall R L, Borchers R, Bailey J L 1975 Reduced nicotinamide adenine dinucleotide–nitrate reductase of *Chlorella vulgaris*. Purification, prosthetic groups and molecular properties. *J. Biol. Chem.* **250**: 4120–7

Solomonson L P, Spehar A M 1977 Model for the regulation of nitrate assimilation. *Nature* (Lond.) **265**: 373–5

Spiller H, Latorre C, Hassan M E, Shanmugam K T 1986 Isolation and characterisation of nitrogenase-derepressed mutant strains of cyanobacterium *Anabaena variabilis*. *J. Bact.* **165**: 412–19

Stacey G, Tabita F R, Van Baalen C 1977 Nitrogen and ammonia assimilation in the cyanobacteria. Purification of glutamine synthetase from *Anabaena* sp. strain C A. *J. Bact.* **132**: 596–603

Stacey G, Van Baalen C, Tabita F R 1979 Nitrogen and ammonia assimilation in the cyanobacteria: regulation of glutamine synthetase. *Arch. Biochem. Biophys.* **194**: 457–67

Stevens S E, Van Baalen C 1974 Control of nitrate reductase in a blue–green alga. The effects of inhibitors, blue light and ammonia. *Arch. Biochem. Biophys.* **161**: 146–52

Stewart W D P 1963 Liberation of extracellular nitrogen by two nitrogen-fixing blue–green algae. *Nature* (Lond.) **200**: 1020–1

Stewart W D P 1980 Some aspects of structure and function in N_2-fixing cyanobacteria. *A. Rev. Microbiol.* **34**: 497–536

Stewart W D P, Haystead A, Pearson H W 1969 Nitrogenase activity in heterocysts of blue–green algae. *Nature* (Lond.) **224**: 226–8

Stewart W D P, Lex M 1970 Nitrogenase activity in the blue–green alga *Plectonema boryanum* strain 594. *Arch. Mikrobiol.* **73**: 250–60

Stewart W D P, Rowell P 1975 Effects of L-methionine-DL-sulphoximine on the assimilation of newly fixed NH_3, acetylene reduction and heterocyst production in *Anabaena cylindrica. Biochem. Biophys. Res. Commun.* **65**: 846–57

Stewart W D P, Rowell P 1977 Modifications of nitrogen-fixing algae in symbioses. *Nature* (Lond.) **265**: 371–2

Stewart W D P, Rowell P 1986 Biochemistry and physiology of nitrogen fixation with particular emphasis on nitrogen-fixing phototrophs. *Pl. Soil* **90**: 167–91

Stewart W D P, Rowell P, Cossar J D, Kerby N W 1985 Physiological studies on N_2-fixing cyanobacteria. In Ludden P W, Burris J E (eds) *Nitrogen Fixation and CO_2 Metabolism,* Elsevier, New York, pp 269–79

Stewart W D P, Rowell P, Hawksford M J, Sampaio M J A M, Ernst A 1982 Nitrogenase and aspects of its regulation. *Israel J. Bot.* **31**: 168–89

Stewart W D P, Rowell P, Kerby N W, Reed R H, Machray G C 1987 N_2-fixing cyanobacteria and their potential applications. *Phil. Trans. R. Soc. Lond. Series B* **317**: 245–58

Stewart W D P, Rowell P, Ladha J K, Sampaio M J A M 1979 Blue–green algae (cyanobacteria) – some aspects related to their role as sources of fixed nitrogen in paddy soils. *Proceedings of Nitrogen and Rice Symposium,* IRRI, Manilla, pp 263–83

Stewart W D P, Rowell P, Rai A N 1983 Cyanobacteria–eukaryotic plant symbioses. *Ann. Microbiol. (Institut Pasteur)* **134B**: 205–28

Subramanian G, Shanmugasundaram S 1986 Uninduced ammonia release by the nitrogen-fixing cyanobacterium *Anabaena. FEMS Mirobiol. Lett.* **37**: 151–4

Syrett P J 1981 Nitrogen metabolism of microalgae. In Platt T (ed) *Physiological Bases of Phytoplankton Ecology,* Canadian Bulletin of Fisheries and Aquatic Sciences, Bull 210, Ottawa, pp 182–210

Syrett P J 1987 Nitrogen assimilation by eukaryotic algae. In Ullrich W R, Apparicio P J, Syrett P J, Castillo F (eds) *Inorganic Nitrogen Metabolism,* Springer-Verlag, Berlin, pp 25–31

Talley D J, White L H, Schmidt R R 1972 Evidence for NADH- and NADPH-specific isozymes of glutamate dehydrogenase and continuous inducibility of the NADH-specific isozyme throughout the cell cycle of the eukaryotic *Chlorella. J. Biol. Chem.* **247**: 7927–35

Thacker A, Syrett P J 1972 Disappearance of nitrate reductase activity from *Chlamydomonas reinhardii. New Phytol.* **71**: 435–41

Thomas J, Wolk C P, Shaffer P W 1975 The initial organic products of fixation of ^{13}N-labelled nitrogen gas by the blue–green alga *Anabaena cylindrica. Biochem. Biophys. Res. Commun.* **67**: 501–7

Thomas J, Meeks J C, Wolk C P, Shaffer P W, Austin S M, Chien W-S 1977 Formation of glutamine from [^{13}N] ammonia, [^{13}N] dinitrogen and [^{14}C] glutamate by heterocysts isolated from *Anabaena cylindrica. J. Bact.* **129**: 1545–55

Tischner R, Hütterman A 1978 Light mediated activation of nitrate reductase in synchronous *Chlorella. Pl. Physiol.* **62**: 284–6

Tischner R, Hütterman A 1980 Regulation of glutamine synthetase by light and during nitrogen deficiency in synchronous *Chlorella sorokiniana. Pl. Physiol.* **66**: 805–8

Tischner R, Lorenzen H 1979a Nitrate uptake and nitrate reduction in synchronous *Chlorella. Planta* **146**: 287–92

Tischner R, Lorenzen H 1979b Changes in the enzyme pattern in synchronous *Chlorella sorokiniana* caused by different nitrogen sources. *Z.Pflanzenphysiol.* **100**: 333–41

Tischner R, Schmidt A 1982 A thioredoxin mediated activation of glutamine synthetase in synchronous *Chlorella sorokiniana*. *Pl. Physiol.* **70**: 113–6

Tischner R, Schmidt A 1984 Light mediated regulation of nitrate assimilation in *Synechococcus leopoliensis*. *Arch. Microbiol.* **137**: 151–4

Tucker J B 1985 Biotechnology goes to sea. *High Technology* February 34–44

Tuli R, Thomas J 1980 Regulation of glutamine synthetase in the blue–green alga *Anabaena* L-31. *Biochim. Biophys. Acta* **613**: 526–33

Tumer E N, Robinson SJ, Haselkorn R 1983 Different promoters for the *Anabaena* glutamine synthetase gene during growth using molecular or fixed nitrogen. *Nature* (Lond.) **306**: 337–42

Turpin D H, Edie S A, Canvin D T 1984 *In vivo* nitrogenase regulation by ammonium and the effect of MSX on ammonium transport in *Anabaena flos-aquae*. *Pl. Physiol.* **74**: 701–4

Vaishampayan A 1982 Amino acid nutrition in the blue–green alga *Nostoc muscorum*. *New Phytol.* **90**: 545–9

Vanlerberghe G C, Brown L M 1987 Proline overproduction in cells of the green alga *Nannochloris bacillaris* resistant to azetidine- 2-carboxylic acid. *Pl. Cell Env.* **10**: 251–7

Vega J M, Gotor, C, Menacho A 1987 Enzymology of the assimilation of ammonia by the green alga *Chlamydomonas reinhardtii*. In Ullrich W R, Apparicio P J, Syrett P J, Castillo F (eds) *Inorganic Nitrogen Metabolism*, Springer-Verlag, Berlin, pp 132–6

Venkataraman G S 1979 Algal inoculation in rice fields. *Proceedings of Nitrogen and Rice Symposium*, IRRI, Manilla, pp 311–21

Venkataraman G S 1981 *All India Co-ordinated Project on Algae.* Annual Report (1980–81). Division of Microbiology, Indian Agricultural Research Institute, New Delhi

Venkataraman G S, Saxena H K 1963 Studies on nitrogen fixation by blue–green algae. 4. Liberation of free amino acids in the medium. *Ind. J. Agric. Sci.* **33**: 21–4

Venkataraman L V 1986 Blue–green algae as biofertilizers. In Richmond A (ed) *CRC Handbook of Microalgal Mass Culture*, CRC Press Inc., Boca Raton, Florida, pp 455–71

Vincent W F 1979 Uptake of [^{14}C]methylammonium by plankton communities: a comparative assay for ammonium transport systems in natural waters. *Can. J. Microbiol.* **25**: 1401–7

Vincenzini M, De Philippis R, Ena A, Florenzano G 1986 Ammonia photoproduction by *Cyanospira rippkae* cells 'entrapped' in dialysis tube. *Experientia* **42**: 1040–3

Walker N A, Smith F A, Beilby M J 1979 Amine uniport at the plasmalemma of Charophyte cells 2. Ratio of matter to charge transported and permeability of free base. *J. Mem. Biol.* **49**: 283–96

Walsby A E 1965 Biochemical studies on extracellular polypeptides. *Br. Phycol. Bull.* **2**: 514–15

Watanabe A 1951 Production in cultural solution of some amino acids by the atmospheric nitrogen-fixing blue–green algae. *Arch. Biochem. Biophys.* **34**: 50–5

Wheeler P A 1980 Use of methylammonium as an ammonium analogue in nitrogen transport and assimilation studies with *Cyclotella cryptica* (Bacillariophyceae). *J. Phycol.* **16**: 328–34

Wheeler P A 1983 Phytoplankton nitrogen metabolism. In Carpenter E J, Capone D G (eds) *Nitrogen in the Marine Environment*, Academic Press, New York, pp 839–68

Wheeler P A, Hellebust J A 1981 Uptake and concentration of alkylamines by a marine diatom. *Pl. Physiol.* **67**: 367–72

Wolk C P, Thomas J, Shaffer P W, Austin S M, Galonsky A 1976 Pathway of nitrogen metabolism after fixation of ^{13}N-labelled nitrogen gas by the cyanobacterium *Anabaena cylindrica*. *J. Biol. Chem.* **251**: 5027–34

Wright S A, Syrett P J 1983 The uptake of methylammonium and dimethylammonium by the diatom *Phaeodactylum tricornutum*. *New Phytol.* **95**: 189–202

Yamada K, Kinoshita S, Tsunoda T, Aida K 1972 *The Microbial Production of Amino Acids*, Halsted Press, New York

Yoch D C, Gotto J W 1982 Effect of light intensity and inhibitors of nitrogen assimilation on NH$_4^+$ inhibition of nitrogenase activity in *Rhodospirillum rubrum* and *Anabaena* sp. *J. Bact.* **151**: 800–6

Zimmerman W J, Boussiba S 1987 Ammonia assimilation and excretion in an asymbiotic strain of *Anabaena azollae* from *Azolla filiculoides* Lam. *J. Pl. Physiol.* **127**: 443–50

Zumft W G 1972 Ferredoxin: nitrite oxidoreductase from *Chlorella*. Purification and properties. *Biochim. Biophys. Acta* **276**: 363–75

4

The biotechnology of mass culturing *Dunaliella* for products of commercial interest

A. Ben-Amotz and M. Avron

Introduction

Photosynthesis is the most abundant energy-storing and life-supporting process on earth. It is not surprising, therefore, that utilization of the photosynthetic machinery for the production of energy, chemicals and food by mass-culturing microalgae has had particular appeal. The early trials of algal biomass production concentrated on the freshwater green alga *Chlorella* as a potential source of single cell protein (Burlew, 1953; Tamiya, 1957) and were pioneered by scientists at the Carnegie Institution in California and at the Tokugawa Institute of Biology in Tokyo. Since then, many other potential applications for large-scale cultures have been advanced, including the production of food, feed, extractable chemicals, wastewater treatment, aquaculture, and health promoting algal preparations. Several reviews and books on the progress of outdoor mass cultivation of microalgae have appeared (Soeder and Binsack, 1978; Goldman, 1979a, b; Shelef and Soeder, 1980; Becker, 1985; Barclay and McIntosh, 1986; Oswald, 1986; Richmond, 1986a, b; Benemann *et al.*, 1987). Most of the literature deals with freshwater microalgae grown autotrophically or heterotrophically. Serious attempts to utilize mass culture systems for marine microalgae were initiated only in the last ten years, prompted by requirements in the mariculture food chain biotechnology, the search for lipid producing microalgae as a means for large-scale biological energy storing systems (SERI, 1985) and the investigation of specific algae which accumulate industrially interesting products (Parkinson, 1987).

Several drawbacks limit the large-scale expansion and utilization of mass-cultured algae:

(a) Very few products which economically justify commercial production have been identified in microalgae.
(b) Only limited biological know-how is available on techniques to control the chemical composition of algae for enhancement of a selected product or products.

(c) The technology for mass-cultivation of algae is still under development.

(d) Growing a particular algal species in an outdoor pond is greatly hampered by contamination with other organisms. Creating conditions under which a desired algal species predominates is not trivial, and is essential for successful commercial cultivation.

The alga *Dunaliella*

General

The green alga *Dunaliella* is probably the most successful microalga for outdoor cultivation described so far. This halotolerant alga has several features which have made it a favourite for mass-cultivation (Ben-Amotz and Avron, 1983a):

(a) Mass-cultivation of algae in open ponds requires considerable land, in areas where solar light is plentiful and temperatures are moderate throughout the year. Such conditions exist essentially in arid areas, where freshwater is normally scarce but salty water, including seawater, is often available. *Dunaliella* thrives under such conditions since it requires media containing about 6–12% NaCl for optimal growth.

(b) *Dunaliella* is one of very few microorganisms which can thrive in media containing such high salt concentrations. This provides a great selective advantage which allows the cultivation of a relatively pure culture.

(c) The high salt concentration also minimizes the number of predatory species which need to be treated.

(d) Since *Dunaliella* thrives in a simple inorganic medium, growth of non-photosynthetic organisms, such as bacteria and fungi which cannot digest the algae themselves, is severely limited.

(e) Under appropriate growth conditions *Dunaliella* accumulates massive amounts of one highly priced product, β-carotene (over 10% of the algal organic weight), in addition to glycerol (around 20–40%) and the remaining algal meal.

(f) Lacking a cell wall, dried *Dunaliella* is easily and fully digestible by animals and humans.

The genus *Dunaliella* belongs to the class Chlorophyceae and the order Volvocales. There are several ill-defined species. All are unicellular, ovoid in shape, 8–25 µm long, 5–15 µm wide, and motile with two equal long flagellae (Butcher, 1959). The alga contains one large cup-shaped chloroplast with a single pyrenoid embedded in the basal portion (Fig. 4.1). The pyrenoid is surrounded by polysaccharide granules – the storage product of green algae. Like most green algae, *Dunaliella* contains typical organelles: chloroplast, pyrenoid, membrane-bound nucleus, mitochondria, small vacuoles, Golgi bodies and an eyespot. The chief morphological character-

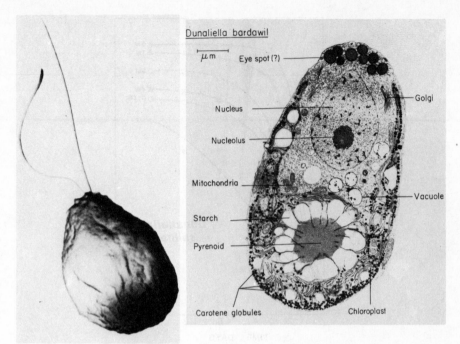

Fig. 4.1 *Dunaliella bardawil.* A scanning electron-micrograph and an electron micrograph.

istic of *Dunaliella*, in contrast to other algae, is the lack of a rigid polysaccharide cell wall. The cell is enclosed by a thin elastic plasma membrane and by a mucus 'surface coat'. It therefore responds rapidly to osmotic changes by changes in cell shape and volume. The lack of a rigid cell wall increases the sensitivity of *Dunaliella* to tension forces by different mechanical means and imposes some limitations on the treatment of cultures.

Dunaliella is found naturally in many aquatic marine habitats and predominates in salt water bodies which contain more than 10% salt. Typical examples of natural unialgal cultures of *Dunaliella* are the Dead Sea in Israel (Nissenbaum, 1975), the Pink Lake in Australia (Borowitzka, 1981) and the Great Salt Lake in Utah, USA (Felix and Rushforth, 1979; Post, 1981). *Dunaliella* shows a remarkable degree of adaptation to a variety of salt concentrations and can be grown in media containing 0.2% to saturated salt (about 35%; Fig. 4.2).

Osmoregulation

The mechanism by which an organism adapts to various salt concentrations is termed 'osmoregulation'. In *Dunaliella*, it was shown to function through the ability of the alga to vary its intracellular concentration of mostly a single substance, glycerol (Ben-Amotz and Avron, 1973; Brown, 1976; Avron, 1986). When the alga is adapted to grow in media containing widely

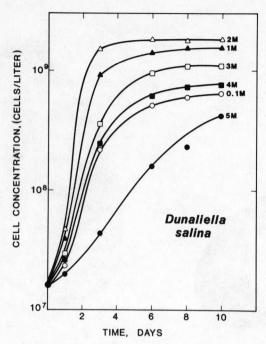

Fig. 4.2 Growth of *Dunaliella salina* adapted to media containing the indicated salt concentrations (1 м NaCl ≃ 6% salt).

different salt concentrations, the intracellular glycerol concentration is directly proportional to the extracellular salt concentration and is sufficient to account for most of the required osmotic pressure (Fig. 4.3). The intracellular sodium and potassium concentrations are around 50 and 200 mм, and are independent of the salt concentration at which the algae are cultivated (Pick *et al.*, 1986).

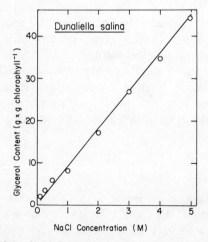

Fig. 4.3. Intracellular glycerol concentration in *Dunaliella* as a function of the medium salt concentration in which the algae were cultivated.

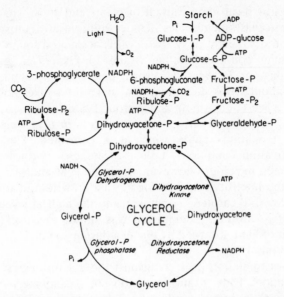

Fig. 4.4 Postulated pathways accounting for the synthesis of glycerol from CO_2 or stored polysaccharides during growth or a hyperosmotic shock; and synthesis of polysaccharides from glycerol during a hypoosmotic shock.

When exposed to higher or lower salt concentrations in their media, the algae behave like perfect osmometers, rapidly shrinking or swelling, respectively. Within minutes after the rapid volume change, synthesis or elimination of glycerol is initiated and continues for a few hours until the cell volume returns to its original ovoid size (Ben-Amotz and Avron, 1973). Glycerol synthesis under hypertonic conditions (increase in salt concentration) and glycerol elimination under hypotonic conditions (decrease in salt concentration) are independent of protein synthesis and occur in the light or in the dark. Synthesis of glycerol in the dark requires, of course, the presence of a sufficient store of polysaccharides.

The biochemical mechanism responsible for the synthesis or elimination of glycerol in *Dunaliella* has been studied in detail, and is believed to involve a unique 'glycerol cycle' including several novel enzymes (Fig. 4.4).

β-Carotene production

In hypersaline lakes, the *Dunaliella* strain which predominates is often red, rather than green, in colour due to massive accumulation of a single pigment, β-carotene. Of the many strains of the genus *Dunaliella* described only two, *Dunaliella bardawil* and *Dunaliella salina* Teod., have been shown to possess the capacity to produce large amounts of β-carotene when cultivated under appropriate conditions (Ben-Amotz *et al.*, 1982; Loeblich, 1982). Electron micrographs of *Dunaliella bardawil* (Fig. 4.1) indicate that the massive amounts of β-carotene are located in a large number of chloroplastic, lipoidal globules located in the interthylakoid space. This is in contrast to the low β-carotene content of plants and algae which is located within the chloroplast membranes. Under appropriate cultivation, more than 10% of the dry weight of *D. bardawil* is accounted for by β-carotene. Other, β-carotene non-accumulating strains of *Dunaliella*, grown under similar conditions, or *D. bardawil* grown under non-accumulating conditions, contain about 0.3% β-carotene and have no visible globules in the interthylakoid space (Ben-Amotz and Avron, 1983a).

The rate and extent of β-carotene accumulation in *D. bardawil* is determined by the conditions under which it is cultivated. Under standard laboratory cultivation conditions, little β-carotene is synthesized and the algae appear green in colour. However, when the light intensity is increased much beyond the intensity required for normal growth, and when the rate of growth is limited, β-carotene accumulates to the highest levels (Fig. 4.5). The amount of β-carotene accumulated was shown to correlate with the integral amount of light absorbed by the alga during one division cycle (Ben-Amotz and Avron, 1983b).

β-Carotene accumulation protects against the deleterious effects of high intensity irradiation. Thus, strains which do not accumulate β-carotene, or β-carotene-poor *D. bardawil* produced by the addition of photobleaching herbicides, die when exposed to high irradiation under limiting growth conditions, while β-carotene rich *D. bardawil* survives under the same conditions (Ben-Amotz *et al.*, 1987). This may explain the previously mentioned predominance of *D. bardawil* or *D. salina* Teod. over green strains of *Dunaliella* in some saline lakes in nature, where low concentrations of algae are exposed to high solar irradiation under nutrient limiting conditions.

The β-carotene accumulated in *Dunaliella bardawil* has been shown to be composed of mostly two isomers, all-*trans* and 9-*cis*, in approximately equal

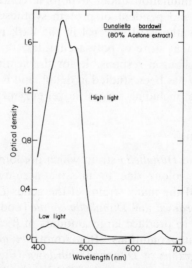

Fig. 4.5 The absorption spectrum of 80% acetone extracts from *Dunaliella bardawil* grown under conditions that do or do not induce high β-carotene accumulation.

amounts (Ben-Amotz *et al.*, 1982, 1988a). The ratio of the 9-*cis* to the all-*trans* isomer increases with increase in the light intensity to which the algae are exposed (Ben-Amotz *et al.*, 1988).

The biotechnology of growing *Dunaliella*

Pond construction

Optimal design of large-scale commercial systems for mass culturing *Dunaliella*, should address the following subjects: size, shape and depth, mixing, liner, aeration and CO_2 supply.

Design

Most of the large-scale commercial ponds of *Dunaliella* consist of growth channels in the form of oblong raceways constructed of repeated units, with each unit an independent production system (Fig. 4.6). This follows the basic design developed for growing *Chlorella* and *Spirulina* (Richmond, 1986b). The pond walls are generally constructed with bricks, concrete or simply earth beams. Ponds constructed entirely of fibreglass have also been successfully employed. Large-scale sloping culture systems were not employed for *Dunaliella* as have been designed and operated for freshwater microalgae in Czechoslovakia (Seltik *et al.*, 1970), Bulgaria (Vendlova,

Fig. 4.6 A typical commercial *Dunaliella* pond.

1969) and Peru (Heussler *et al.*, 1978). The size of the pond is generally determined by the area which can conveniently be well mixed by operation of one paddle-wheel. Thus, a commercial production pond of *Dunaliella* has an average surface area of 1000–4000 m^2 (see Dodd, 1985).

Depth

The light which impinges upon an algal culture is absorbed, mostly by the algae themselves, as it penetrates into the pond. Since solar irradiation provides the sole energy source for the autotrophic *Dunaliella*, it is the amount of light absorbed which generally limits the productivity of algae that can be attained in an outdoor horizontal pond. In a standard commercial pond, containing $3–10 \times 10^8$ cells l^{-1} (200–600 mg algal dry weight l^{-1}), effective light (400–700 nm) is essentially fully absorbed in less than 5 cm of medium. Therefore, from a productivity point of view, deeper cultures will not result in a higher yield. Indeed, experiments carried out with small outdoor cultures of *Dunaliella bardawil* indicated that the productivity, expressed as algae m^{-2} day^{-1}, was identical when the depth of the ponds was varied between 5 and 30 cm (the concentration of the algae, of course, was higher the lower the depth). Nevertheless, in the operation of large-scale open water bodies of more than 1000 m^2, engineering restrictions make it essentially impossible to maintain a uniformly smooth pond bottom and, therefore, it is impractical to circulate the culture in an average pond depth of less than 10 cm. In practice, most commercial *Dunaliella* ponds have a depth of between 10 and 20 cm.

Mixing

Continuous mixing of the microalgal cultures is required to prevent cell sinking and thermal stratification, to maintain even nutrient distribution (including dissolved inorganic carbon) and to remove excess oxygen. The latter may be of greater significance than hitherto considered, since it has become clear that the productivity of such ponds can be limited by the deleterious effects of photoinhibition (Powles, 1985; Vonshak, 1987). Since photoinhibition is a function of both light intensity and the ambient O$_2$ concentration, and since the latter can reach values of three-fold supersaturation in a poorly stirred pond, the removal of excess dissolved oxygen may be of paramount importance.

Several mixing techniques have been tested in *Dunaliella* ponds. These include paddle-wheels, pumps and air-lifts. The lack of a rigid cell wall in *Dunaliella*, with the resulting cellular fragility and sensitivity to mechanical forces mandated the exclusive use of paddle-wheels. Attempts to circulate the algae in the raceways by airlifts or centrifugal pumps resulted in cell breakage and eventually in culture collapse. Paddle-wheels constructed with long arms so as to run at slow revolution, avoiding cell damage, but maintaining a linear liquid velocity of at least 10 cm s^{-1} proved most efficient in commercial production units of *Dunaliella*. The motile behaviour of

Dunaliella and its positive phototactic response allow slower mixing than that commonly needed to keep the filamentous *Spirulina* and the non-motile *Chlorella* in suspension.

More involved methods of mixing (Laws *et al.*, 1983, 1986), which were reported to increase the productivity of *Phaeodactylum tricornutum* by two-fold, have not been tested in cultures of *Dunaliella*.

Liner

Algal production ponds must be essentially impermeable and have a hydraulically smooth bottom surface in order to prevent loss of media, entrainment of dirt and clay during mixing and the establishment of denitrifying bacteria. To achieve this goal commercial microalgal ponds have been lined by plastic sheathing or constructed of fibreglass, concrete, or asphalt concrete. Concrete has been used for many years in *Chlorella* ponds in Taiwan. It requires a high initial investment but lasts for many years. The salty water employed in *Dunaliella* ponds requires attention to the type of concrete employed, and special coating to avoid deterioration of the concrete and corrosion of the metal reinforcing the concrete. The use of asphaltic concrete or asphaltic spray is not recommended. In one case, where rubber asphaltic spray was used in 0.4 ha commercial *Dunaliella* ponds the asphalt gradually disintegrated and asphalt pieces floated on the medium ponds following inoculation of the ponds in a medium containing 15% NaCl. Since the same asphalt in freshwater remained intact, it was suggested that salt penetration disintegrated the asphalt.

Liners of plastic sheathings are most common at present for pond lining. UV-resistant polyvinylchloride (PVC) and white reinforced UV-resistant polyethylene were found to be most suitable for lining *Dunaliella* ponds. Extensive production units, which do not use mixing and do not require culture flow, can be used with natural clay lining, as found in many salt ponds such as those utilized by Betatene Ltd, in Whyalla, Australia.

CO₂ supply

Large-scale intensive algal production ponds require a continuous supply of soluble inorganic carbon to maintain maximal growth. *Dunaliella* is apparently unable to grow heterotrophically, i.e. utilizing organic carbon compounds as its sole carbon source. We were unable to grow the alga with acetate or glucose in the dark, and it seems that the algae are obligate photoautotrophs. The required inorganic carbon is most commonly supplied in the form of gaseous CO_2. Assuming a maximal productivity of 20 g *Dunaliella* m^{-2} day^{-1}, the CO_2 demand for maintaining this productivity (assuming that carbon amounts to 50% of the algal ash-free dry weight, see Goldman, 1979b) will be $20 \times 0.5 \times 44/12 = 37$ g CO_2 m^{-2} day^{-1}, or 37 kg day^{-1}, for a single commercial production unit of 1000 m^2. It is clear, therefore, that an efficient carbonation system is essential to minimize CO_2 loss. Since CO_2 is most commonly introduced into the system in the form of bubbles of pure CO_2, methods have been sought that will retain the bubbles until they are completely absorbed by the medium.

The total solubility of inorganic carbon (Ci) is the sum of the solubility of CO_2, HCO_3^- and CO_3^{2-}. The solubility of CO_2 is essentially independent of pH, only weakly dependent on salt concentrations, but strongly temperature dependent (Millero and Morse, 1982). Thus, whereas the solubility of CO_2 at 15 °C and 1 atm. in water and 2 M NaCl is 1.02 and 0.95 ml CO_2 per ml of solution, the values at 40 °C are 0.53 and 0.51, respectively. When the pH is lower than 5.0 the total solubility of Ci is essentially equal to the solubility of CO_2. When the pH is between 5 and 9 the solubility of HCO_3^- need also be considered and when the pH is higher than 9 the solubility of CO_3^{2-} should also be taken into account. In practice, between pH 6.5–9.5 where HCO_3^- accounts for most of the dissolved Ci the following equation (which is correct for 1 atm.) applies:

$$\log (HCO_3^-) = pH - pK_1 + \log [\alpha(\%CO_2)\ 4.46 \times 10^{-4}]$$

where (HCO_3^-) is expressed in moles l^{-1}, α is the solubility of CO_2 (ml CO_2 ml solution^{-1}) and $(\%CO_2)$ is the precentage of CO_2 in the gas phase. The value of the apparent pK in the equation decreases slightly with temperature and rather strongly with an increase in the salt concentration. Thus, the apparent pK_1 for HCO_3^- in water changes from 6.39 at 20 °C to 6.31 at 40 °C, but decreases to around 5.8 in 3 M NaCl solution. The apparent pK_2 is affected even stronger, changing from 10.3 in water to around 8.5 in 3 M NaCl (see Krumgalz, 1980; Lazar et al., 1983).

From these data, it is clear that the maximal amount of Ci which can be dissolved in an algal growth medium is a function of pH, the concentration of CO_2 in the gas phase, the salt concentration and the temperature.

In laboratory cultures of D. salina, provided with 10 mM NaHCO₃ the algae grow best between pH 7.5 and 9.5, and growth rate decreases sharply at higher pH. However, the algae will grow well up to pH 11 if provided with a higher initial NaHCO₃ concentration (Fig. 4.7), presumably because at the higher pH range, the supply of soluble CO_2 becomes limited.

In practice, the rate limiting factor in providing adequate supply of Ci to the algae with gaseous CO_2 is the rate of dissolution. The rate of solution of gaseous CO_2, provided as bubbles, into the liquid depends on contact time of the bubble with the medium, its size and the state of saturation of the medium for CO_2. In experiments which we ran to measure the solubility of CO_2 gas in a vertical column it was found that CO_2 bubbles, with an average diameter of 1 mm, rose at a rate of about 20 cm s^{-1} and were completely dissolved at about 1 m. The solubility characteristics were essentially the same in water, 3 M NaCl or a growth medium containing 3 M NaCl and about 300 mgl^{-1} Dunaliella cells. Special equipment is required to achieve high transfer rates through long residence times and small bubble size. Sprayers with counter-current flow carbonation chambers provide highly efficient carbonation but often cause algal breakdown when used with intact algae (Zenz, 1972). At the other extreme, plastic floating domes containing pure CO_2 provide slow exchange of CO_2 with long residence

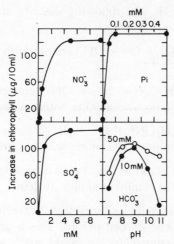

Fig. 4.7 Requirement of *D. salina* for major nutrients. Reaction mixtures contained in a total volume of 10 ml, except as otherwise noted: NaCl, 1.5 M; NaNO$_3$, 5 mM; NaH$_2$PO$_4$, 0.1 mM; MgSO$_4$, 5 mM; KCl, 1 mM; NaHCO$_3$, 10 mM; CaCl$_2$, 0.3 mM; FeCl$_3$, 1.5 μM; EDTA, 30 μM; MnCl$_2$, 7 μM; ZnCl$_2$, 0.8 μM; CoCl$_2$, 20 nM; CuCl$_2$, 0.2 nM; TRIS-CAPS, 40 mM, pH 7.5. Samples were incubated at 25 °C with continuous illumination of white light ($\sim$ 3000 lux) for 72 h. The difference in chlorophyll content between the time of inoculation (about 1 μg/10 ml) to 72 h is plotted. Initial values of nutrient concentrations supplied and pH are indicated.

time. The use of domes is losing favour, mostly due to the gradual increase of photosynthetic O$_2$ partial pressure under the dome. In between are a variety of gas bubbling devices (glass, ceramic, plastic) which produce small bubbles passed either through a special depression within the pond or through columns by the pond with a continuous flow of the pond content.

pH Control

The pH of the medium in which the cells grow affects both the algae and the liquid chemistry. *Dunaliella* has a wide pH optimum between 7 and 9, but even at pH 6 or 10 the algae maintain a good growth rate, about two-thirds of the rate at the optimum pH. For optimal growth a higher Ci concentration is required when the pH exceeds 9.

In natural fresh or sea water (Ca^{2+} content about 10 mM) several calcium salts, most notably carbonates, phosphates and sulphates can precipitate causing algal flocculation and a reduction in growth rate (Engster *et al.*, 1980; Lazar *et al.*, 1983; Sukenik and Shelef, 1984). It is, therefore, important to maintain the pH in the algal culture within the optimal growth range (7–9) and to avoid an increase of the pH above 8 when the growth medium contains more than 1 mM Ca^{2+}.

In autotrophic algal cultures the pH rises mostly due to the light-dependent fixation of CO$_2$ with NO$_3^-$ uptake contributing to a further increase in pH. The conversion of NH$_4^+$ to an amino acid within a protein involves net acidification, while the use of NH$_3$ as the sole source of nitrogen causes no net change in pH. The following describe the overall reactions at neutral pH:

$HCO_3^- + H_2O \rightarrow [CH_2O] + O_2 + OH^-$ (photosynthetic fixation of CO_2 to carbohydrate)

$NO_3^- + [CH_2O]_3 \rightarrow [C_3H_5ON] + 2O_2 + OH^-$ (conversion of nitrate and a carbohydrate to the amino acid alanine within a protein)

$NH_4^+ + [CH_2O]_3 \rightarrow [C_3H_5ON] + H^+ + 2H_2O$ (conversion of NH_4^+ and a carbohydrate to the amino acid alanine within a protein).

$NH_3 + [CH_2O]_3 \rightarrow [C_3H_5ON] + 2H_2O$ (conversion of NH_3 and a carbohydrate to the amino acid alanine within a protein).

Uncontrolled autotrophic algal cultures show, therefore, an increase in pH during the day which can exceed pH 10. This is generally avoided by the use of commercial pH controllers, i.e. a pH meter which has the ability to transmit an on/off signal to a valve at predetermined pH levels. Normally, as the pH increases beyond a predetermined value (e.g. 7.5), a solenoid valve is opened and introduces a flow of CO_2 gas. It closes when the pH is thereby reduced to a predetermined value (e.g. 7.3). There are two major advantages to the use of CO_2 as a pH controlling acid: (a) it is safe in that the pH can never reach dangerously low levels, and (b) it maintains the Ci concentration in the pond approximately constant. Its major disadvantages are the need for a special solubility device and the slow build-up of excess bicarbonate in the medium to compensate for the inorganic nitrogen uptake. The latter can be easily controlled by the occasional addition of HCl. Alternatively, when added HCO_3^- serves as the major source of CO_2 for photosynthesis, addition of HCl through a pH controller can be used to maintain the required pH in the pond. In this case, special safety devices need to be installed to avoid an overdose which can be lethal to the algae (pH lower than 5).

Salt concentration and predators

Dunaliella bardawil grows optimally in a medium containing 1–2 M NaCl (Ben-Amotz and Avron, 1981), but can grow well at higher salt concentrations up to saturated brine. Most organisms cannot tolerate NaCl concentrations which exceed 2 M. The few that are adapted to these elevated salinities include halotolerant and halophilic bacteria, a few ciliates, *Artemia salina*, a few amoebae and certain fungi (Post *et al.*, 1983). The heterotrophic nature of the bacteria does not allow their proliferation in the *Dunaliella* ponds unless the medium contains an organic load in sufficient quantities. Bacterial blooms in *Dunaliella* culture may occur at high temperatures (>35 °C) when the algal membrane loses its integrity and glycerol leaks into the medium (see Wegmann *et al.*, 1980). Similar blooms may occur when the culture is treated by vigorous mechanical means (certain centrifuges or pumps) which break the algae and return broken fragments and glycerol into the pond. A well maintained pond at >2 M NaCl will generally contain a very small number of halophilic bacteria. Few amoebae and zooplankton ciliates survive in media containing more than 2 M NaCl (Post *et al.*, 1983). Some of these organisms can be very dangerous predators of *Dunaliella*, mainly in hot periods when the temperature of the medium rises above 38 °C.

Most commercial *Dunaliella* ponds employ freshwater or seawater, augmented with commercial salt, to reach the desired concentration in the medium. Thereafter the salt medium is used in a closed controlled system by recycling the medium back to the pond after harvesting. Betatene Ltd in Australia, on the other hand, harvests *Dunaliella* from an open salt water pond where the algae grow naturally as the major autotrophic organism. Following harvesting, the effluent is returned to the lagoon and not to the pond, thus minimizing organic contamination of the pond.

Nutrient supply

Dunaliella grows well in a strictly inorganic medium, with no requirement for any organic factors. The major nutrients that need to be considered in mass-culture are discussed in the following sections (see also Kaplan *et al.*, 1986).

Nitrogen

Dunaliella can use either nitrate, ammonia or urea as a source of nitrogen. Since ammonia is toxic when used in concentrations exceeding 5 mM, particularly when the pH exceeds 8 (see Abeliovich and Azov, 1976; Azov and Goldman, 1982), and urea supplies an organic carbon source which stimulates the growth of heterotrophs, nitrate is the preferred source of nitrogen in mass cultures. As can be seen in Fig. 4.7, about 5 mM nitrate was required for maximal growth of *D. salina* under the conditions specified. Nitrogen is not stored to a significant degree and therefore growth ceases as soon as the source of nitrogen is exhausted or removed. The value of a few millimolar, as shown in Fig. 4.7, reflects mostly the quantitative requirement for nitrogen rather than the affinity for nitrate. From shorter-term experiments it is clear that the affinity for nitrate is considerably smaller, below 0.1 mM. Indeed, no detectable nitrate can be determined in algae ponds which ceased growing due to exhaustion of the nitrogen source.

As previously noted (see p. 96), maximal β-carotene production in *D. bardawil* is observed when light intensity is maximal and growth rate is limited. One convenient way to limit the growth rate for maximizing β-carotene production is by limiting the nitrogen supply. The gross composition of the alga is, of course, also changed by such limitation leading to a decrease in protein and lipid content and an increase in total carbohydrates. Thus, *D. bardawil* grown on a complete medium, with 5 mM nitrate, contained on a dry weight basis: 40% proteins, 20% carbohydrates, 18% lipids, 20% glycerol, 1.5% chlorophyll and 0.5% β-carotene; whereas algae grown under the same conditions with 0.5 mM nitrate contained: 24% proteins, 36% carbohydrates, 15% lipids, 20% glycerol, 0.5% chlorophyll and 5% β-carotene (Ben-Amotz, 1987).

Phosphorus

As can be seen in Fig. 4.7, relatively low concentrations of inorganic phosphate, below 0.1 mM, are required for optimal growth of *D. salina*. *Dunaliella* accumulates phosphate when provided with adequate supply, until its total phosphorus content exceeds 0.5 M (L. Karni and M. Avron, unpublished). From NMR analysis it is clear that most of the intracellular phosphate is in the form of polyphosphates (Bental *et al.*, 1988). When transferred to a medium lacking phosphate the cells utilize the stored polyphosphates and will undergo several divisions before phosphate limitation becomes apparent.

The combined presence of calcium and phosphate in cultivation ponds, particularly when the pH exceeds 8, can lead to Ca_3PO_4 precipitation and algal flocculation (see Sukenik and Shelef, 1984), and thus to a serious decrease in algal growth. Phosphate concentration is therefore kept relatively low in·commercial ponds and its level monitored frequently.

Sulphate

Dunaliella requires relatively high sulphate concentration for maximal growth, around 2 mM (Fig. 4.7). When sulphate is limiting cell division is hampered, and the cell size markedly increases. Since it is relatively abundant in seawater (around 30 mM) and tap-water, it seldom needs to be added in commercial cultivation. When present in high concentrations, and in the presence of Ca^{2+} it can precipitate as $CaSO_4$ (solubility around 20 mM in water, and somewhat less in salt solutions), and lead to decreased growth.

Magnesium

Magnesium is required for optimal growth at about 1 mM but is not toxic even at 150 mM. It is accumulated by *Dunaliella* and its intracellular content reaches about 300 mM. An analysis of the ion content of *D. salina* grown under laboratory conditions indicates that magnesium is the major cation which electrically balances the large accumulation of intracellular polyphosphates (Karni and Avron, unpublished).

Magnesium is present in seawater in a concentration around 50 mM. If the pH rises above 9 both $Mg(OH)_2$ and $MgCO_3$ tend to precipitate.

Others

Potassium is required for optimal growth at about 1 mM. *Dunaliella* accumulates potassium intracellularly to a level of about 200 mM (Pick *et al.*, 1986). Seawater contains about 10 mM K^+, and therefore no supplementation of potassium is usually required in commercial cultivation ponds. Nevertheless, some supplementation is usually effected by the addition of the nitrogen source in the form of KNO_3

Calcium is required for optimal growth at about 0.1 mM. Seawater contains about 10 mM calcium, and when the pH exceeds 8, it can precipitate as the following salts: $Ca(HCO_3)_2$, $CaCO_3$, $CaSO_4$ and $Ca_3(PO_4)_2$ (Lazar *et al.*, 1983). These calcium precipitates are a major factor which makes pH control in *Dunaliella* ponds an essential element for a successful operation. To avoid precipitations pH should be maintained below 7.5.

The medium of *Dunaliella* usually contains $FeCl_3$ in a chelated complex, usually with EDTA at around 1 mM. The basic need for iron in algae for different metabolic processes was summarized by O'Kelley (1974), and similar considerations apply to *Dunaliella*. Excess of $FeCl_3$ up to 50 μM was found to be non-toxic to *Dunaliella* under different NaCl concentrations ranging from 0.5–4 M.

Four micronutrients were found to be essential for growth of *Dunaliella*: manganese, zinc, cobalt and copper. These are normally added in the micromolar range (around 5, 1, 0.1 and 0.01, respectively). In most media containing technical salt or seawater, addition of these micronutrients is not needed, as they are present in excess for the algal requirement. The β-carotene rich alga *D. bardawil* requires higher concentrations of around 0.1 μM copper for optimal growth.

Temperature

The effect of temperature on microalgae has been summarized (Soeder and Stengel, 1974; Richmond, 1986a, b). *Dunaliella* has a very wide range of temperature tolerance from below freezing to around 45 °C. Indeed, it is naturally found in areas experiencing low winter temperatures in the Ukraine (Masyuk, 1961), and high summer temperatures in southern California (Klausner, 1986) or in the Dead Sea in Israel (Nissenbaum, 1975). Under laboratory controlled conditions, the optimal growth temperature for *Dunaliella* is around 32 °C with a broad optimum between 25 °C and 35 °C. In medium containing high salt concentrations the temperature tolerance increases by a few degrees.

Glycerol retention by the cells is strongly temperature-dependent. Below 25 °C little or no glycerol is found in the growth medium. Above 25 °C, the rate of glycerol release into the medium gradually increases and above 40 °C it increases dramatically, so that by 50 °C the alga loses all its glycerol into the medium within a few minutes (Wegmann *et al.*, 1980). This leak of glycerol in *Dunaliella* seems to be related to the effect of temperature on membrane organization. It was shown recently (Lynch and Thompson, 1982) that the composition of the phospholipids of the photosynthetic membranes in *Dunaliella* changes during adaptation from growth at 30 °C to 12 °C.

The fact that high culture temperatures cause leakage of glycerol imposes a difficulty in outdoor cultivation of *Dunaliella*. The free extracellular glycerol can serve as an organic carbon substrate for bacteria and fungi. Indeed, bacterial and filamentous fungal contamination is quite common in *Dunaliella* ponds whenever the temperature exceeds 36 °C, particularly when the salinity is maintained below 2.5 M. Thus, the most suitable areas for growing *Dunaliella* are in arid zones where the hot air is dry leading to a significant evaporation rate, which in turn reduces the pond temperature. Indeed, in Eilat, Israel, where the summer air temperature can exceed 43 °C, the high evaporation rate of around 1 cm day^{-1} maintains the temperature in the commercial *Dunaliella* ponds below 36 °C.

Harvesting

The techniques employed for harvesting microalgae in general were recently summarized (Benemann *et al.*, 1980; Richmond, 1986b). *Dunaliella* grown outdoors reaches the same low level of biomass as most other microalgae (around 0.5 g l^{-1}), and therefore the economics and efficiency of the technique employed to separate the product from the over 99% of medium presents a major limiting factor in the commercial utilization of the alga. Of the many methods tried for efficient harvesting of microalgae, only a few were found to be effective at a reasonable operational cost.

Dunaliella possesses some specific features which should be considered in relation to harvesting: (a) the lack of a rigid cell wall and the resulting fragility of the cell membrane; (b) an outer cell layer, termed a surface coat, which possesses mucus properties; (c) the high salinity and therefore density of the growth medium; and (d) the motility of the alga and its rapid phototactic response. These characteristics impose certain limitations on the application of the common harvesting methods used with other microalgae to *Dunaliella*.

In general, filtration of *Dunaliella* through sand filters, cellulose fibres, and other filterable materials has not proved practical. The algae clog the filter rapidly by forming a layer of mucous material which prevents further filtration unless backwashing is performed frequently. Repeated backwashing breaks the algae and releases glycerol and organic matter to the medium (Naghavi and Malone, 1986).

Pressure filtration (Mohn, 1980) has also been tried unsuccessfully with *Dunaliella*. Different commercial apparatuses are available for pressure filtration. They employ low pressure in a closed flowing filter system where the free medium passes through the filter while the algae are gradually concentrated. Units consisting of hollow fibres, filter trays, porous tubings, etc., made of different materials, have not provided continuous filtration for *Dunaliella*. Shortly after a pressure filtration is initiated, *Dunaliella* clogs the filter and an increase of pressure or backwashing provide only a temporary relief.

Large-scale filtration of naturally growing *Dunaliella* in salt ponds by passing the diluted culture through diatomaceous earth has been patented in Australia (Ruane, 1974a). The filtered algae are extracted from the diatomaceous earth with an organic solvent, yielding an extract containing β-carotene which can be further concentrated and purified (Ruane, 1974b). The system suffers from the need to wash the diatomaceous earth from the algal debris, and the low efficiency of filtration.

Attempts to concentrate *Dunaliella* by microstrainers operating continuously with medium backwashing were also rejected, probably due to the small size of *Dunaliella*. Screens with pores around 1 μm are required to obtain efficient harvesting, while much larger pores are sufficient to collect filamentous algae such as *Spirulina* or colonial algae such as *Scenedesmus* (Kormanik and Cravens, 1979).

Centrifugation has been most widely employed for harvesting microalgae (Mohn, 1980). It has been used for many years as the sole harvesting method to collect *Chlorella* in Japan and Taiwan for the preparation of algal health food. Continuous-flow batch-type centrifuges have been most popular. They are very efficient but involve a relatively high initial investment, energy and labour cost for maintenance and discharging. Continuous-flow automatic-discharge (self cleaning) centrifuges were only partially successful in harvesting *Dunaliella*. The major problem noted was a high degree of cell destruction possibly related to the air intake during discharge. In addition, the discharged algal paste (up to 40% algal dry weight) does not flow out unless its water content is increased to about 90%.

Flocculation combined with sedimentation and flotation have been used extensively in waste water treatments (Golueke and Oswald, 1965; McGary, 1971). Flocculation may be spontaneous, or it may be induced by adding chemicals which promote the formation of algal aggregates or flocs which can be floated or sedimented. *Dunaliella* cells flocculate on the addition of flocculants such as aluminium sulphate (alum), ferrous and ferric sulphate, ferric chloride, lime, commercially available polyelectrolytes, etc. As with most other microalgae, alum at concentrations about 150 mg l^{-1} was found to be the most efficient chemical agent for flocculation of *Dunaliella*. The required concentration of flocculant depends on many variables: NaCl concentration, temperature, pH, dissolved organic matter, bacteria, phosphate concentration, etc., and the exact amount to be added should be tested experimentally. The flocced algae cannot be used directly for the food market unless the flocculant is safe or is completely released from the algae prior to utilization. β-Carotene can be extracted from the flocced *Dunaliella*, leaving behind the aluminium contaminated cell debris. The use of alum flocculation in *Dunaliella* ponds which operate in a closed system (i.e. the medium is returned to the pond following harvesting), mandates special care to avoid recycling of the chemical flocculant back into the growth ponds. 'Safe flocculants', such as CaO, Ca(OH)$_2$ and CaCO$_3$ have not been used so far in large-scale operations.

A new method of harvesting *Dunaliella*, hydrophobic binding, was developed recently in Australia (Curtain and Snook, 1983; Curtain *et al.*, 1983). Harvesting is based on the observation that *Dunaliella* grown at salinities higher than 4 M acquires a hydrophobic surface coat. When the algal suspension is in contact with a hydrophobic surface, the algae are adsorbed, and can later be extracted for β-carotene. This process does not seem to be effective with actively growing *Dunaliella*, presumably because the development of a hydrophobic cell coat may be related to the transition from a motile cell to a non-motile cyst, as is common in salt evaporation ponds.

The motility of *Dunaliella* prevents effective use of simple sedimentation for harvesting. However, the features of motility and positive phototaxis can be used to attract the algae by light to migrate onto a smooth 70° angle plate, slide down in the form of a thin layer and be collected as an algal concentrate. This process of phototactic sedimentation was applied to *Dunaliella* in our laboratory with high efficiency. A small-scale continuous counter flow system (Forsell and Hedstrom, 1975) with weakly illuminated clear plastic plates, yielded a consistent production of an algal concentrate of about 5–10% solids (a 100-fold concentration factor). Scaling-up of the phototactic sedimentation units has not been attempted.

Dehydration

β-Carotene can be extracted directly from wet *Dunaliella* paste and this is the process employed in several production facilities. The need for dry *Dunaliella* powder as a product for the health food market has led to the development of several techniques for dehydrating *Dunaliella* paste while protecting its β-carotene. Several methods have been employed to dry microalgae such as *Chlorella*, *Scenedesmus* and *Spirulina*. The most common are spray drying, drum drying, freeze drying and sun drying. The first three have been applied to the β-carotene-rich *Dunaliella* and yielded satisfactory results in terms of uniformity of the powder and stability of the β-carotene. Some grinding is necessary to complete powderizing in drum drying and lyophilization, mainly in algae with a high glycerol content. The sensitivity of β-carotene to oxidation and photodestruction may lead to a rapid loss of β-carotene during the drying process and storage period, unless precautions are taken by the inclusion of antioxidants, working with inert gas and exclusion of light. Similar problems were encountered during the drying and processing of alfalfa, where the necessity to protect and maintain carotenoids was important to enhance the market value of the final product. The information available on the production of dry stable alfalfa (Bauernfeind, 1981), can be applied to the production of *Dunaliella* powder.

The cost of dehydration, when added to the cost of algal growth and harvesting, has formed an economic barrier to the mass production of algal powder produced for low-cost products, such as protein or feed. However, high market value algae such as the β-carotene-rich *Dunaliella* can be economically produced by the presently available technology for microalgal production.

Productivity

All attempts to use microalgae for outdoor production of biomass or chemicals are subjected to yield limitations dictated by the nature of the available light and the photosynthetic process (see Goldman, 1979b; Avron, 1989).

Productivity depends, among others, on the availability of sunlight, the type of photosynthetic machinery of the alga, nutrient availability, temperature, and design characteristics of the culture. Maximal productivity calculations under natural outdoor conditions set an upper limit to the light conversion efficiency in the order of 3% (or 25 g biomass m^{-2} day^{-1}). Paramount in such considerations are losses due to the limited light absorption range (400–700 nm) which decrease the available solar energy by about 50%. The photosynthetic energy conversion efficiency of the absorbed light results in only about 25% of this energy being stored in organic products. Further losses are due to light scattering and refraction ($\sim$ 10%), light saturation ($\sim$ 60%), respiration and photorespiration ($\sim$ 5%), photoinhibition ($\sim$ 10%) and temperature limitation ($\sim$ 20%).

Dunaliella represents a special case in optimizing productivity. The aim is not to increase the productivity of algal biomass but to maximize the production of β-carotene even at the expense of algal biomass. As indicated above, β-carotene accumulation in *Dunaliella* increases the higher the light intensity and the slower the rate of growth. However, under these conditions productivity is rather low due to the slow rate of growth. Thus, in practice, maximal β-carotene productivity is found in areas having high light intensity when the growth rate is somewhat limited. Maximal β-carotene productivity of around 500 mg m^{-2} day^{-1} has been observed, but long-term average yearly yields are around 250 mg m^{-2} day^{-1}. Since such algae contain about 5% β-carotene, these yields correspond to 10 g and 5 g algal dry weight m^{-2} day^{-1}, respectively. Higher biomass productivity can be obtained by growing the algae under conditions promoting a lower accumulation of β-carotene. Cultures of *Dunaliella bardawil* which grow under such conditions (i.e. low salinity, high nitrate concentration) can attain maximal productivities of around 20 g dry algae (containing $\sim$ 0.5% β-carotene) m^{-2} day^{-1}.

Markets

There are three existing avenues for marketing high-β-carotene-*Dunaliella*: for food colouring, as feed and in the health food industry.

β-Carotene has been used for many years as a food colouring agent and the availability of a natural source provides an extra attraction in this area.

β-Carotene has also been widely used as a source of vitamin A in animal feed. Adding dry *Dunaliella* or an extract of the algae high in β-carotene to a vitamin A-deficient chick diet (Ben-Amotz et al., 1986) indeed showed that it is an excellent source of the vitamin and provides in addition, a yolk colour enhancing agent (the latter is most likely due to the lutein present in the algal meal). Similar responses were also shown by vitamin A-deficient rats fed on a diet supplemented with *Dunaliella bardawil* (Ben-Amotz et al., 1988b).

Recent epidemiological and oncological studies suggest that normal to high levels of β-carotene in the body protect against cancer development in humans. These studies provide evidence that humans and animals who maintain above average β-carotene levels have a lower incidence of several types of cancer (Peto *et al.*, 1981; Suda *et al.*, 1986). A comprehensive long-term study, funded by the US National Cancer Institute, is presently being carried out to determine whether a daily intake of β-carotene supplements would reduce the occurrence of cancer in humans. A first approach to the study of the use of *Dunaliella* β-carotene in humans was reported recently (Jensen *et al.*, 1987) through observations on the level of β-carotene isomers in human sera. These observations provide an extra impetus for the use of *Dunaliella* products in the health food market.

Natural β-carotene, as found in *Dunaliella bardawil* and in most fruits and vegetables, contains a mixture of all-*trans* β-carotene together with a few other isomers, predominantly 9-*cis* β-carotene (Ben-Amotz *et al*, 1982, 1988a). As previously noted, the proportion of the 9-*cis* isomers in *Dunaliella* can reach 50% of the total β-carotene. A controlled transformation of the 9-*cis* to the all-*trans* isomer seems to take place in humans (Jensen *et al.*, 1987). It has not been resolved at present whether the beneficial effects of natural β-carotene in cancer prevention are correlated with the isomer composition. However, it is clear that the β-carotene of *Dunaliella* and that of fruits, vegetables, and other light-exposed plant parts is not identical to the synthetic, >99% all-*trans*, β-carotene which presently accounts for most of the commercially available product.

In comparison to the high market value of β-carotene, the other possible products of *Dunaliella* have a lower value even though their market size might be larger than that of β-carotene. The two major secondary commercial products of *Dunaliella* are glycerol and protein. Glycerol is an important commercial organic chemical which is currently produced mostly from petrochemical sources. As previously noted the amount of glycerol in *Dunaliella* is directly proportional to the extracellular salt concentration. Thus, for example, algae grown in a 3 M NaCl medium contain about 40% glycerol (in addition to β-carotene). Of course, growth conditions optimal for maximal glycerol productivity are not identical to those which provide maximal β-carotene production.

The protein of *Dunaliella* is similar in composition to other common plant proteins such as soybean meal, being low in the sulphur amino acids, cysteine and methionine, but is relatively high in lysine (Ben-Amotz and Avron, 1980).

The algal paste of *Dunaliella*, harvested from commercial cultivation ponds may become a source of several other biochemicals, including enzymes, but little work performed thus far points to a specific, economically viable product.

Commercial producers

At present, at least five companies are actively engaged in cultivating *Dunaliella* for commercial purposes. These are:

1. Koor-Foods Ltd, 2 Ibn-Gvirol St., Tel-Aviv 64077, Israel

2. Microbio Resources, Inc., Calipatria, CA 92233, USA
3. Cyanotech Corp., Woodinville, WA 98072, USA
4. Western Biotechnology Ltd, Bayswater, W.A. 6053, Australia
5. Betatene Ltd, Cheltenham, Victoria 3192, Australia

References

Abeliovich A, Azov Y 1976 Toxicity of ammonia to algae in sewage oxidation ponds. *Appl. Env. Microbiol.* **31**: 801–6

Avron M 1989 The efficiency of biosolar energy conversion by aquatic photosynthetic organisms. In Cohen Y, Rosenberg E (eds) *Microbial Mats*, ASM Press, Baltimore, pp 385–7

Avron M 1986 The osmotic component of halotolerant algae. *Trends Biochem. Sci.* **11**: 5–6

Azov Y, Goldman J C 1982 Free ammonia inhibition of algal photosynthesis in intensive culture. *Appl. Env. Microbiol.* **43**: 735–9

Barclay W R, McIntosh R P 1986 *Algal Biomass Technologies. An Interdisciplinary Perspective*, J. Cramer (Beiheffe zur Nova Hedwigia, 83)

Bauernfeind J C (ed) 1981 *Carotenoids as Colorants and Vitamin A Precursors: Technological and Nutritional Applications*, Academic Press, New York

Becker E W (ed) 1985 *Production and Use of Microalgae*, E. Schweizerbart'sche Verlags, Stuttgart, (Ergebnisse der Limnologie 20)

Ben-Amotz A 1987 Effect of irradiance and nutrient deficiency on the chemical composition of *Dunaliella bardawil* Ben-Amotz and Avron (Volvocales, Chlorophyta). *J. Pl. Physiol.* **131**: 479–87

Ben-Amotz A, Avron M 1973 The role of glycerol in the osmotic regulation of the halophilic alga *Dunaliella parva*. *Pl. Physiol.* **51**: 875–8

Ben-Amotz A, Avron M 1980 Glycerol, beta-carotene and dry algal meal production by commercial cultivation of *Dunaliella*. In Shelef G and Soeder C J (eds) *Algae Biomass: Production and use*, Elsevier, Amsterdam, pp 603–10

Ben-Amotz A, Avron M 1981 Glycerol and beta-carotene metabolism in the halotolerant alga *Dunaliella*: a model system for biosolar energy conversion. *Trends Biochem. Sci.* **6**: 297–9

Ben-Amotz A, Avron M 1983a Accumulation of metabolites by halotolerant algae and its industrial potential. *A. Rev. Microbiol.* **37**: 95–119

Ben-Amotz A, Avron M 1983b On the factors which determine massive beta-carotene accumulation in the halotolerant alga *Dunaliella bardawil*. *Pl. Physiol.* **72**: 593–7

Ben-Amotz A, Edelstein S, Avron M 1986 Use of the beta-carotene rich alga *Dunaliella bardawil* as a source of retinol. *Br. Poultry Sci.* **846**: 313–23

Ben-Amotz A, Gressel J, Avron M 1987 Massive accumulation of phytoene induced by norflurazon in *Dunaliella bardawil* (Chlorophyceae) prevents recovery from photoinhibition. *J. Phycol.* **23**: 176–81

Ben-Amotz A, Katz A, Avron M 1982 Accumulation of beta-carotene in halotolerant algae: purification and characterization of beta-carotene rich globules from *Dunaliella bardawil* (Chlorophyceae). *J. Phycol.* **18**: 529–37

Ben-Amotz A, Lers A, Avron M 1988a Stereoisomers of β-carotene and phytoene in the alga *Dunaliella bardawil*. *Pl. Physiol.* **86**: 1286–91

Ben-Amotz A, Mokady S, Avron M 1988b The β-carotene rich alga *Dunaliella bardawil* as a source of retinol in a rat diet *Br. J. Nutr.* **56**: 443–9

Benemann J R, Koopman B, Weissman J C, Eisenberg D, Goebel R 1980 Development of microalgae harvesting and high rate pond technologies in California. In Shelef G, Soeder C J (eds) *Algal Biomass: Production and Use*, Elsevier North Holland Biomedical Press, pp 457–95

Benemann J R, Tillett D M, Weissman J C 1987 Microalgae biotechnology. *Trends Biotechnol.* **5**: 47–53

Bental M, Oren-Shamir M, Avron M, Degani H 1988 ^{31}P and ^{13}C NMR studies of the phosphorus and carbon metabolites in the halotolerant alga *Dunaliella salina*. *Pl. Physiol.* **87**: 320–4

Borowitzka L J, 1981 The microflora: adaptation to life in extremely saline lakes. *Hydrobiologia* **81**: 33–46

Brown A D 1976 Microbial water stress. *Bact. Rev.* **40**: 803–46

Burlew J W (ed) 1953 *Algal Culture from Laboratory to Pilot Plant*, Carnegie Institution of Washington, Publ 600, Washington, DC

Butcher R W 1959 An introductory account of the smaller algae of British coastal waters. I. Introduction and Chlorophyceae. *Fish. Invest. Ser.* **31**: 175–91

Curtain C C, Snook H 1983 *Method for Harvesting Alga*, International Patent WO 83/01257

Curtain C C, Looney F D, Regan D L, Ivancic N M 1983 Changes in the ordering of lipids in the membrane of *Dunaliella* in response to osmotic pressure changes. *Biochem. J.* **213**: 131–6

Dodd J C 1985 Some design approaches to microalgae culture ponds. *Appl. Phycol. Forum* **2**: 4–5

Engster H P, Harrie C E, Weare J H 1980 Mineral equilibria in a six-component seawater system. *Geochim. Cosmochim. Acta* **44**: 1335–47

Felix E A, Rushforth S R 1979 The algal flora of the Great Salt Lake, Utah, USA. *Nova Hedwigia* **31**: 163–95

Forsell B O, Hedstrom B 1975 Lamella sedimentation: a compact separation technique *J. Wat. Pollut. Control Fed.* **47**: 834–42

Goldman J C 1979a Outdoor algal mass cultures. I. Application. *Wat. Res.* **13**: 1–19

Goldman J C 1979b Outdoor algal mass cultures. II. Photosynthetic yield limitations. *Wat. Res.* **13**: 119–36

Golueke C G, Oswald W J 1965 Harvesting and processing sewage-grown planktonic algae. *J. Wat. Pollut. Control Fed.* **37**: 471–98

Heussler P, Castillo S J, Merino M F, Vasquez V V 1978 Improvements in pond construction and CO_2 supply for the mass production of algae. In Soeder C J, Binsack R (eds) *Microalgae for Food and Feed: A Status Analysis*, E Schweizerbart'sche Verlags, Stuttgart, pp 254–58 (Ergebnisse der Limnologie 11)

Jensen C D, Howes T W, Spiller G A, Pattison T S, Whittan J H, Scala J 1987 Observations on the effects of ingesting *cis*- and *trans*-beta-carotene isomers on human serum concentrations. *Nutr. Rep. Int.* **35**: 413–22

Kaplan D, Richmond A, Dubinsky Z, Aaronson S 1986 Algal nutrition. In Richmond A (ed) *Handbook of Microalgal Mass Culture*, CRC Press, pp 147–63

Klausner A 1986 Alga culture: food for thought. *Biotechnology* **4**: 947–53

Kormanik R A, Cravens J B 1979 Removal of algae through microscreening. *Wat. Wastes Eng.* Aug, 31–5

Krumgalz B 1980 Salt effect on the pH of hypersaline solutions. In Nissenbaum A (ed) *Hypersaline Brines and Evaporative Environments*, Elsevier, Amsterdam, pp 73–83

Laws E A, Terry K L, Wickman J, Chalup M S 1983 A simple algal production system designed to utilize the flashing light effect. *Biotechnol. Bioeng.* **25**: 2319–35

Laws E A, Taguchi S, Hirata J, Pang L 1986 Continued studies of high algal productivities in shallow flume. *Biomass* **11**: 39–50

Lazar B, Starnisky A, Katz A, Sass E, Ben-Yaakov S 1983 The carbonate system in hypersaline solutions: alkalinity and $CaCO_3$ solubility of evaporated sea water. *Limnol. Oceanogr.* **28**: 978–86

Loeblich L 1982 Photosynthesis and pigments influenced by light intensity and salinity in the halophilic *Dunaliella salina* (Chlorophyta). *J. Mar. Biol. Assoc. UK* **62**: 493–508

Lynch D V, Thompson G A 1982 Low temperature-induced alterations in the chloroplast and microsomal membranes of *Dunaliella salina*. *Pl. Physiol.* **69**: 1369–75

Masyuk N P 1961 Carotene producing alga *Dunaliella salina* Teod in the salt water reservoir of the Krynsh Oblash. *Ukr. Bot. Zh.* **18**: 100–7

McGary C G 1971 Algal flocculation with aluminium sulfate and polyelectrolites. *Algal Flocc.* **42**: 191–201

Millero F J, Morse J W 1982 The carbonate system in seawater. In Zaborsky D R, Mitsui A, Black C C (eds) *Handbook of Biosolar Resources* Vol. 1 part II CRC Press, pp 517–29

Mohn F H 1980 Experience and strategies in the recovery of biomass from mass cultures and microalgae. In Shelef G, Soeder C J (eds) *Algal Biomass: Production and Use*, Elsevier North Holland Biomedical Press, pp 547–71

Naghavi B, Malone R F 1986 Algae removal by fine sand/silt filtration. *Wat. Res.* **20**: 377–83

Nissenbaum A 1975 The microbiology of the Dead Sea. *Microbiol. Ecol.* **2**: 139–61

O'Kelley J C 1974 Inorganic nutrients. In Stewart W D P (ed) *Algal Physiology and Biochemistry*, Blackwell, pp 610–35

Oswald W J 1988 Large scale algal culture systems (engineering aspects). In Borowitzka L, Borowitzka M (eds) *Micro-algal Biotechnology*, Cambridge University Press, pp 357–94

Parkinson G 1987 New techniques may squeeze more chemicals from algae. *Chem. Eng.* **94**: 19–22

Peto R, Doll R, Buchley J D, Sporn M B 1981 Can dietary beta-carotene materially reduce human cancer rates? *Nature* (Lond.) **290**: 201–7

Pick U, Karni L, Avron M 1986 Determination of ion content and ion fluxes in the halotolerant alga *Dunaliella salina*. *Pl. Physiol.* **81**: 92–6

Post F J 1981 Microbiology of the Great Salt Lake north arm. *Hydrobiologia* **81**: 56–69

Post F J, Borowitzka L J, Borowitzka M A, Mackay B, Moulton T 1983 The protozoa of a Western Australia hypersaline lagoon. *Hydrobiologia* **105**: 95–113

Powles S B 1985 Photoinhibition of photosynthesis induced by visible light. *A. Rev. Pl. Physiol.* **35**: 15–44

Richmond A 1986a Microalga culture. *CRC Crit. Rev. Biotechnol.* **4**: 369–438

Richmond A 1986b *Handbook of Microalgal Mass Culture*, CRC Press

Ruane M 1974a *Recovery of Algae from Brine Suspensions*, Australian Patent 72395/74.

Ruane M 1974b *Extraction of Caroteniferous Materials from Algae*, Australian Patent 72395/74.

Seltik I, Veladmir S, Malek I 1970 Dual purpose open circulation units for large scale culture of algae in temperate zones. I. Basic design and considerations and scheme of pilot plant. *Algal Studies* (Trebon): 11–20

SERI 1985 *Aquatic Species Program Review*, SERI Publication CP 231–2700, US Dept of Energy

Shelef G, Soeder C J (eds) 1980 *Algae Biomass: Production and Use*, Elsevier North Holland Biomedical Press

Soeder C J, Binsack R (eds) 1978 *Microalgae for Food and Feed, a Status Analysis*, E. Schweizerbart'sche Verlags, Stuttgart (Ergebnisse der Limnologie 11)

Soeder C J, Stengel E 1974 Physico-chemical factors affecting metabolism and growth rate. In Stewart W D P (ed) *Algal Physiology and Biochemistry*, Blackwell, pp 714–40

Suda D, Schwartz J, Shklar G 1986 Inhibition of experimental oral carcinogenesis by topical beta-carotene. *Carcinogenesis* **7**: 711–5

Sukenik A, Shelef G 1984 Algal autoflocculation: verification and proposed mechanism. *Biotechnol. Bioeng* **26**: 142–7

Tamiya H 1957 Mass culture of algae. *A. Rev. Pl. Physiol.* **8**: 309–34

Vendlova J 1969 Outdoor cultivation in Bulgaria. *A. Rep. Algol. Lab. Trebon* p 143

Vonshak A 1987 Biological limitations in developing the biotechnology for algae mass cultivation. *Sci. l'eau* **6**: 99–103

Wegmann K, Ben-Amotz A, Avron M 1980 The effect of temperature on glycerol retention in the halotolerant algae *Dunaliella* and *Asteromonas*. *Pl. Physiol.* **66**: 1196–7

Zenz F A 1972 Designing gas absorption. *Chem. Eng.* **79**: 120–38

5

The biotechnological potential of symbiotic algae and cyanobacteria

T. A. V. Rees

Introduction

Definitions of symbiosis include the broad definition of de Bary (1879) which states that it is a relationship between two or more differently named organisms which live together, and that which commonly equates symbiosis with mutualism thereby implying that all organisms in the relationship derive benefit from the association. The former definition encompasses parasitism whereas the latter specifically excludes it. Douglas and Smith (1989) have reviewed the problems associated with these definitions and propose a new and practical definition. This states that symbiosis is a persistent and intimate association between organisms of different size in which the larger organism (host) utilizes novel or enhanced capabilities by the smaller partner(s) (symbiont(s)). For associations with algae or cyanobacteria as symbionts the capability utilized by the host is the release of metabolites. There is considerable information on the metabolites released by algal and cyanobacterial symbionts to the host, and it is this information which forms the basis of this review.

A symbiotic association involves a degree of permanence but the intimacy of the relationship varies. Associations involving algae or cyanobacteria include those where the symbiont is ectosymbiotic or endosymbiotic, i.e. on or within the host tissue; endosymbionts are either extracellular or intracellular. It is to be expected that this variability in the location of the symbionts will affect their degree of structural, biochemical and genetic integration with the host; the extent of this integration has important implications for the biotechnological potential of symbiotic algae and cyanobacteria. Symbiotic algae are, in general, unicellular with the exception of some lichen algae which are filamentous; symbiotic cyanobacteria are predominantly filamentous. The general biology of symbiotic associations has been reviewed recently by Smith and Douglas (1987).

Symbiotic associations with algae and cyanobacteria may be regarded as natural immobilized cell systems. Immobilization of cells in biotechnology involves supporting cells in a porous support material such as alginate, agar, polyurethane or polyvinyl foams with the porosity of the support material allowing diffusion of reagents and products. The major advantages of immobilized cell systems are that they allow increased cell densities, often avoid the necessity of extraction and purification of the products and facilitate continuous production of the products because of their removal by diffusion (Rao and Hall, 1984). Despite these advantages the isolation of commercially important products in a pure form remains a major problem for biotechnology.

In symbiotic associations the host represents the support material. Though movement of algal and cyanobacterial products occurs, it is unlikely that host metabolites are freely available to the symbionts. Symbiont densities are often far greater than occur naturally with non-symbiotic algae and cyanobacteria in the free-living state. Release of metabolites by symbiotic algae and cyanobacteria is continuous, though often at a reduced or negligible rate in darkness, with the released product being stored (e.g. as glycogen) or metabolized by the host. It is this continuous release of metabolites by symbiotic algae and cyanobacteria which makes them of potential interest to biotechnologists, particularly as the metabolite released is often in the form of a single compound, and therefore obviates the necessity for extraction and purification. There is little evidence that host digestion or necrotrophy of symbiotic algae is a common occurrence. Moreover, the mechanism of this release of specific metabolites is also of potential importance because, if understood, it could enable modifications to be made to the membrane properties of non-symbiotic organisms so as to facilitate the release of specific products.

This chapter will review the range of metabolites released by symbiotic algae and cyanobacteria, factors which affect this release, and possible mechanisms involved. Detailed comments are inevitably restricted to the more intensively studied symbiotic associations.

Symbiotic cyanobacteria and ammonia

The majority of symbiotic cyanobacteria are filamentous and possess heterocysts, which are the site of the nitrogen-fixing enzyme, nitrogenase. Nitrogenase is very sensitive to oxygen. Heterocysts are specialized, differentiated cells with thick walls; they lack photosystem II, are unable to fix CO_2 and do not evolve oxygen. These structural and biochemical features of the heterocyst enable the cyanobacterial nitrogenase to operate under aerobic conditions. It is the ability to fix nitrogen which makes cyanobacteria attractive organisms to biotechnologists. Symbiotic cyanobacteria are potentially of particular interest because the product of nitrogen-fixation, ammonia, is released in substantial amounts to the host. Values range from 40% of nitrogen-fixed in the lichen *Peltigera canina* (Rai *et al.*, 1983) to 90% or more in the liverwort *Blasia pusilla* (Stewart and Rodgers, 1977), the tripartite lichen *Peltigera aphthosa* (Rai *et al.*, 1981a) and the liverwort

Anthoceros punctatus (Meeks *et al.*, 1985). In *Peltigera canina* the symbiont, *Nostoc*, also releases glucose (Richardson *et al.*, 1968). Production of ammonia from nitrogen, both chemically (Haber process) and biochemically (nitrogenase), is energetically very costly. The potential advantage of symbiotic cyanobacteria is that ammonia can be produced at a relatively low financial cost.

Symbiotic cyanobacteria form associations with invertebrates (sponges and echiuroid worms) and plants. It is the latter associations which have been more intensively studied.

Cyanobacteria form associations with liverworts (e.g. *Anthoceros* and *Blasia*), mosses (e.g. *Sphagnum*), the fern *Azolla*, about a third of the species of cycads, and the angiosperm *Gunnera*. In the liverworts the cyanobacteria are intercellularly located in cavities on the undersurface of the thallus. Cyanobacteria forming associations with the moss *Sphagnum* may occur intracellularly but most are extracellular. In *Azolla* the symbiont (*Anabaena*) is located in cavities at the base of the dorsal lobe of each leaf; the *Nostoc* symbiont of *Gunnera* is also found at the base of the leaves, but unlike *Anabaena* in *Azolla* the symbionts are intracellularly located in meristematic cells. The cycads are the only plants with symbiotic cyanobacteria located in the roots where they occupy a cylindrical region between the inner and outer cortex. All cyanobacteria which form associations with higher plants are dependent to a considerable extent on the host for their carbon and energy supply (Stewart *et al.*, 1980).

Symbiotic cyanobacteria usually have higher rates of nitrogen-fixation (Stewart *et al.*, 1980) and higher heterocyst frequencies (Table 5.1) than their free-living counterparts. An exception are the cyanobacteria which form bipartite associations in lichens; these cyanobacteria have rates of nitrogen-fixation and heterocyst frequencies which are comparable to free-living cyanobacteria.

The principal reason for release of ammonia by symbiotic cyanobacteria appears to be a combination of high nitrogenase activity and low glutamine synthetase levels. The *Nostoc* symbionts of *Peltigera aphthosa* and *P. canina* possess 3% and 6% the amount of glutamine synthetase protein of their free-living counterparts; the value for *Anabaena* in *Azolla caroliniana* is 5% (Orr and Haselkorn, 1982) or 20% (Stewart *et al.*, 1980). However, there are exceptions. In *Anthoceros punctatus*, grown with N_2, the symbiotic *Nostoc* possesses 32% of the glutamine synthetase activity but 74% of the glutamine synthetase protein level found in free-living *Nostoc*; growth on ammonia gives values of 48% activity and 176% protein level compared with ammonia-grown free-living *Nostoc* (Joseph and Meeks, 1987). The authors suggest that, in this association, post-translational control of cyanobacterial glutamine synthetase is operating and, moreover, that glutamine synthetase and its regulation are not involved directly in control of heterocyst differentiation. *Nostoc* in the cycads *Cycas* and *Zamia* have relatively high glutamine synthetase activities and protein levels suggesting that, if a nitrogenous compound is released, it is unlikely to be ammonia (Lindblad and Bergman, 1986). Comparable results have been obtained with activities

Table 5.1   Heterocyst frequencies in symbiotic cyanobacteria

Association	Heterocyst frequency (%)	Reference
Lichens – cyanobacteria as sole symbionts		
Lobaria scrobiculata	3.9	
Nephroma laetevirens	4.1	
N. parile	4.9	
Peltigera canina	4.9	
P. evansiana	4.7	Kershaw (1985)
P. polydactyla	5.8	
P. praetextata	4.4	
P. venosa	7.8	
Sticta fuliginosa	6.0	
S. limbata	4.9	
Lichens – cyanobacteria located in cephalodia		
Lobaria amplissima	21.6	
L. laetevirens	30.4	
L. pulmonaria	35.6	
Nephroma arcticum	14.1	Kershaw (1985)
Peltigera aphthosa	21.1	
Solorina crocea	17.8	
S. saccata	14.7	
Bryophytes		
Anthoceros punctatus	43	Rodgers and Stewart (1977)
Blasia pusilla	30	Rodgers and Stewart (1977)
Ferns		
Azolla sp.	20–30	Hill (1975)
Gymnosperms		
Cycas circinalis	30	Grilli Caiola (1980)
C. revoluta – root tip	7	Lindblad and Bergman (1986)
C. revoluta – root base	30	Lindblad and Bergman (1986)
C. rumphii	20	Grilli Caiola (1980)
Dioon edule	30	Grilli Caiola (1980)
Encephalartpos altensteinii	20	Grilli Caiola (1980)
E. lebomboensis	15	Grilli Caiola (1980)
E. lehmannii	25	Grilli Caiola (1980)
E. longifolius	30	Grilli Caiola (1980)
E. natalensis	20	Grilli Caiola (1980)
E. villosus	25–30	Grilli Caiola (1980)
E. umbeluziensis	30	Grilli Caiola (1980)
Macrozamia communis	30	Grilli Caiola (1980)
Zamia skinneri – root tip	17	Lindblad *et al.* (1985)
Z. skinneri – root base	45	Lindbald *et al.* (1985)
Angiosperms		
Gunnera arenaria	80	Silvester and McNamara (1976)

of glutamate synthase with decreases in activities in *Nostoc* from *Peltigera* and *Anabaena* from *Azolla caroliniana* (Rai *et al.*, 1981b), but comparable levels to free-living cyanobacteria in *Nostoc* from·*Cycas revoluta* (Lindblad et al., 1987).

The potential use of cyanobacteria in biotechnology is discussed in Ch. 3 and 9.

Lichen algae and cyanobacteria

Lichens are symbiotic associations between fungi (mainly Ascomycete with a few Basidiomycetes) and algae; about 8% of lichen species have cyano-bacterial symbionts. All symbionts are extracellular. Most associations contain one species of symbiont but a few contain both an algal and a cyanobacterial symbiont, with the latter confined to specialized structures called cephalodia. In most lichens the symbionts occur in a distinct layer below the upper cortical layer of fungal hyphae, and constitute 3–10% of the total biomass (Smith, 1975). The exception are some species of lichen with cyanobacterial symbionts which are distributed throughout the lichen, and which may make up to 50% of the total biomass.

There is substantial movement of photosynthate from the algal and cyano-bacterial symbionts to the fungal host with over 90% of fixed carbon being released by symbionts in *Cladonia convoluta* (Tapper, 1981). Photosynthetic products are in the form of a single compound, with cyanobacteria releasing glucose, and algae polyols (Table 5.2). It is interesting to note that lichen cyanobacteria are the only documented instance of symbiotic cyanobacteria

Table 5.2 Photosynthetic products released by lichen algae and cyanobacteria

Symbiont	Number of lichens investigated	Photosynthetic product released
Algae		
Coccomyxa	3	ribitol
Hyalococcus	2	sorbitol
Myrmecia	4	ribitol
Stichococcus	2	sorbitol
Trebouxia	13	ribitol
Trentepohlia	6	erythritol
Cyanobacteria		
Calothrix	1	glucose
Nostoc	10	glucose
Scytonema	1	glucose

(Source: Smith, 1980)

releasing a non-nitrogenous product, though lichen cyanobacteria also release ammonia (see p 116). A major problem for the potential use of lichen algae and cyanobacteria in biotechnology is that there is little information on the biosynthetic pathways (or their control), especially for polyols, or the mechanism of release. Though mannitol is synthesized in appreciable quantities by a number of algae (Lewis and Smith, 1967), the biosynthesis of erythritol, ribitol and sorbitol, in substantial amounts, appears to be an unique characteristic of algae in lichen associations. An exception is sorbitol, which is produced by both symbiotic and free-living algae of the genus *Stichococcus*. In free-living *Stichococcus* sorbitol is produced as a compatible solute during osmoregulation (Brown and Hellebust, 1980). A further problem, which could limit their biotechnological potential, is that release of photosynthate by lichen algae and cyanobacteria decreases rapidly on isolation from the association, such that 6–12 h after isolation of symbionts from *Peltigera aphthosa* only about 5% of their photosynthate is released (Green and Smith, 1974). Cultured lichen algae grow very slowly and release little or no photosynthate (Smith *et al.*, 1969). This decrease in release of photosynthate is accompanied by an increase in incorporation of carbon into the ethanol-insoluble fraction, particularly polysaccharide. The mechanism of polyol and glucose biosynthesis in the intact association presumably prevents this diversion of carbon to polysaccharide. Neither low pH nor host homogenate induces an increase in the release of photosynthate (Smith, 1975).

In a discussion of possible mechanisms involved in the induction of release of photosynthate in the lichen symbiosis, Smith (1974) suggested the following:

(a) Several chemical substances are produced by the fungus to effect release. A number of effectors rather than a single compound would be required, as several metabolic processes are affected, sometimes in both algae and cyanobacteria within the same lichen. He concluded that this production is unlikely.

(b) *Specific* transport mechanisms and enzymes are affected. He suggested that the physical contact between fungus and symbiont is important, with some membrane property such as membrane potential being affected. This might result in either a reversal in the normal direction (i.e. inwards) of nutrient flux, or a reduction in the inward flux relative to the outward flux. Charge might be carried along protein, polysaccharide or glycoprotein links between alga and fungus.

Recent results obtained from symbiotic associations suggest that such mechanisms may have some validity. A number of characteristics associated with rhizobia (symbiotic nitrogen-fixing bacteria present in root nodules of leguminous plants) can be expressed in cultured *Bradyrhizobium* by increasing the concentration of potassium ions (Gober and Kashket, 1987). Growth on low oxygen (0.2%), in the presence of high potassium concentrations (8–12 mM), causes development of the ability to reduce acetylene, induction of the ammonia transport system and the K^+/H^+ antiporter, increased levels of the haem biosynthetic enzymes and repression of glutamine synthetase and capsular polysaccharide synthesis. It would appear that a number of metabolic processes associated with a symbiotic organism can be effected by a single compound. That such changes occur in *Bradyrhizobium* in the presence of high potassium concentrations is of particular interest as Smith (1974) suggests that release of cations by the fungus in the lichen association may effect release of polyols or glucose by the symbiont.

The suggestion that physical contact may be important in facilitating release (Smith, 1974) is also supported by recent evidence. Immobilization of cultures of the symbiotic cyanobacterium *Anabaena azollae* results in the restoration of some of the characteristics associated with the cyanobacterium when in symbiosis with the fern *Azolla* (Shi *et al.*, 1987). These include release of ammonia and increased heterocyst frequency, acetylene reduction, hydrogenase activity and mucilage production. Whether such changes are due to physical contact (i.e. immobilization), and/or the ionic environment of the immobilization matrix, remains to be elucidated. That release of glucose in the lichen *Peltigera polydactyla* continues in the presence of digitonin (which selectively affects fungal membrane permeability) (Chambers *et al.*, 1976), suggests control of the ionic environment by the fungus may not be important. However, it is possible that charged sites in the fungal cell wall, and/or the gelatinous sheath which surrounds the *Nostoc* symbionts (Honegger, 1984), are already occupied. It is noteworthy that immobilization of another symbiotic cyanobacterium, *Anabaena azollae*, causes increased mucilage production (Shi *et al.*, 1987). Use of the immobilization techniques described by Shi *et al.* (1987) with lichen algae and cyanobacteria, to determine whether this induces release of photosynthate, would be of interest. It would be a pleasing irony if biotechnological techniques were to provide important insights into aspects of symbiotic associations!

The release of photosynthate (as a single compound), by symbiotic algae and cyanobacteria from lichens, could be coupled to the *generation* of an electrochemical gradient of protons. In membrane vesicles of *Escherichia coli* an electrical potential across the cell membrane is generated by the carrier-mediated efflux of lactose (Kaczorowski and Kaback, 1979; Kaczorowski *et al.*, 1979), and uptake of amino acids by *E. coli* can be driven by the efflux of gluconate (Bentraboulet *et al.*, 1979). Furthermore, lactate efflux by *Streptococcus cremoris* results in at least a 12% net energy gain, and efflux drives leucine uptake which is sensitive to the protonophore carbonyl cyanide-4-trifluoromethoxy-phenylhydrazone (FCCP) (Otto *et al.*, 1980). The ribitol concentration in the symbiont *Trebouxia* of *Hypogymnia physodes* is at least 260 mM (Farrar and Smith, 1976); in principle a concentration high enough to generate an electrochemical gradient of protons (Michels *et al.*, 1979).

Symbiotic marine algae

Symbiotic dinoflagellates

Symbiotic dinoflagellates form associations with marine invertebrates, in particular coelenterates and molluscs. Of these, the most ecologically important are reef-building corals. Most symbiotic dinoflagellates belong to the genus *Symbiodinium*, the exception being *Amphidinium klebsii* which forms associations with turbellarians. Though *Symbiodinium*, from a range of marine associations, has been classified as a single species, *S. micro-adriaticum*, recent evidence suggests that symbiotic marine invertebrates harbour separate and distinct species of the genus *Symbiodinium* (Trench and Blank, 1987). This apparent taxonomic diversity is reflected in biochemical (Table 5.3) and ultrastructural (Trench and Blank, 1987) differences between symbiotic dinoflagellates from different associations.

Dinoflagellates are unicellular motile algae with the peculiar feature of permanently condensed chromosomes. In symbiotic associations, *Symbiodinium* displays the same features, except that the two flagella are lost. However, cultured symbiotic dinoflagellates display normal (i.e. free-living) morphology. In most symbiotic associations the dinoflagellates are located intracellularly in the gastrodermal layer and are surrounded by a membrane of host origin. An exception are the symbionts of the Tridacnidae (giant clams) where the symbionts are intercellular and are located in the haemal sinuses of the enlarged siphonal tissue (Trench *et al.*, 1981).

Freshly isolated symbionts of marine alga–invertebrate symbioses release a proportion of their photosynthate (Table 5.3a). There are a number of products, the proportion of which depends on the host from which the symbionts were isolated and the experimental conditions (Table 5.3b). However, the intracellular and extracellular distributions of ^{14}C-labelled compounds are sufficiently different to suggest that the release is not due to cell lysis. The major compounds released are glycerol (21–95%), organic acids (0–69%), glucose (0.5–21%) and alanine (1–9%). Differences between *Symbiodinium* isolated from different hosts are exemplified by *Zoanthus pacifica* and the Scyphozoan *Rhizostoma*. Symbionts from *Zoanthus* release 42% of their photosynthate, of which 95% is glycerol and 3% organic acids, whereas symbionts of *Rhizostoma* release 20% of their photosynthate of which 21% is glycerol and 69% organic acids (Trench, 1971b).

In common with many symbiotic algae the proportion of photosynthate released declines with time after isolation. For example, symbionts of the anemone *Anthopleura elegantissima* release 31% of their photosynthate when freshly isolated, but 10 h after isolation this value declines to 6% (Trench, 1971c). However, the proportion of photosynthate released immediately after isolation, and 10 h after isolation, is increased by the addition of host homogenate (Table 5.3b). Moreover, the proportion of photosynthate released, in the presence of host homogenate, by symbionts 10 h after isolation is the same as for freshly isolated symbionts (Trench,

1971c). In the absence of host homogenate, the mean percentage release
for symbionts from a range of alga–invertebrate symbioses is 10%, whereas
in the presence of host homogenate it increases to about 30% (Hinde,
1983). Host homogenate increases the photosynthetic rate of freshly isolated
symbionts (Trench, 1971c; Muscatine et al., 1972), and also alters the

Table 5.3 Photosynthetic products released by marine symbiotic algae.
(a) Effect of host homogenate on photosynthate release by freshly isolated
symbionts

Host	Percentage of total fixed carbon released		Reference
	Seawater	Seawater + host homogenate	
COELENTERATA			
Scleractinia:			
Agaricia agaricites	21	44	Muscatine et al. (1972)
Fungia scutaria		25	Trench (1971b)
Plesiastrea versipora	5	9–42	Hinde (1983)
Pocillopora damicornis	4	38	Muscatine (1967) Muscatine and Cernichiari (1969)
Alcyonaria:			
Capnella gaboensis	4		Hinde (1983)
Zoanthidea:			
Zoanthus pacifica	10	42–47	Trench (1971 b,c)
Zoanthus pulchellus	7		von Holt and von Holt (1968)
Zoanthus sociatus	32	45–48	von Holt and von Holt (1968) Trench (1974)
Actiniaria:			
Aiptasia pulchella		35	Trench (1971b)
Anemonia sulcata	2	3	Gallop (1974)
Anthopleura elegantissima	31	49–58	Trench (1971b,c)
Macranthea cookei		40	Trench (1971b)
Rhizostomae:			
Cassiopeia frondosa		23	Trench (1971b)
Rhizostoma sp.		20	Trench (1971b)
PLATYHELMINTHES			
Acoela:			
Convoluta roscoffensis	1–5	5–16	Muscatine et al. (1974) Boyle and Smith (1975)
MOLLUSCA			
Heterodonta:			
Tridacna crocea	2	37	Muscatine (1967)
Nudibranchia:			
Pteraeolidia ianthina	2	1–10	Hinde (1983)

Table 5.3 Cont.

(b) Compounds released by freshly isolated symbionts in the presence of host
homogenate

Host	Percentage distribution of released photosynthate						Reference
	Gly-cerol	Glu-cose	Ala-nine	Organic acids	Gly-collate	Lipid	
COELENTERATA							
Agaricia agaricites	62.5	13.4	7.2			13.2	Muscantine *et al.* (1972)
Fungia scutaria	61.2	5.0	4.5	13.7	8.5		Trench (1971b)
Pocillopora damicornis	82.0	0.7	1.7		6.4		Muscatine and Cernichiari (1969)
Zoanthus pacifica	95.0	0.9	1.1	3.0			Trench (1971b)
Zoanthus sociatus	75.0	16.0	1.0				Trench (1974)
Aiptasia pulchella	24.8	1.3	5.4	65.3	3.2		Trench (1971b)
Anthopleura elegantissima	43.0	0.5	1.2	53.0	2.3		Trench (1971b)
Macranthea cookei	42.6	20.8	5.6	26.0			Trench (1971b)
Cassiopeia frondosa	43.5	14.7	7.3	30.9	3.6		Trench (1971b)
Rhizostoma sp.	20.7	1.8	4.3	69.2	4.0		Trench (1971b)
MOLLUSCA							
Tridacna crocea	80.0	4.0	9.0				Trench (1979)

composition of the released photosynthate. For freshly isolated symbionts
from *Anthopleura elegantissima*, incubated without host homogenate, 19%
of the photosynthate released is glycerol and 80% organic acids. In the presence of host homogenate 43% is glycerol and 55% organic acids.
Furthermore, the proportions of the release products change with time
after isolation. Of the photosynthate released by freshly isolated symbionts
from *Anthopleura* 43% is as glycerol, but this declines with time after
isolation, and there is a corresponding increase in the proportion of alanine
released. There is also a decrease in the amount of organic acids, particularly fumarate and succinate, released; this may reflect utilization of organic
acid carbon for the biosynthesis of alanine. Release of organic acids may
reflect an absence of nitrogen in the medium, i.e. in the presence of nitrogen
the release products would be alanine and/or other amino acids. The presence of host homogenate has no stimulatory effect on the proportion of
photosynthate released as alanine up to 5 h after isolation of symbionts, but
markedly increases it thereafter (Trench, 1971c). It is possible that this

reflects increasing nitrogen deficiency and the presence of ammonia in the host extract (Hinde, 1983). However, there have been no experiments conducted on the effects of either nitrogen deficiency or ammonia on release of photosynthate by symbiotic dinoflagellates (see below).

These observations have led to the development of the concept of the 'host factor', i.e. the host in symbiosis produces a compound (or compounds) which stimulates release of photosynthate by the symbionts. The identity of the 'host factor' is unknown, but if elucidated could prove of considerable biotechnological potential. The factor is not present in either aposymbiotic (i.e. animals rid of their symbionts) or non-symbiotic animals, but is present in aposymbiotic animals reinfected with symbiotic algae (Hinde, 1983). Furthermore the 'host factor' is often not species specific. For example, host homogenate from the coral *Pocillopora damicornis* will stimulate release of photosynthate by freshly isolated symbionts from the giant clam *Tridacna crocea* (Muscatine, 1967). The factor is soluble (Muscatine *et al.*, 1972), is not lost on dialysis (Hinde, 1983) and is thermolabile (Muscatine, 1967; Trench, 1971c), suggesting the possibility that the factor is proteinaceous. Moreover, Hinde (1983) has suggested that denaturation of bound 'host factor' may account for the decrease in the amount of photosynthate released with time of isolation. A low molecular weight (2–10 KDa) fraction from the host homogenate of the jellyfish *Cassiopea xamachana* markedly inhibits the transport of alanine by freshly isolated symbionts, and is effective at concentrations as low as 5 μg protein ml^{-1} (Carroll and Blanquet, 1984a,b). Unfortunately there is no information on the effect of this fraction on release of photosynthate by the symbionts.

A problem with all investigations on the effect of the 'host factor' is that freshly isolated symbionts are contaminated with host protein (Steen, 1986). It remains a distinct possibility that the results obtained with 'host factor' are an artefact of interactions between these surface 'contaminants' and soluble 'host factor'. Alternatively, the surface 'contaminants' may be an integral component of the release mechanism (Mews, 1980; Hinde, 1983; Douglas, 1987, 1988).

The foregoing discussion implies that symbiotic dinoflagellates release only low molecular weight metabolites. However, evidence is accumulating that suggests that this may be misleading. Indeed, early work with the coral *Agaricia agaricites* showed that up to 13% of fixed ^{14}C released by symbionts is lipid (Muscatine *et al.*, 1972). Subsequent work, mainly by Patton's group, has suggested that in the intact association, release of photosynthate is primarily in the form of lipid and glycerol. Release of lipid by symbiotic dinoflagellates would account for the high levels of lipid in the host; 34% of the dry weight of surface tissue of the coral *Pocillopora capitata* (Patton *et al.*, 1977) and 43% in *Condylactis gigantea* (Kellogg and Patton, 1983). In the coral *Pocillopora capitata* the proportion of ^{14}C which appears in host lipid is greatest in the triglyceride fraction, with 80% of the lipid being composed of myristic acid, palmitic acid, palmitoleic acid and stearic acid (Patton *et al.*, 1977). The observations that freshly isolated symbionts from the sea anemone *Condylactis gigantea* and the coral *Stylophora pistillata* release fat droplets, which are predominantly triglyceride (Kellogg and Patton, 1983; Patton and Burris, 1983), provides considerable support for the suggestion that lipid is a major release product of symbiotic dinoflagellates. There is, however, an apparent discrepancy in the distribution of ^{14}C within host lipid. It can be present exclusively in the glycerol moiety (Muscatine and Cernichiari, 1969; Trench, 1971a; Schmitz and Kremer, 1977; Schlichter *et al.*, 1983) or predominantly in the non-glycerol (fatty acyl) moieties (Patton *et al.*, 1983; Battey and Patton, 1984). However, a recent report (Battey and Patton, 1987) suggests that both glycerol and lipid are released by symbionts in the sea anemone *Condylactis gigantea*, with glycerol being utilized by the host as a respiratory substrate and lipid as a carbon storage product. The potential use of algal-derived lipid as an alternative source of fuel is described by Calvin and Taylor (Ch. 6). The effect of 'host factor' on the release of lipid is not known.

A further biotechnological potential of symbiotic dinoflagellates is in the production of amino acids. Marine alga–invertebrate symbioses occur in a nutrient-poor environment and are considerably more productive than the surrounding waters. It has been suggested (Yonge, 1936; Lewis and Smith, 1971; Muscatine and Porter, 1977) that, in part, this is due to nitrogen recycling within symbiotic associations, with symbionts assimilating the host nitrogenous excretion product (ammonia) and releasing amino acid(s). Evidence for this assertion has been criticized by Rees (1986, 1989), and in the absence of experiments demonstrating enhanced release of amino acids by freshly isolated symbionts incubated with ammonia, it is difficult to evaluate the potential use of symbiotic dinoflagellates as a source of amino acids. Furthermore, it is difficult to reconcile substantial release of glycerol and lipid with nitrogen recycling, but photosynthetically-fixed ^{14}C does appear in host protein (Trench, 1971a; Hinde, 1983). However, the appearance of label in host protein cannot be taken as evidence for amino acid release by symbionts. Catabolism of glycerol and lipid yields acetyl-CoA. Label from acetyl-CoA will remain in the dicarboxylic acids of the Krebs cycle, and therefore will appear in amino acids via transamination.

Other symbiotic marine algae

Unicellular marine algae which form symbiotic associations with invertebrates represent a surprising taxonomic diversity with representatives from all the major classes (Taylor, 1982; Smith and Douglas, 1987). However, for most of these associations, the paucity of information precludes evaluation of their biotechnological potential. The exception is the Prasinophycean alga *Tetraselmis convolutae* (previously *Platymonas convolutae*) which forms an association with the acoel turbellarian *Convoluta roscoffensis*.

Adult *Convoluta* always contain symbionts; aposymbiotic juveniles do not mature and die if they are not infected with *Tetraselmis*. The symbionts are intracellular and the four flagella and eyespot, which are characteristic of cultured symbionts, are lost. The cell wall is also lost and this probably accounts for the ease with which freshly isolated symbionts lyse. However, *Tetraselmis convolutae* has been successfully cultured but, in common with most symbiotic algae, loses the ability to release photosynthate (Gooday, 1970). Freshly isolated symbionts release photosynthate as amino acids (alanine (mainly), glycine, serine and glutamine). There is no stimulation of release by low pH (Muscatine *et al.*, 1974) and though host homogenate does increase the amount of photosynthate released, the increase is only from 1–5% to 5–16% of ^{14}C fixed (Table 5.3a). However, evidence suggests that in the intact association about 50% of photosynthate is released to the host (Boyle and Smith, 1975). The suggestion (Keeble, 1910) that the symbionts are responsible for recycling the nitrogenous excretion product (uric acid) of the host has been criticized recently (Douglas, 1989). In common with symbiotic dinoflagellates there is some evidence that lipid, principally in the form of fatty acids and sterols, is released by the symbionts to the host (Meyer *et al.*, 1979).

What is the mechanism of photosynthate release by marine symbionts?

The limited information available on the mechanism of photosynthate release by marine symbiotic algae is that it is stimulated by a 'host factor' which is only produced by symbiotic invertebrates (Hinde, 1983). However, it is noteworthy that *Cassiopea xamachana* produces a 'host factor' which inhibits sodium-dependent alanine uptake by its symbionts (Carroll and Blanquet, 1984a,b).

Though there is no evidence that these two 'host factors' are the same compound it is suggested that they are. Moreover, it is suggested, by analogy with the stimulation of maltose release from symbiotic *Chlorella* by low pH, that increased sodium ion concentrations are involved in the mechanism for stimulation of photosynthate release by symbiotic marine algae, particularly dinoflagellates.

Topologically the perialgal vacuolar space is external to the host. Regulation of the intracellular sodium concentration by Na^+, K^+-ATPases, located in the perialgal vacuolar membrane, would increase the perialgal vacuolar sodium concentration. It is suggested, therefore, that glycerol is produced by symbiotic dinoflagellates as an osmoregulatory compatible solute. Glycerol is the major compatible solute produced by a number of marine and freshwater algae including *Asteromonas gracilis*, *Chlamydomonas* and *Dunaliella* (Ben-Amotz and Avron, 1983). There is no information on osmoregulation in either symbiotic or free-living dinoflagellates. The chloroform-extractable fraction (mainly triglycerides and free fatty acids) increases with increasing sodium chloride concentration in *Botryococcus braunii* (2% to 13% organic dry weight) and *Dunaliella salina* (1.6% to 5.2% organic dry weight) (Ben-Amotz *et al.*, 1985). In *Dunaliella salina* the glycerol content increases from 9.4% to 27.7% organic dry weight, and the combined glycerol and chloroform-extractable fraction from 11% to 33%, when the sodium chloride concentration is increased from 0.5

to 2 M. In symbiotic dinoflagellates there may be a further enhancement of lipid production due to nitrogen deficiency.

A possible role for the 'host factor, is as an ionophore. A 'host factor' is present in *Cassiopea xamachana* which inhibits sodium-dependent alanine uptake by freshly isolated symbionts (Carroll and Blanquet, 1984a,b), but it is not known how specific this effect is. It is envisaged that the ionophore would act on the perialgal vacuolar membrane rather than the algal cell membrane, though it is possible that the latter could also occur. The effect would be to decrease the sodium concentration in the perialgal vacuolar space and facilitate release of lipid and glycerol by the symbionts.

Symbiotic *Chlorella*

Chlorella is the major algal symbiont in freshwater alga–invertebrate symbioses. Hosts include ciliate protozoans, sponges, coelenterates, neo-rhabdocoel turbellarians and bivalve molluscs. However, despite the taxonomic diversity of its hosts, evidence suggests that the symbionts are restricted to the *Chlorella vulgaris* group (Reisser, 1984; Douglas and Huss, 1986; Douglas, 1987) which comprises *C. vulgaris, C saccharophila* and *C. sorokiniana* (Kessler, 1982).

In the protozoa, sponges and the coelenterate hydra the symbionts are intracellular. Each algal cell resides within a vacuole (perialgal vacuole) which is delimited by a membrane of host origin. In turbellarians and molluscs it is not known whether the symbionts are intracellular or inter-cellular. In common with other algal (but not cyanobacterial; see p. 117) symbionts their location within the host tissue is sufficiently peripheral to enable them to photosynthesize.

The major photosynthetic product released by symbiotic *Chlorella* is the disaccharide maltose, but algae isolated from some hosts release glucose, fructose or xylose (Table 5.4). Substantial release of photosynthate only occurs at low pH (Muscatine, 1965; Muscatine *et al.*, 1967; Cernichiari *et al.*, 1969; Douglas, 1987). Experiments in which freshly isolated symbiotic *Chlorellas* are allowed to fix $^{14}CO_2$ and the ^{14}C-labelled products identified by chromatography, show that trace amounts of alanine and glycollate are also released. Moreover, below pH 4.5 ammonia is also released (Rees, 1989), but this could be removed easily by the addition of alkali. Given that only trace amounts of ^{14}C-maltose are detectable in the cell (Cernichiari *et al.*, 1969; Ziesenisz *et al.*, 1981) it is presumed that release must occur against a considerable concentration gradient (but see below). This, together with the fact that transport of 6-deoxyglucose in *Chlorella* is driven by the protonmotive force (the electrochemical gradient of protons across the cell membrane) (Komor and Tanner, 1974, 1976), suggests that release of maltose may be achieved by a similar mechanism. It is possible that the protein (if it exists) responsible for maltose release behaves in a similar manner to other *Chlorella* transport proteins, but the electrochemical gradient of protons into the cell drives the efflux of maltose against a

Table 5.4 Photosynthetic products released by symbiotic *Chlorella*

Host	Photosynthetic products released	Percentage of total photosynthetically fixed carbon	Reference
Anodonta cygnea	Maltose		
Climacostomum virens	Glucose, fructose, xylose	8	Reisser (1984)
Coleps hirtus	Xylose	4	Reisser (1984)
Dalyellia viridis	Maltose		Douglas (1987)
Ephydatia fluviatilis	Glucose	9–16	Wilkinson (1980)
Euplotes daidaleos	Fructose, xylose	6	Reisser (1984)
Hydra viridis	Maltose	45–85	Muscatine (1965) Cernichiari *et al.* (1969) Douglas (1987)
Paramecium bursaria	Maltose	5–87	Muscatine *et al.* (1967) Reisser (1984) Douglas (1987)
Spongilla sp.	Glucose	4	Muscatine *et al.* (1967)
Spongilla fluviatilis	Glucose	15	Reisser (1984)
Stentor polymorphus	Maltose	6	Reisser (1984)
Typhloplana viridata	Maltose		Douglas (1987)

concentration gradient. In support of this contention is the acidic pH optimum for release and the sensitivity of release to the protonophore carbonyl cyanide *m*-chlorophenylhydrazone (CCCP) (Smith and Douglas, 1987).

An alternative hypothesis is that maltose release is coupled to the *generation* of an electrochemical gradient of protons (as outlined in p. 121). This hypothesis is unlikely for two reasons. Lactose efflux in *E. coli* is sensitive to external pH and increases with increasing pH (Kaczorowski and Kaback, 1979), whereas the opposite is found for maltose release by symbiotic *Chlorella*. Secondly, there is no convincing evidence for a substantial outwardly-directed gradient of maltose concentration across the cell membrane.

Does maltose release occur against a substantial concentration gradient? ^{14}C-maltose is only detected at very low levels intracellularly in freshly isolated and cultured symbionts (Cernichiari *et al.*, 1969; Ziesenisz *et al.*, 1981). Irrespective of the fact that the specific activity of intracellular maltose is unknown, the interpretation that this involves release *against* a substantial *concentration* gradient is unlikely. The gradient is simply that of counts per minute. With a release rate of 1.19 fmol cell^{-1} h^{-1} for freshly isolated symbionts of the European strain of green hydra (Douglas and Smith, 1984) and a cell density of 10^7 cells ml^{-1}, the concentration of maltose in the medium after 1 h incubation is 11.9 μM. With a cell volume of 71 μm^3 (Douglas and Smith, 1984) and the same cell density the total algal cell volume is 7.1×10^8 μm^3 compared with a total volume of 10^{12} μm^3 (i.e.

the algae occupy 0.07% of the total volume). Assuming an identical specific activity for intracellular and extracellular maltose, the concentrations would be identical when the external counts are 1400× the intracellular counts. Moreover, this presupposes an equal distribution of maltose within the cell; the available evidence (Cernichiari et al., 1969) suggests otherwise.

The only evidence for movement of maltose *against* a concentration gradient is that of Cernichiari et al., (1969). ^{14}C-Maltose continues to be released in the presence of up to 10% w/v ^{12}C-maltose. However, this probably represents exchange diffusion (Smith, 1974) and not release against a substantial concentration gradient. The ratio of counts per minute for ^{14}C-maltose in the medium to that in the cells is 307 (Cernichiari et al., 1969; Table 1b). In this experiment 0.1 ml wet packed volume of algae was suspended in 4.0 ml medium. Therefore maltose was released against an apparent 7.7-fold gradient (assuming the specific activity is constant). A release rate of 600 μg maltose ml^{-1} wet packed algae h^{-1} is given by Cernichiari et al., (1969). For the above experiment the medium concentration of maltose would be 21 μM, and the intracellular concentration, assuming a 7.7-fold gradient would be 2.7 μM. If it is assumed that maltose is only present in the cytosol (Cernichiari et al., 1969) and that the cytosol of *Chlorella* is 40% of the total cell volume (Atkinson et al., 1974) then the cytosolic concentration becomes 6.8 μM and the gradient three-fold. Moreover, *in vivo*, maltose released by the symbionts will be hydrolysed by host maltase (Mews, 1980) suggesting that maltose is released *down* a concentration gradient.

The above intracellular concentrations of maltose appear to be very low and experiments are required to determine the precise size and location of the maltose pool. The intracellular concentration may indeed be low, either because the mechanism of release is so efficient that no accumulation occurs, or because biosynthesis is very sensitive to feedback inhibition. That sucrose, as opposed to maltose, is the normal disaccharide end product of photosynthesis in *Chlorella* suggests that there is a diversion from one cytosolic biosynthetic pathway to the other, which is effected by a decrease in external pH. The intracellular signal for this change is unknown.

Hypotheses for the mechanisms of maltose release include modification of surface polysaccharide synthesis (Cernichiari et al., 1969), alteration of the substrate specificity of a glucosyl transferase located on the cell surface (Jolley and Smith, 1978) and modification of a cell surface protein by host maltase (Mews, 1980) or a 'host factor' (Douglas, 1987, 1988). Further experiments are required to elucidate the mechanisms involved in release as they could provide valuable insights into ways of inducing release of metabolites from plant cells.

A major problem with symbiotic *Chlorella* from some associations is that, on isolation into culture, the ability to release maltose at low pH is lost (Jolley and Smith, 1978; Douglas, 1987). Notable exceptions are strains of *Chlorella*, isolated from the ciliate *Paramecium bursaria*, which continue to release maltose after many years in culture. Why there should be this difference between symbiotic *Chlorella* strains is unclear, but it has been suggested that in those strains which lose the ability to release maltose on isolation, the presence of a 'host factor' is necessary for the induction of release (Douglas, 1987, 1988). The identity of the 'host factor' is unknown, and that some symbiotic *Chlorella* strains, in particular those isolated from *Paramecium bursaria*, continue to release maltose in its absence suggests that different release mechanisms may be involved.

The host in symbiosis

The question of how the host is altered biochemically by the presence of symbionts has received scant attention. It would be surprising if the changes were limited; in parasitic associations they are obvious and deleterious.

Lichens produce antibiotics (in particular, usnic acid) and dyes (Hale, 1967), which are probably fungal in origin. Symbiotic *Paramecium bursaria* produces a 'factor' which discourages predation by the protozoan *Didinium nasutum* (Berger, 1980). The identity of the 'factor', and whether it is produced by the host or symbionts, is unknown.

Conclusions

This chapter has attempted to review the biotechnological potential of symbiotic algae and cyanobacteria. So far, only the symbiotic cyanobacteria have attracted serious attention (Chs. 3 and 9). The major theme of the chapter has been that the potential of symbiotic organisms may be indirect, namely that the mechanisms of release of metabolites may be of greater consequence than the metabolites themselves. However, in the absence of experimental investigations into the mechanisms of release, the role of symbiotic algae and cyanobacteria in biotechnology is likely to remain potential.

Acknowledgements

I thank Dr A. E. Douglas, Prof. D. O. Hall and Sir David Smith FRS for their constructive comments.

References

Atkinson A R, John P C L, Gunning B E S 1974 The growth and division of the single mitochondrion and other organelles during the cell cycle of *Chlorella* studied by quantitative stereology and three dimensional reconstruction. *Protoplasma* **81**: 77–109

de Bary A 1879 Die Erscheinung der Symbiose. *Naturfoschung Versammlung Cassel*, L I, Tagebl. p 121

Battey J F, Patton J S 1984 A reevaluation of the role of glycerol in carbon trans-location in zooxanthellae–coelenterate symbiosis. *Mar. Biol.* **79**: 27–38

Battey J F, Patton J S 1987 Glycerol translocation in *Condylactis gigantea. Mar. Biol.* **95**: 37–46

Ben-Amotz A, Avron M 1983 Accumulation of metabolites by halotolerant algae and its industrial potential. *A. Rev. Microbiol.* **37**: 95–119

Ben-Amotz A, Tornabene T G, Thomas W H 1985 Chemical profile of selected species of microalgae with emphasis on lipids. *J. Phycol.* **21**: 72–81

Bentraboulet M, Robin A, Kepes A 1979 Artificially induced active transport of amino acids driven by the efflux of a sugar via a heterologous transport system in de-energized *Escherichia coli. Biochem. J.* **178**: 103–7

Berger J 1980 Feeding behaviour of *Didinium nasutum* on *Paramecium bursaria* with normal or apochlorotic zoochlorellae. *J. Gen. Microbiol.* **118**: 397–404

Boyle J E, Smith D C 1975 Biochemical interactions between the symbionts of *Convoluta roscoffensis. Proc. R. Soc. Lond.* B **189**: 121–35

Brown L M, Hellebust J A 1980 The contribution of organic solutes to osmotic balance in some green and Eustigmatophyte algae. *J. Phycol.* **16**: 265–70

Carroll S, Blanquet R S 1984a Alanine uptake by isolated zooxanthellae of the mangrove jellyfish, *Cassiopea xamachana.* I Transport mechanisms and utilization. *Biol. Bull.* **166**: 409–18

Carroll S, Blanquet R S 1984b Alanine uptake by isolated zooxanthellae of the mangrove jellyfish, *Cassiopea xamachana.* II Inhibition by host homogenate fraction. *Biol. Bull.* **166**: 419–26

Cernichiari E, Muscatine L, Smith D C 1969 Maltose excretion by the symbiotic algae of *Hydra viridis. Proc. R. Soc. Lond.* B **173**: 557–76

Chambers S, Morris M, Smith D C 1976 Lichen physiology. XV The effect of digitonin and other treatments on biotrophic transport of glucose from alga to fungus in *Peltigera polydactyla. New Phytol.* **76**: 485–500

Douglas A E 1987 Experimental studies on symbiotic *Chlorella* in the neorhabdocoel turbellaria *Dalyellia viridis* and *Typhloplana viridata. Br. Phycol. J.* **22** 157–61

Douglas A E 1988 The influence of host contamination on maltose release by symbiotic *Chlorella. Limnol. Oceanogr.* **33**: 295–7

Douglas A E 1989 Alga–invertebrate symbiosis. A. *Proc. Phytochem. Soc. Eur.* **29**: (In Press)

Douglas A E, Huss V A R 1986 On the characteristics and taxonomic position of symbiotic *Chlorella. Arch. Microbiol.* **145**: 80–4

Douglas A E, Smith D C 1984 The green hydra symbiosis. VII Mechanisms in symbiont regulation. *Proc. R. Soc. Lond.* B **221**: 291–319

Douglas A E, Smith D C 1989 A working definition of symbiosis that avoids the concept of mutual benefit. *Biol. J. Linn. Soc.* (In Press)

Farrar J F, Smith D C 1976 Ecological physiology of the lichen *Hypogymnia physodes.* III The importance of the rewetting phase. *New Phytol.* **77**: 115–25

Gallop A 1974 Evidence for the presence of a 'factor' in *Elysia viridis* which stimulates photosynthate release from its symbiotic chloroplasts. *New Phytol.* **73**: 1111–17

Gober J W, Kashket E R 1987 K^+ regulates bacteroid-associated functions of *Bradyrhizobium. Proc. Nat. Acad. Sci. USA* **84**: 4650–4

Gooday G W 1970 A physiological comparison of the symbiotic alga *Platymonas convolutae* and its free-living relatives. *J. Mar. Biol. Assoc. UK* **63**: 419–34

Green T G A, Smith D C 1974 Lichen physiology. XIV Differences between lichen algae in symbiosis and in isolation. *New Phytol.* **73**: 753–66

Grilli Caiola M 1980 On the phycobionts of the cycad coralloid roots. *New Phytol.* **85**: 537–44

Hale M E 1967 *The Biology of Lichens,* Edward Arnold, London
Hill D J 1975 The pattern of development of *Anabaena* in the *Azolla–Anabaena* symbiosis. *Planta* **122**: 179–84
Hinde R 1983 Host release factors in symbiosis between algae and invertebrates. In Schwemmler W, Schenk H E A (eds) *Endocytobiology, Endosymbiosis and Cell Biology,* Vol 2, Walter de Gruyer, Berlin, pp 709–26
Honegger R 1984 Cytological aspects of the mycobiont–phycobiont relationship in lichens. *Lichenologist* **16**: 111–27

Jolley E, Smith D C 1978 The green hydra symbiosis. I Isolation, culture and characteristics of the *Chlorella* symbiont of 'European' *Hydra viridis. New Phytol.* **81**: 637–45
Joseph C M, Meeks J C 1987 Regulation of expression of glutamine synthetase in a symbiotic *Nostoc* strain associated with *Anthoceros punctatus. J. Bact.* **169**: 2471–5

Kaczorowski G J, Kaback H R 1979 Mechanism of lactose translocation in membrane vesicles from *Escherichia coli.* I Effect of pH on efflux, exchange, and counterflow. *Biochemistry* **18**: 3691–7
Kaczorowski G J, Robertson D E, Kaback H R 1979 Mechanism of lactose translocation in membrane vesicles from *Escherichia coli.* II Effect of imposed $\Delta\Psi$, ΔpH, and $\Delta\mu_{H+}$. *Biochemistry* **18**: 3697–704
Keeble F 1910 *Plant–Animals. A Study in Symbiosis,* Cambridge University Press
Kellogg R B, Patton J S 1983 Lipid droplets, medium of energy exchange in the symbiotic anemone *Condylactis gigantea:* a model coral polyp. *Mar. Biol.* **75**: 137–49
Kershaw K A 1985 *Physiological Ecology of Lichens,* Cambridge University Press
Kessler E 1982 Chemotaxonomy in the Chlorococcales. *Prog. Phycol. Res.* **1**: 111–35
Komor E, Tanner W 1974 The hexose–proton symport system of *Chlorella vulgaris.* Specificity, stoichiometry and energetics of sugar-induced proton uptake. *Eur. J. Biochem.* **44**: 219–23
Komor E, Tanner W 1976 The determination of the membrane potential of *Chlorella vulgaris.* Evidence for electrogenic sugar transport. *Eur. J. Biochem.* **70**: 197–204

Lewis D H, Smith D C 1967 Sugar alcohols (polyols) in fungi and green plants. I Distribution, physiology and metabolism. *New Phytol.* **66**: 143–84
Lewis D H, Smith D C 1971 The autotrophic nutrition of symbiotic marine coelenterates with special reference to hermatypic corals. I Movement of photosynthetic products between the symbionts. *Proc. R. Soc. Lond.* B **178**: 111–29
Lindblad P, Bergman B 1986 Glutamine synthetase: activity and localization in cyanobacteria of the cycads *Cycas revoluta* and *Zamia skinneri. Planta* **169**: 1–7
Lindblad P, Hallbom L, Bergman B 1985 The cyanobacterium–*Zamia* symbiosis: C_2H_2-reduction and heterocyst frequency. *Symbiosis* **1**: 19–28
Lindblad P, Rai A N, Bergman B 1987 The *Cycas revoluta–Nostoc* symbiosis: enzyme activities and carbon metabolism in the cyanobiont. *J. Gen. Microbiol.* **133**: 1695–9

Meeks J C, Enderlin C S, Joseph C M, Chapman J S, Lollar M W L 1985 Fixation of [^{13}N]N_2 and transfer of fixed nitrogen in the *Anthoceros–Nostoc* symbiotic association. *Planta* **164**: 404–14
Mews L K 1980 The green hydra symbiosis. III The biotrophic transport of carbohydrate from alga to animal. *Proc. R. Soc. Lond.* B **209**: 377–401

Meyer H, Provasoli L, Meyer F 1979 Lipid biosynthesis in the marine flatworm *Convoluta roscoffensis* and its algal symbiont *Platymonas convolutae. Biochim. Biophys. Acta* **573**: 464–80

Michels P A M, Michels J P J, Boonstra J, Konings W N 1979 Generation of an electrochemical proton gradient in bacteria by excretion of metabolic end products. *FEMS Microbiol. Lett.* **5**: 357–64

Muscatine L 1965 Symbiosis of hydra and algae. III Extracellular products of the algae. *Comp. Biochem. Physiol.* **16**: 77–92

Muscatine L 1967 Glycerol excretion by symbiotic algae from corals and *Tridacna* and its control by the host. *Science* **156**: 516–19

Muscatine L, Boyle J E, Smith D C 1974 Symbiosis of the acoel flatworm *Convoluta roscoffensis* with the alga *Platymonas convolutae. Proc. R. Soc. Lond.* B **187**: 221–34

Muscatine L, Cernichiari E 1969 Assimilation of photosynthetic products of zooxanthellae by a reef coral. *Biol. Bull.* **137**: 506–23

Muscatine L, Karakashian S J, Karakashian M W 1967 Soluble extracellular products of algae symbiotic with a ciliate, a sponge and a mutant hydra. *Comp. Biochem. Physiol.* **20**: 1–12

Muscatine L, Pool R R, Cernichiari E 1972 Some factors influencing selective release of soluble organic material by zooxanthellae from reef corals. *Mar. Biol.* **13**: 298–308

Muscatine L, Porter J W 1977 Reef corals: mutualistic symbioses adapted to nutrient-poor environments. *BioScience* **27**: 454–60

Orr J, Haselkorn R 1982 Regulation of glutamine synthetase and synthesis in free-living and symbiotic *Anabaena* spp. *J. Bact.* **152**: 626–35

Otto R, Sonnenberg A S M, Veldkamp H, Konings W N 1980 Generation of an electrochemical proton gradient in *Streptococcus cremoris* by lactate efflux. *Proc. Nat. Acad. Sci. USA* **77**: 5502–6

Patton J S, Abraham S, Benson A A 1977 Lipogenesis in the intact coral *Pocillopora capitata* and its isolated zooxanthellae: evidence for a light-driven carbon cycle between symbiont and host. *Mar. Biol.* **44**: 235–47

Patton J S, Battey J F, Rigler M W, Porter J W, Black C C, Burris J E 1983 A comparison of the metabolism of bicarbonate ^{14}C and acetate 1-^{14}C and the variability of species lipid compositions in reef corals. *Mar. Biol.* **75**: 121–30

Patton J S, Burris J E 1983 Lipid synthesis and extrusion by freshly isolated zooxanthellae (symbiotic algae). *Mar. Biol.* **75**: 131–6

Rai A N, Rowell P, Stewart W D P 1981a $^{15}N_2$ incorporation and metabolism in the lichen *Peltigera aphthosa* Willd. *Planta* **152**: 544–52

Rai A N, Rowell P, Stewart W D P 1981b Glutamate synthase activity in symbiotic cyanobacteria. *J. Gen. Microbiol.* **126**: 515–8

Rai A N, Rowell P, Stewart W D P 1983 Interactions between cyanobacteria and fungus during $^{15}N_2$-incorporation and metabolism in the lichen *Peltigera canina. Arch. Microbiol.* **134**: 136–42

Rao K K, Hall D O 1984 Photosynthetic production of fuels and chemicals in immobilized systems. *Trends Biotechnol.* **2**: 124–9

Rees T A V 1986 The green hydra symbiosis and ammonium. I The role of the host in ammonium assimilation and its possible regulatory significance. *Proc. R. Soc. Lond.* B **229**: 299–314

Rees T A V 1989 The green hydra symbiosis and ammonium. II Ammonium assimilation and release by freshly isolated symbionts and cultured algae. *Proc. R. Soc. Lond.* B **235**: 365–82

Reisser W 1984 The taxonomy of the green algae endosymbiotic in ciliates and a sponge. *Br. Phycol. J.* **19**: 309–18

Richardson D H S, Hill D J, Smith D C 1968 Lichen physiology. XI The role of the alga in determining the pattern of carbohydrate movement between lichen symbionts. *New Phytol.* **67**: 469–86

Rodgers G A, Stewart W D P 1977 The cyanophyte–hepatic symbiosis. I Morphology and physiology. *New Phytol.* **78**: 441–58

Schlichter D, Svoboda A, Kremer B P 1983 Functional autotrophy of *Heteroxenia fuscescens* (Anthozoa: Alcyonaria): carbon assimilation and translocation of photosynthates from symbionts to host. *Mar. Biol.* **78**: 29–38

Schmitz K, Kremer B P 1977 Carbon fixation and analysis of assimilates in a coral–dinoflagellate symbiosis. *Mar. Biol.* **42**: 305–13

Shi D J, Brouers M, Hall D O, Robins R J 1987 The effects of immobilization on the biochemical, physiological and morphological features of *Anabaena azollae*. *Planta* **172**: 298–308

Silvester W B, McNamara P J 1976 The infection process and ultrastructure of the *Gunnera–Nostoc* symbiosis. *New Phytol.* **77**: 135–41

Smith D C 1974 Transport from symbiotic algae and symbiotic chloroplasts to host cells. *Symp. Soc. Exp. Biol.* **28**: 485–520

Smith D C 1975 Symbiosis and the biology of lichenised fungi. *Symp. Soc. Exp. Biol.* **29**: 373–405

Smith D C 1980 Mechanisms of nutrient movement between lichen symbionts. In Cook C B, Pappas P W, Rudolph E D (eds) *Cellular Interactions in Symbiosis and Parasitism*, Ohio State University Press, Columbus, pp 197–227

Smith D C, Douglas A E 1987 *The Biology of Symbiosis*, Edward Arnold, London

Smith D C, Muscatine L, Lewis D H 1969 Carbohydrate movement from autotrophs to heterotrophs in parasitic and mutualistic symbiosis. *Biol. Rev.* **44**: 17–90

Steen R G 1986 Evidence for heterotrophy by zooxanthellae in symbiosis with *Aiptasia pulchella*. *Biol. Bull.* **170**: 267–78

Stewart W D P, Rodgers G A 1977 The cyanophyte–hepatic symbiosis. II Nitrogen fixation and the interchange of nitrogen and carbon. *New Phytol.* **78**: 459–71

Stewart W D P, Rowell P, Rai A N 1980 Symbiotic nitrogen-fixing cyanobacteria. A. *Proc. Phytochem. Soc. Eur.* **18**: 239–77

Tapper R 1981 Direct measurement of translocation of carbohydrate in the lichen *Cladonia convoluta*, by quantitative autoradiography. *New Phytol.* **89**: 429–37

Taylor F J R 1982 Symbioses in marine microplankton. *Ann. Inst. Océanogr.* **58**(S): 61–90

Trench R K 1971a The physiology and biochemistry of zooxanthellae symbiotic with marine coelenterates. I The assimilation of photosynthetic products of zooxanthellae by two marine coelenterates. *Proc. R. Soc. Lond. B* **177**: 225–35

Trench R K 1971b The physiology and biochemistry of zooxanthellae symbiotic with marine coelenterates. II Liberation of fixed ^{14}C by zooxanthellae *in vitro*. *Proc. R. Soc. Lond. B* **177**: 237–50

Trench R K 1971c The physiology and biochemistry of zooxanthellae symbiotic with marine coelenterates. III The effect of homogenates of host tissues on the excretion of photosynthetic products *in vitro* by zooxanthellae from two marine coelenterates. *Proc. R. Soc. Lond. B* **177**: 251–64

Trench R K 1979 The cell biology of plant–animal symbiosis. *A. Rev. Pl. Physiol.* **30**: 485–531

Trench R K 1974 Nutritional potentials in *Zoanthus sociatus* (Coelenterata, Anthozoa). *Helg. Meer.* **26**: 174–216

Trench R K, Blank R J 1987 *Symbiodinium microadriaticum* Freudenthal, S. *goreauii* sp. nov., *S. kawagutii* sp. nov. and *S. pilosum* sp. nov.: gymnodinioid dinoflagellate symbionts of marine invertebrates. *J. Phycol.* **23**: 469–81

Trench R K, Wethey D S, Porter J W 1981 Observations on the symbiosis with zooxanthellae among the Tridacnidae (Mollusca, Bivalvia). *Biol. Bull.* **161**: 180–98

von Holt C, von Holt M 1968 The secretion of organic compounds by zooxanthellae isolated from various types of *Zoanthus. Comp. Biochem. Physiol.* **24**: 83–92

Wilkinson C R 1980 Nutrient translocation from green algal symbionts to the freshwater sponge *Ephydatia fluviatilis. Hydrobiologia* **75**: 241–50

Yonge C M 1936 Mode of life, feeding, digestion and symbiosis with zooxanthellae in the Tridacnidae. *Scientific Reports of the Great Barrier Reef Expedition* **1**: 283–321

Ziesenisz E, Reisser W, Wiessner W 1981 Evidence of *de novo* synthesis of maltose excreted by the endosymbiotic *Chlorella* from *Paramecium bursaria. Planta* **153**: 481–5

6

Fuels from algae

M. Calvin and S. E. Taylor

Introduction

Most of the world's energy supply is derived from non-renewable fossil fuels; in the United States 95% of energy utilized in 1982 was obtained from petroleum, coal and natural gas (Calvin 1984). Though there is much debate on the quantity of these fuels still available, and on their longevity, it is unarguable that the reserves of these resources are finite. It is, therefore, essential that alternative sources of energy be identified and developed.

There are also economic and environmental problems associated with the continued use of fossil fuels. The energy cost in extracting a barrel of oil is approaching the total energy content of a barrel of oil. At some point in the near future it will require more energy to extract the oil than is contained within the oil. Combustion of the fossil fuels has resulted in a sharp increase in the global carbon dioxide concentration over the last 100 years. Increases in the mean sea level have also been observed during this period, providing evidence that increased global temperatures are associated with higher carbon dioxide concentrations. The climatic effects of this warming trend could be catastrophic, with vast areas of agricultural land becoming barren (Calvin 1984).

One attractive alternative to petroleum and coal is photosynthetically-generated biomass. It is renewable, the technology for its production and utilization is within our grasp, and its use has no effect on the carbon dioxide concentration of the atmosphere. Carbon utilization is a closed cycle when biomass-derived fuels are used: any carbon dioxide that is liberated by combustion was just recently incorporated by the photosynthetic organism, resulting in no net change in carbon dioxide concentration.

Can algae be utilized as a source of energy? There are actually two parts to this question and we will address each separately. First, can enough algae be obtained to provide significant amounts of material, and can this be done so that the energy input is kept low? Second, can the total algal biomass or specific components of the algae be economically utilized as fuel? Both questions must be answered in the affirmative before algal-derived fuels can be considered as an alternative to fossil fuels.

Algal biomass production

Ryther (1959) calculated that the production of algae in the sea is comparable to that of agricultural productivity on land. Under optimum conditions, algal production can compete with that of sugarcane, the most productive of the land plants. Both can generate biomass at a rate of 20–30 g dry weight $m^{-2} d^{-1}$, which represents a photosynthetic conversion efficiency of about 2% (Whitesides and Elliott, 1984; Hanisak and Ryther, 1986). In some instances algal systems have even higher photosynthetic conversion efficiencies than land-based agriculture (Ben-Amotz, 1980; Laws *et al.*, 1986; Arad, 1987). In addition, aquatic biomass production has the advantages of an availability of area without the use of agricultural land, water availability, buffered temperature variations due to the high heat capacity of water, abundant light availability, a continuous growth cycle, and in many instances low climatic fluctuations (Chynoweth, 1980; Indergaard, 1983; Whitesides and Elliott, 1984).

Studies of algal biomass production have utilized a wide variety of species, with representatives from the brown algae (Phaeophyceae), the red algae (Rhodophyceae), the green algae (Chlorophyceae) the diatoms (Bacillariophyceae) and the cyanobacteria, or blue–green algae (Cyanophyceae). All of the species studied can be grouped by size into one of two classes: the macroalgae and the microalgae. The macroalgae are large multicellular organisms, with sizes reaching 200 kg per plant (Indergaard, 1983). The microalgae are microscopic organisms, and are usually either unicellular or arranged in small filaments or colonies. The methods for propagation, harvesting and utilization differ enough between these two size classifications that we will treat them as two separate topics.

Macroalgae

The macroalgae, also called kelp or seaweed, are currently utilized for both food and chemical extraction. Species of the brown alga genus *Laminaria* and the red alga genus *Porphyra* constitute a significant portion of the diet of the Far East. The polysaccharides of the brown and red algae, called phycocolloids, are utilized in the pharmaceutical, chemical and food industries. About 40 000–50 000 metric tons of phycocolloids are used worldwide each year, at a value of $US300–400 million (Indergaard, 1983; Whitesides and Elliott, 1984). Because of this demand, techniques for the propagation and harvesting of species from genera like *Ascophyllum, Laminaria, Macrocystis* (all brown algae), *Chondrus* and *Gelidium* (both red algae) have been established. Current yields of kelp harvests range from 5 to 39 dry ash free metric tons per hectare per year (DAFMT $ha^{-1} yr^{-1}$) (Bird, 1986). It has been estimated that for algal biomass to be competitive as a biomass source for the gasification to methane, it will require yields of 50–170 DAFMT ha^{-1} yr^{-1} (Bird, 1986; Snyder and Brehany, 1987). Therefore, before macroalgae can be utilized economically as an energy alternative to natural gas, it will be necessary to build on the current technology and increase both the area under propagation and the biomass yields per unit area.

Many different environmental factors control yield (Jackson, 1980), but only three can be readily manipulated: nutrition, physical constraint, and competition/predation. While other factors, such as light intensity and water temperature, play a key role in regulating productivity, little can be done to manipulate these parameters outside of choosing species that are tolerant to the local conditions.

Nitrogen is the key limiting factor in most natural situations. Ammonia and nitrate are the only forms readily utilized by the macroalgae. Providing additional nitrogen almost always results in higher photosynthetic rates and higher yields (Jackson, 1980). In some situations phosphorus may also be limiting. Compounding the nitrogen and phosphorus limitation is the poor efficiency of uptake of these nutrients. At best 5–20% of the available nitrogen and phosphorus are utilized by the macroalgae.

The application of exogenous fertilizer (e.g. ammonium nitrate, ortho-phosphate) to algae is currently used in commercial operations, but it adds a significant amount to the cost of the biomass. Since the economic return on biomass for energy will be less than that for food or chemicals, alternative and hopefully less expensive sources of nutrients have been proposed. The residue of the biomass to fuel conversion process has been reapplied to algal cultures, and it works adequately as a nutrient source (Ryther and Hanisak, 1982). Problems do arise though, for progressive batches of the residue-treated algae may have higher nitrogen and ash contents which, as we will discuss later, can affect fuel yields. Sewage and agricultural run-offs may also be utilized, though the availability of these items is probably not enough to supply the nutrients necessary for the scale envisioned for marine energy farms (Indergaard, 1983). In addition, all of these sources have application problems; how are they to be supplied uniformly to a large-scale algal farm?

Another source of nutrients can be found in the ocean. Productivity of the brown seaweed *Sargassum muticum* was as high in seawater undergoing rapid mixing as in nutrient-amended water (Gellenbeck and Chapman, 1986). The deep (and cool) waters are enriched in nitrogen and phosphorus, and recirculating these waters could supply the algae with the nutrients needed. In some areas natural upwelling takes care of this problem, but in others pumping may be necessary (Chynoweth, 1980). The cost of pumping the water up to the surface can be defrayed by combining the algal farms with either a nuclear power plant or thermal gradient-driven energy plant, both of which require large quantities of cool water. One problem with this method is that it combines algal farming with energy technologies that are either not in favour at the moment (nuclear), or that are still in the developmental phase (thermal gradient). There is also a report suggesting that deep waters are deficient in one or more of the essential trace elements

(Jackson, 1980). In addition, such systems are confined to a limited number of areas as they require deep waters very near to a land mass for easily accessible cold water.

Algal farming also requires a means of physically supporting and confining the seaweed. Many of the seaweeds can adhere to the shallow bottom, but restricting farming to these shallow coastal waters will reduce the area available for biomass production, will restrict the number of species that can be considered, and will spread the farms over a thin, elongated area requiring decentralization of the biomass to fuel conversion process. Artificial supports have been constructed and tried, but have yet to prove successful on a large scale, and will also add to the overall cost of biomass production (Chynoweth, 1980; Indergaard, 1983; Snyder and Brehany, 1988). Another approach is to build ponds on land near the ocean, and to pump seawater through these ponds (Gellenbeck and Chapman, 1986; Hanisak and Ryther, 1986). This allows protection from wave and storm damage, and facilitates harvesting. The disadvantage with such systems is the added cost of pumping and the limited amount of available coastal land.

Establishment of the algal community and controlling the predators and competitors of the domesticated species is the third problem. Seaweeds are at the bottom of the food chain, and in some instances can be decimated by an efficient grazer. For example, maintenance of *Macrocystis* populations off the coast of California can require the control of the sea urchins by poisoning with calcium oxide (Indergaard, 1983). Pathogens and fish can also reduce biomass yields (Jackson, 1980).

Some species of seaweed are more competitive than others, and dominate the natural community. If the biomass conversion process is fine-tuned to handle only one species this may cause difficulties. Application of herbicides to kill existing species before introduction of the domesticated species (Indergaard, 1983) or manual weeding (Jackson, 1980) may be necessary. In addition some seaweeds, for example *Porphyra* (Waaland et al., 1986) and *Sargassum* (Gellenbeck and Chapman, 1986), require annual seeding so many of these problems have to be confronted year after year. Establishment of land-based ponds could eliminate all of these problems.

The techniques for harvesting the macroalgae are well established. The brown algae are usually harvested mechanically by boat, and the red algae by hand. Yields as high as 300 tons fresh weight per boat per day are obtainable with the brown algae (Indergaard, 1983). In terms of rate of production the harvesting of the algal biomass would not be a bottleneck in the biomass to energy conversion process, though it is still projected as the most costly step, accounting for 44% of the total cost of the fuel produced (Bird, 1986).

Microalgae

Though the microalgae share many of the same problems and limitations with the macroalgae, the strategies for their cultivation and the resulting energy products differ. Microalgae cannot be cultivated in the open sea, but must be propagated within small lakes, ponds, or tanks. Their productivity is higher than that of the macroalgae, with yields of 30–60 DAFMT ha$^-$ yr^{-1} reported for the cyanobacterium *Spirulina* and the green alga *Chlorella* (Benemann *et al.*, 1986). Microalgae have been contemplated as a biomass source for the conversion to methane (for example, see Samson and LeDuy, 1982; Matsunaga and Izumida, 1984), but because of the increased nutritional demands and harvesting costs, they are usually considered for their direct biosynthetic production of high energy fuels (i.e. glycerol, lipids, isoprenoids, and hydrogen) (Benemann *et al.*, 1986; Johnson, 1988).

Controllable variables that influence microalgal yields include nutrition, design and operation of the pond system, and harvesting procedures. Nitrogen and phosphorus are limiting factors in microalgal productivity, and the application of these nutrients by addition of sewage or the recycling of processing wastes has been proposed. But unlike the macroalgae, carbon availability also plays a key role in microalgal cultivation. This arises because closed pond systems are normally used in microalgal production, resulting in low mixing rates and low inorganic carbon availability. Though carbon limitation can directly affect photosynthetic rates, it is the rise in pH with the drop in carbon dioxide concentration that has the greatest effect on productivity (Jackson, 1980). To overcome a carbon limitation, either rapid mixing of the cultures is required, or carbon dioxide must be added to the system. This amendment of carbon dioxide can be done by the addition of bicarbonate, or by pumping in carbon dioxide directly. One possibility is to construct the microalgae farms near industrial sources of carbon dioxide (Benemann *et al.*, 1986; Johnson, 1988).

Proponents of microalgal fuel production envision large ponds in the desert regions and tropics (Johnson, 1988), with mechanical means of circulating the water/algae slurry through the ponds. For economic reasons ponds would probably be unlined, and the large water loss due to leakage and evaporation would be made up with saline water supplies buried beneath the surface.

The rate and degree of mixing of the cultures in these ponds is a crucial variable. Available irradiance can be controlled by mixing rates, and irradiance has been found to be a key factor in determining algal productivity (Weissman and Goebel, 1987). The use of specially shaped bottoms to enhance mixing has resulted in 25–30% increases in algal productivity (Dugan *et al.*, 1988). The use of shallow raceways can also enhance biomass productivity. Yields of 40 g ash free dry weight m^{-2} d^{-1} and photosynthetic conversion efficiencies of 8–11% have been reported for *Tetraselmis suecica* grown in vertical mixing flumes (Laws *et al.*, 1986). For comparison, if enlarged to a full-scale operation this would represent a production rate of 146 DAFMT ha^{-1} yr^{-1}.

One of the largest problems with microalgal biomass is the difficulty of harvesting. In most full-scale operations the algal solids will comprise only 0.06–0.015% of the total weight of the water/algae slurry, so the microalgae have to be collected and concentrated at least 1000× before they can be utilized (Shelef *et al.*, 1986). Various strategies have been tested using a combination of biological, chemical and physical processes to acquire the necessary concentration of cells.

The most direct approach for concentration of the algae is by centrifugation, but the cost of handling such vast quantities of solution in this manner is prohibitive (Benemann *et al.*, 1986). Microalgae can also be concentrated while still in large ponds by inducing an aggregation of the cells, either through chemical additives or by activation of natural biological mechanisms (bioflocculation). Chemical flocculation can be initiated by: (a) the addition of organic polyelectrolytes followed by the introduction of inorganic agents, such as alum or ferric chloride; (b) the exposure to low levels of ozone followed by treatment with organic polyelectrolytes (Shelef *et al.*, 1986); or (c) the proper selection of water hardness and pH (autoflocculation) (Benemann *et al.*, 1986). Bioflocculation is a natural aggregation of microalgae caused by the production of extracellular macromolecules which cause cellular agglutination. Easily controllable changes in environmental conditions will induce bioflocculation in many species including some from the *Micractinium* and *Synechocytis* genera (Lincoln and Koopman, 1986).

Once the algae have been aggregated by flocculation, they can be harvested from the ponds by either natural sedimentation, dissolved air flotation, or filtration by screens with micropores. At the present time all of these procedures are too costly to make the microalgae competitive with other sources of biomass (Benemann *et al.*, 1986), but research is currently underway to decrease the expense of harvesting.

Methane production from biomass

Methanization processes

There are three methods currently utilized for the conversion of biomass to methane: thermal gasification, catalytic hydrogenation and biological gasification. In thermal gasification the biomass is exposed to high temperatures at reduced oxygen levels, resulting in the conversion of the biomass to gas, oil, char, carbon and ash. The major components of the gas are methane, carbon dioxide and carbon monoxide. This procedure is best suited for biomass with a relatively low water content. Most algae have a water content of greater than 80%, and would require a drying process before thermal gasification could be carried out. The pre-gasification drying step(s) makes this procedure economically unfavourable for use with algal biomass (Chynoweth, 1980).

The second method of converting biomass to fuel is by catalytic hydrogenation, also referred to as direct liquefaction. In this process biomass is exposed to hydrogen gas under high pressures (1200 p.s.i.g.) and elevated temperatures (400 °C) in the presence of a suitable catalyst (cobalt molybdate). Under these conditions the long-chain macromolecules are cracked and the heteroatoms (oxygen, nitrogen, sulphur) are removed resulting in the production of gaseous and liquid hydrocarbons. This has been attempted with the green alga *Chlorella pyrenoidosa* by Chin and Engel (1981), and resulted in the successful generation of gas (25%), oil (38%) and asphaltenes (17%). The gas contained C_1 to C_4 gases, with methane the major component, carbon dioxide, carbon monoxide and water. The oils were C_{10}–C_{26} alkanes and alkenes, with the major components C_{15}, C_{17}, C_{20} and C_{22} alkanes.

Though this technique will generate both gases and liquid hydrocarbons from algal biomass, it is not economically feasible at this point. Most of the water must be removed before the algae can be fed into the reactor vessel, and this adds an enormous cost to the production and harvesting of the biomass. But the most crucial limitation to this technique is acquiring an inexpensive source of hydrogen gas. If such a source was available, research monies would be better spent developing the technology to use it directly instead of consuming it in hydrogenation reactions.

The third procedure available, biological gasification, is the most likely candidate for large-scale utilization in the near future. Biological gasification is an anaerobic digestion of the biomass by a mixed microbial population, and it generates methane and carbon dioxide as its major end-products. The digestion of the biomass is a two step process (Fig. 6.1), consisting of an 'acid-forming' stage followed by the 'methanization' stage (Golueke, 1980).

In the acid-forming step, the complex biological macromolecules of the biomass are first hydrolysed to their simple soluble components, and these molecules are then converted to small fatty acids, carbon dioxide, hydrogen

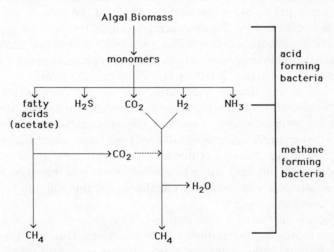

Fig. 6.1 Degradation of algal biomass to methane and other products by biological gasification. The biomass is processed in two steps by distinct heterogeneous microbial populations. In the first phase, the acid-forming stage, the complex biological macromolecules are cleaved to their monomeric components and then degraded further to fatty acids, carbon dioxide, etc. In the second phase, the methane-forming stage, these degradative products are transformed into methane, carbon dioxide and water.

gas, hydrogen sulphide, and ammonia. The fatty acids formed range in size from formic acid to caproic acid (C_1–C_6), but it is acetic acid that is predominant. This step results in a drop in the pH of the digestion mixture, giving rise to the title 'acid-forming' stage.

The acid-forming step is catalysed by a mixed population of 'acid-formers' or 'non-methanogenic' bacteria. These microbes are fast growing with a wide tolerance to a range of environmental conditions, and thus this step is usually not limiting in the conversion of biomass to methane (Chynoweth, 1980; Golueke, 1980).

In the second step of the digestion, the methane-forming step, a different population of bacteria converts the products of the acid-formers to methane and carbon dioxide. This occurs through one of two processes: (a) the conversion of the fatty acids to methane and carbon dioxide; or (b) a reaction between carbon dioxide and hydrogen gas forming methane and water. These reactions are carried out by a heterogeneous population of 'methanogenic' bacteria. These microbes are slow growing, and are very intolerant to oxygen and to variations in pH. They will function only between pH 6.0 and 7.5, with pH 7 the optimum for methane production (Chynoweth, 1980). Because of the slow growth rate and sensitivity to environmental changes of these bacteria, the methanization step is usually rate limiting.

The entire conversion process is sensitive to the composition of the starting material. The carbon to nitrogen ratio (C/N) of the biomass is of extreme importance. If the ratio is too high, acid formation occurs at a higher rate, growth of the microbes is limited by lack of nitrogen, and the methanogenic bacteria present cannot utilize the acids fast enough, resulting in a decrease in pH to below the optimal range. If the ratio is too low, all the ammonia produced by the acid-formers cannot be used for growth, so the ammonia accumulates to levels that are lethal for both the acid-forming bacteria and the methanogenic bacteria. Reports of the optimum C/N ratio range from 35/1 (Golueke, 1980) to 11/1 (Chynoweth, 1980).

In addition to the overall C/N ratio, the specific composition of the biomass may also affect the methanization process. Certain constituents may be indigestible by the microbes. Lignin, a polyaromatic component of wood, is an important example. As we will discuss later, high sulphur content and high ash content may also affect methane production. Pretreatment of the biomass with heat and acid and alkaline hydrolysis will increase the availability of substrates for conversion to methane, but this will add to the cost of the gas produced.

The process is also temperature sensitive. There appear to be two distinct temperature niches for populations of the microbes. One set, termed the mesophilic bacteria, have a temperature optimum from 30 to 40 °C, while a second set, the thermophilic bacteria, have an optimum range from 55 to 65 °C. Though the thermophilic bacteria may have higher rates and yields of methane production, the heat generated by the fermentation is not sufficient to maintain the higher temperatures, so additional heat must be provided. This would add to the cost of the process, and place economic restraints on the use of higher fermentation temperatures (Golueke, 1980).

The two steps in the biogasification process, acid-formation and methanization, can be temporally separated. Separate microbial populations can be maintained, which will allow for the conversion of the biomass to be halted at the formation of the fatty acids. These can be stored in open containers and later converted to methane by introduction of the methanogenic bacteria. This is important if the supply of biomass is seasonal but the demand for the methane is constant (Roos and Ristroph, 1987).

Composition of algal biomass

The procedures for biological gasification of terrestrial biomass and municipal solid wastes have been established. The use of these procedures with aquatic biomass is dependent upon the molecular composition of the algal biomass, and the ability of the microbial populations to digest any components unique to the algae.

After water, the major components of macroalgae are the polysaccharides. Though similar to terrestrial plants in their percentage of polysaccharides, macroalgal species are quite different in the composition of these structural carbohydrates. In terrestrial plants cellulose is the major structural component, but in macroalgae it makes up only a small proportion of the total algal polysaccharide, at most about 7%. The most abundant components in macroalgal species are alginic acid and mannitol in the brown algae and the partially-sulphated galactans (agar and carrageenan) of the red algae. The green algae contain varying amounts of mannans, xylans, glucomannans, pectic acids and hemicellulose in addition to the small amounts of cellulose (Indergaard, 1983; Whitesides and Elliott, 1984).

The next most abundant component of macroalgae is made up of inorganic salts, and is referred to as non-volatile solids or ash. Terrestrial plants have ash contents as low as 5% of the dry weight, but in macroalgal species it is generally higher, and can reach levels of 45% of the total dry weight (Ryther and Hanisak, 1982; Indergaard, 1983). Ash represents cellular components that are not convertible to methane, and will thus reduce the yield per unit weight. Show (1981) has calculated that the potential biomass yield of the brown alga genus *Macrocystis* is greater than that of the red alga *Chondrus* (17.4 kg m^{-2} yr^{-1} vs 16.4 kg m^{-2} yr^{-1}) but because of the higher ash content of *Macrocystis* (Indergaard, 1983), its potential caloric yield is less than that of *Chondrus* (49 000 kcal m^{-2} yr^{-1} for *Macrocystis*, 52 300 kcal m^{-2} yr^{-1} for *Chondrus*, Show (1981)). This factor is important to remember when comparing published values for methane production. Methane production is often reported in terms of methane produced per unit of volatile solids, with no mention of the relative concentration of the volatile and non-volatile components.

Another major component is protein, which composes about 1–7% of the dry weight and contains 90% of the total nitrogen found in macroalgae (Show, 1981). It follows that protein is the major nitrogen component in the C/N ratio. This ratio will vary with nitrogen availability. When supplemental nitrogen is provided to the algae, the C/N ratios can be reduced to the

4–8/1 range (Jackson, 1980; Ryther and Hanisak, 1982; Samson and LeDuy, 1982). In natural situations, during periods of nitrogen deficiency, the C/N ratios can range as high as 50/1 (Jackson, 1980).

A major factor in the consideration of algae for biomass production is the absence of any lignin. Lignin can account for up to 20% of the dry weight of land plants, and as it is not digestible it represents a loss in carbon available for conversion to methane. In addition, it may actually interfere with the conversion process. Thus this lack of lignin makes algae an attractive alternative to terrestrial biomass for use as feedstock for the biological gasification process.

Biological gasification of algae

Biomethanization studies have been carried out with both macroalgae and microalgae utilized as starting material. In most studies methane yields from the algae have been greater than those reported from terrestrial species (Chynoweth, 1980; Indergaard, 1983). Conversion of the available carbon to methane ranges from 47% to 65% (Ryther and Hanisak, 1982; Indergaard, 1983; Matsunaga and Izumida, 1984), and the gas produced ranges from 60% to 72% methane, with the remainder made up of carbon dioxide and trace levels of nitrogen (Samson and LeDuy, 1982; Yang, 1982; Matsunaga and Izumida, 1984; Hanisak and Ryther, 1986).

The algae provided all the necessary nutrients to support the microbial culture, and it was not necessary to add supplements to obtain bioconversion of the biomass (Indergaard, 1983). A wide range of bacterial cultures can utilize the biomass, with origins ranging from pig manure sludge (Yang, 1982) to estuary sediments (Ryther and Hanisak, 1982) to cultures derived from heavily saline environments (Matsunaga and Izumida, 1984). Mesophilic populations of the bacteria produced higher methane yields than did the thermophilic cultures (Chynoweth, 1980).

It was not necessary to wash the macroalgae before digestion; making an algal slurry with fresh water was sufficient. When fresh water is used during digestion, methane yields from unwashed kelp have often been higher than those from freshwater-washed kelp (Chynoweth, 1980). This has been attributed to natural buffering capacity (probably carbonates) found on the surface of the kelp. In one study *Eucheuma* sp. biomass was used in anaerobic fermentation, and the pH of the algal/microbial mixture was measured during digestion. A pH of 6.5 was found when washed kelp was used, but unwashed kelp resulted in a pH of 7 (Yang, 1982), which is the optimum for biomethanization. In a study that may lead to further reductions in the cost of processing the biomass, Matsunaga and Izumida (1984) reported that the digestions can be carried out in saline solution if salt-tolerant microbial cultures are utilized for the inoculum.

A variation in the C/N ratio of the algae can effect methane production, and addition of nitrogen to the digester is sometimes necessary to obtain maximum yields (Chynoweth, 1980), though addition of nitrogen does not always have a significant effect (Yang, 1982). It is important to remember that the overall C/N ratio is not critical, but it is the level of carbon that is actually available for digestion that determines the performance of the biomethanization process. Samson and LeDuy (1982) have reported on biomethanization experiments with the cyanobacterium *Spirulina maxima* as the substrate for biomethanization. The biogasification reactions produced yields of methane that were comparable to other biomass yields, despite the algal biomass having a C/N ratio of 4.2/1. They attributed this result to the ability of the microbes to utilize the carbon present, produce large amounts of the organic acids, and thus compensate for the high ammonia levels. So a prediction of digester performance cannot be made on overall C/N ratios, but must instead be based on both the quality and quantity of the carbon source.

The composition of the carbohydrate has also been found to influence methane yields. Increased mannitol levels in the brown algae have been correlated with an increase in methane yield (Chynoweth, 1980). Gellenbeck and Chapman (1986) have found that mannitol levels increased and alginic acid levels decreased in the brown seaweed *Sargassum muticum* over time. It may be possible to maximize algal conversion to methane by planning the harvest of the biomass in terms of polysaccharide composition.

The polysaccharides of the red algae are also readily utilized by the bacteria. Hanisak and Ryther (1986) found that agar, which makes up 25% of the dry weight of *Gracilaria tikvahie*, was easily converted to methane. The carrageenan levels in *Eucheuma cottonii* biomass decreased at the same rates as the rest of the polysaccharides during digestion, again illustrating the ease of utilization of these unique algal polysaccharides. Because of its high levels of the sulphated glucan carrageenan, *E. cottonii* did cause some problems during digestion. Large amounts of sulphate were produced during the digestion, which decreased methane production. The authors speculate that this restriction was caused by a depletion of the available reducing equivalents by the reduction of the sulphate, resulting in a decrease in the reduction of acetate to methane (King *et al.*, 1986).

To be competitive with other sources of synthetic natural gas, the cost of algal biomass-derived methane must be in the range of $5–6 per MM BTU (Benemann *et al.*, 1986). Current estimates using present biomass and methane yields put the price at $14–80 per MM BTU (Bird, 1986; Snyder and Brehany, 1988) for near-shore kelp farming. If the best case studies can be believed, then reducing the price by half would make algae-derived methane competitive. The best target for increased efficiency is in the production and harvesting of the biomass, which Bird (1986) has estimated to account for 70% of the total cost of the gas. Increased methane yields and reduced need for gas clean-up will also affect the overall cost, so study of all parts of the biomass to methane system should be continued.

Usable area may also limit the overall contribution of algal biomass: with a productivity of 50 DAFMT ha^{-1} yr^{-1}, it would take 127 000 square miles of coastal waters to produce enough gas to satisfy the current needs of the United States. This represents 14% of the total United States coastal waters, including Hawaii and Alaska, that would be available for marine biomass farming (Chynoweth, 1980). The environmental and political costs for dedicating this much coastline for algal farming may be too high.

Direct production of energy-rich fuels by algae

Ethanol and gasoline from glycerol

Dunaliella, a genus of unicellular green algae, is found in saline environments with NaCl concentrations from 0.4 M to greater than 5 M (Ben-Amotz, 1980), with optimum growth rates occurring in NaCl concentrations between 1 and 2 M (Nakas *et al.*, 1986). Amazingly, *Dunaliella* does this without the benefit of a supportive cell wall to compensate for changes in osmotic pressure and turgor pressure; the organism is bounded only by a membrane. To counterbalance the osmotic stress of their environment, these algae synthesize large amounts of glycerol as an alternative osmoticum.

Under optimum growth conditions glycerol production in *Dunaliella* can reach levels of 8 g m^{-2} d^{-1}, and can account for up to 85% of the algal dry weight. This production represents a photosynthetic conversion efficiency of about 4%, which is comparable to or even greater than that of the most productive terrestrial plants (Whitesides and Elliott, 1984). Under conditions of high irradiance the isoprenoid β-carotene will also accumulate, reaching levels as high as 8% of the dry weight of the cell (Ben-Amotz, 1986).

Glycerol is not a good candidate for use as a liquid fuel. It has a relatively high oxygen content and thus a low energy value. The heat of combustion for glycerol is 4.3 kcal g^{-1}, about three times lower than the average for petroleum-derived gasoline of 11.4 kcal g^{-1}. It also is quite viscous, making it difficult to handle and transport and difficult to deliver to the combustion chambers of an engine. But glycerol can be converted to more useful products by a combination of bacterial fermentation and catalytic conversion.

At least three different bacteria are capable of fermenting the glycerol in a *Dunaliella* biomass preparation to other less viscous, higher energy compounds: a *Bacillus* sp.; a *Klebsiella* sp.; and *Clostridium pasteurianum* (Nakas *et al.*, 1986). The products of the *Bacillus* fermentation are ethanol (92% by weight of the total product yield), and acetate (8%) (Fig. 6.2). Fermentation with *Klebsiella* results in the production of 1,3-propanediol (88%), acetate (9%) and ethanol (3%). A fermentation with *C. pasterianum* generates *n*-butanol (72%), ethanol (21%), acetate (1%) and trace amounts of 1,3-propanediol, and results in slightly lower overall yield when compared to the other two microbes. Unamended biomass required a 300-fold concentration to generate the highest yields, and the bacteria were not halophytic, so some expense would be added to the processing due to concentration of the algae and salt removal via freshwater washing. Halophilic bacteria are not needed with washed *Dunaliella* cells because the algae

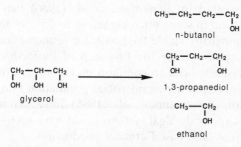

Fig. 6.2 Possible conversions of glycerol to liquid fuels via bacterial fermentations (from Nakas *et al.*, 1986).

do not accumulate large amounts of NaCl internally; the high levels of glycerol allow the algae to exclude salt.

Ethanol, *n*-butanol, and possibly 1,3-propanediol can be used as liquid fuels. They have higher energy contents than glycerol (e.g. ethanol has a heat of combustion of 7.1 kcal g^{-1}), and flow easily at normal temperatures. Ethanol is currently used in Brazil as a substitute for gasoline, and in the midwestern United States as an additive to gasoline (gasohol).

The use of ethanol as a gasoline additive requires the costly removal of water by co-distillation with benzene to a level of less than 1% of the total weight. This has led Costa *et al.* (1985) to propose the conversion of ethanol to gasoline as an economically viable alternative to its use in gasohol. When ethanol (or butanol or 1,3-propanediol) is introduced to a column containing a zeolite shape-selective catalyst (Kokotailo *et al.*, 1978) it will be converted to a mixture of alkanes, alkenes, and aromatics similar to that of gasoline (Weisz *et al.*, 1979; Costa *et al.*, 1985). A water content of 5% has no effect on the yield nor on the stability of the catalyst, so ethanol derived from a simple distillation can be utilized (Costa *et al.*, 1985). Recent results have indicated that with a modified zeolite catalyst it may be possible to generate fuels and chemical feedstocks from dilute ethanol without prior distillation

(Dao and Mao, 1988). Though glycerol could possibly be used directly in this catalytic process, the best results are obtained with starting materials that have higher hydrogen to carbon ratios, and that have reduced amounts of heteroatoms (i.e. oxygen) (Weisz et al., 1979). In comparison to glycerol, the products of the above mentioned bacterial fermentations have both increased H/C ratios and reduced oxygen content.

The biomass remaining after the glycerol fermentation can be further utilized as a fuel source. The β-carotene can be isolated and converted into gasoline-like fuels by catalytic cracking as described in a later section of this chapter. The remaining residue could then be used as a starting material for biomethanization, though the prior removal of a large portion of its carbon may result in C/N ratio problems.

Ethanol from carbohydrates

It has been suggested by Benemann et al. (1986) that the high level of carbohydrates found in some microalgae, up to 75% of the dry weight, may provide an economically viable feedstock for fermentation to ethanol. Since the microbes responsible for the first step of biomethanization can break down these polysaccharides to common metabolic intermediates, it should be possible to develop a mixed microbial population that could ferment the algae to ethanol. Benemann et al. (1986) have estimated that ethanol produced from large-scale algal systems could have up to ten times the yield per area of current land-based ethanol production.

Algal hydrocarbons

Algae are capable of producing large quantities of highly reduced, hydrocarbon-like molecules. These compounds can be classified by their biosynthetic origin: they are all derived from either the isoprenoid biosynthetic pathway or from fatty acid/lipid biosynthesis. They are of value due to their high energy content, about double per unit mass of any other biological compound. The smaller isoprenoids ($<C_{16}$) can be used directly as liquid fuels, and the larger molecules can easily be converted to liquid fuels by use of the proper catalysts. The technology for economically converting lipids to liquid fuels still needs to be developed.

Most of the hydrocarbon producing species are microalgae. Some of the most productive hydrocarbon synthesizers are found in the green alga Botryococcus (Fig. 6.3), where up to 86% of the dry weight is hydrocarbon (Maxwell et al., 1968). This alga is thought to exist in one of two metabolic states, either in the rapidly growing 'active' state or in the quiescent 'resting' state. In the 'active' state between 0.1–36% of the dry weight are long chain linear alkenes, mostly C_{27}, C_{29} and C_{31} dienes (Wolf, 1986). These molecules are probably derived from elongation and decarboxylation of fatty acids. The hydrocarbons found in the 'resting' state usually constitute 25–40% of the total dry weight and are branched alkenes called botryococcenes. These molecules are C_{30}–C_{37} isoprenoids, with a general formula of C_nH_{2n-10}

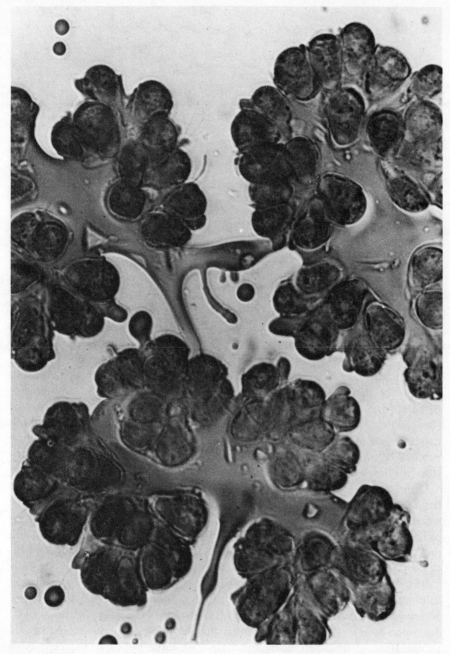

Fig. 6.3 *Botryococcus braunii* algae showing isoprenoid production. (Photograph by F. Wolf).

(Fig. 6.4) (Wolf *et al.*, 1985a,b). Recent work (Wolf *et al.*, 1985b) has suggested that the growth rate in the 'resting' state is quite significant, and that the difference between the two states is not growth but a regulation of car-

botryococcene

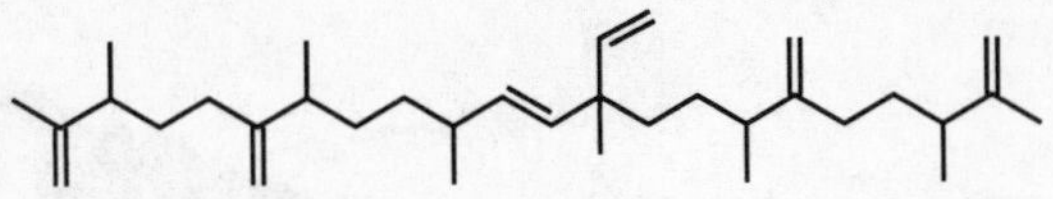

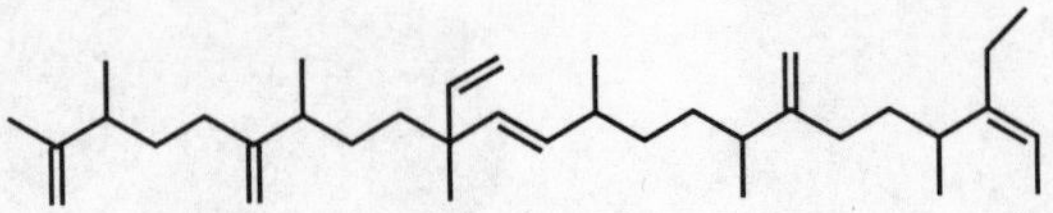

darwinene

Fig. 6.4 Structure of two of the isoprenoids found in *Botryococcus*, the C_{34} botryococcene and the C_{36} darwinene. These compounds are probably formed via a condensation of two C_{15}-isoprenoid molecules followed by further methylations (Wolf *et al.*, 1985).

bon partitioning between the fatty acid biosynthetic pathway and the isoprenoid pathway (Wolf, 1986).

There is no clear understanding of the physiological mechanism that regulates the shift from the linear alkenes to the branched isoprenoids, and many laboratories are unable to obtain isoprenoid production in culture. This has made study of this alga difficult and its economic exploitation impossible at this point in time. But if mass production were possible, and large quantities of isoprenoids could be obtained, the technology is available to convert them directly into liquid fuel. Work with $>C_{30+}$ isoprenoids isolated from terrestrial plants has shown that these materials can be solvent (hexane) extracted and then passed over a zeolite catalyst. The resulting product has a carbon distribution pattern almost identical to gasoline derived from petroleum (Weisz *et al.*, 1979), and could be used as a substitute in combustion engines.

Quite a few species of microalgae will divert a large proportion of their fixed carbon into the production of lipids; values as high as 72% of the dry weight have been reported (Shifrin and Chisholm, 1981). The lipids are either neutral triacylglycerides or polar diacylphospholipids, consisting mostly of combinations of saturated C_{14}, C_{16}, and C_{18} fatty acids and unsaturated C_{18} and C_{20} fatty acids (Pohl, 1982; Tillett and Benemann, 1988).

The identification and isolation of these high-yielding lipid-producing microalgae has been the target of a project directed by the Solar Energy Research Institute (SERI, Golden, Colorado) (Barclay *et al.*, 1987; Johnson, 1988). The ultimate goal of this programme is the selection and propagation of lipid-producing microalgae in large ponds in the deserts of the southwestern United States, utilizing saline water that is available in the region. They have been successful in selecting strains that are tolerant of high temperatures and high salinities, and have found that between 19–36% of their ash free dry weight is lipid under non-stress conditions (Barclay *et al.*, 1987; Tadros, 1987). Most of the favourable algae are diatoms (e.g. *Amphora*, *Navicula*, *Nitzschia*).

Lipid production can be further increased if the microalgae are stressed by nutrient deficiency. It has been known for quite a while that nitrogen stress will increase lipid production in *Chlorella* (Spoehr and Milner, 1949). This has been applied to current studies, and nitrogen stress, along with silicon deficiency in the diatoms (Shifrin and Chisholm, 1981; Roessler, 1987), has been found to cause large increases in algal lipid content. Many of the species isolated by Barclay *et al.* (1987) have shown two- to three-fold increases in lipid content when exposed to N or Si deficiencies. Management of outdoor ponds to induce these nutrient deficiencies should be possible to maximize yields.

Outdoor field-scale pond studies have been carried out over a period of 45 days with *Isochrysis*. Lipid production was measured at $6 \text{ g m}^{-2} \text{ d}^{-1}$, comprising 25% of the total ash free dry weight (Arad, 1987). If this project could be scaled up to year round production, this would represent a lipid yield of 21 tons $\text{ha}^{-1} \text{ yr}^{-1}$. But even with these high yields Benemann *et al.* (1986) have calculated the price of extracted lipids to be at least \$5.50 per gallon, and further processing is necessary before usable liquid fuels could be obtained from the lipids.

The technology for the conversion of lipids to liquid fuels has not been fully developed for the simple reason that currently triglycerides are more valuable to the food industry than as an energy source. It has been found that some triglycerides can be used as diesel fuel substitutes or in diesel blends, but long-term engine problems usually result (Kaufman, 1984). It is possible to trans-esterify the fatty acid components of the lipids, and methyl-esters of the smaller fatty acid chains (C_{18} and smaller) can be used as diesel fuel substitutes without the deposition of carbon observed with triglycerides (Kaufman, 1984; Pryde, 1984). Standard biomass and petroleum technologies, such as pyrolytic conversion to gas and oils or cracking on catalytic columns, could also be adapted for use with lipids to generate gasoline-type fuels.

Another approach that should be considered is the genetic modification of the algae to divert carbon flow from fatty acids to true acyclic hydrocarbons. Both macroalgae (e.g. *Macrocystis*) and the cyanobacteria (e.g. *Phormidium luridum* and *Anacystis nidulans*) have small amounts of alkanes and alkenes, predominantly heptadecane and pentadecane (Han *et al.*, 1968, 1969; Show, 1981). They are derived from the enzymatic decarboxylation or the reduction/decarbonylation of fatty acids (Han *et al.*, 1969; Cheesbrough and Kolattukudy, 1984). The physical characteristics of these compounds make them easier to handle and manipulate than the lipids, and they also have higher energy values being devoid of any oxygen. Therefore, the manipulation of algal metabolism to increase the production of these alkanes instead of lipids could be a beneficial project for the genetic engineers. The same consideration should be given to increasing carbon allocation to the production of isoprenoids, which are found in varying amounts in a wide range of algal species (Fenical, 1982), and have the same advantages as alkanes over lipids. If lipids are chosen as the desired product, it may be possible to enhance their quality as a fuel by reducing the level of unsaturation. Changes in lipid saturation have been accomplished in terrestrial plants (rapeseed) by direct genetic modification.

Hydrogen production

In addition to carbon-based fuels, it may be possible to use algae for the production of hydrogen (Benemann *et al.*, 1980). Hydrogen can be produced through one of two mechanisms: (a) using the nitrogenase of cyanobacteria; or (b) utilizing the hydrogenase activity found in most algae.

Nitrogenase activity has been reported in over 125 strains or species of cyanobacteria, and is found in both heterocystous and non-heterocystous forms (Stewart, 1980). The enzyme normally catalyses the ATP-driven reduction of molecular nitrogen to ammonia, but will also catalyse the production of hydrogen gas (Benemann and Weare, 1974). Under the proper culture conditions it may be possible to exploit this enzymatic system and utilize solar energy for the generation of large volumes of hydrogen gas.

There are two problems preventing the application of this technique. First, the synthesis of ammonia and hydrogen gas is competitive, so the presence of nitrogen will reduce the hydrogen yields. The logical approach of exposing the algal cultures to N_2-free solution (by argon or helium purging) cannot be applied on a long-term basis. Growth requires the presence of nitrogen, but other nitrogen sources, namely ammonium ion or nitrate, repress the expression of the nitrogenase gene. The second problem arises from the presence of a hydrogen uptake enzyme. This enzyme is usually present when nitrogenase is present, and acts as a scrubbing mechanism by trapping any H_2 produced and reoxidizing it. This probably evolved as a means of recapturing the reducing power that is lost by the organism in the production of hydrogen. Needless to say, this activity results in a decrease in hydrogen yield.

These difficulties may be solved through genetics. Spiller and Shanmugam (1986) are attempting to isolate mutant strains of *Anabaena variabilis* that have expression of nitrogenase in the presence of ammonium ion or nitrate, and have reduced levels of hydrogenase uptake activity. They have isolated strains that exhibit one or the other of these traits, and are using them to develop an understanding of the physiology of this alga to identify the proper targets for genetic manipulation to increase hydrogen yields.

Hydrogen production can also be accomplished via a hydrogenase system found in the vegetative cells of the algae. This enzyme uses available reductant power to produce hydrogen gas, and the reaction can occur in either the dark or the light. It is the use of the algae for the conversion of solar energy into hydrogen gas that has received the most attention (Mahro and Grimme, 1983).

This photoproduction of hydrogen utilizes the algal cell's photosynthetic light harvesting and electron transport system. The overall reaction involves splitting water to produce oxygen and hydrogen, though the actual sites of oxygen and hydrogen production are physically separate (Greenbaum,

1977). The reaction is inhibited by the presence of carbon dioxide, which consumes most of the available reduction potential via carbon fixation, and by oxygen, which either inactivates the hydrogenase or oxidizes the electron carrier (Rosenkrans and Krasna, 1984). To overcome this limitation, hydrogen must be generated under conditions of low carbon dioxide concentrations, and the oxygen removed by either a flow of inert gas (Greenbaum, 1977) or by the presence of reversible oxygen-chelating agents (Rosenkrans and Krasna, 1984). A low concentration of oxygen (and carbon dioxide) is stressful for aerobic organisms, so selection of anaerobic-tolerant hydrogen-producing strains has been attempted and accomplished with *Chlamydomonas reinhardtii* (Ward *et al.*, 1985).

The photoproduction of hydrogen is found primarily in the green algae (e.g. *Chlorella, Chlamydomonas*). Though the hydrogenase enzyme is present in the macroalgae, the seaweeds are unable to produce hydrogen in the light. Greenbaum and Ramus (1983) hypothesize that hydrogenase in the macroalgae is used primarily as a hydrogen uptake system and is not involved in the photoproduction of hydrogen.

Genetic manipulation

It will be necessary to increase the productivity, the tolerance range and the quality of algal biomass before it can be seriously considered as an energy source. Such improvements can be accomplished by combining the selection and breeding techniques of classical genetics with the revolutionary gene manipulation methods of recombinant DNA technology. This work is already underway in laboratories throughout the world.

By selecting for outstanding individuals, and cross-breeding the desired traits into the population, increases in *Porphyra* yields of three to five times have been accomplished. Varieties of *Laminaria japonica* have been cloned that have increased tolerance ranges to light and temperature compared to the wild types (Indergaard, 1983). SERI has collected and cultured over 3000 different microalgae strains, and has selected at least 39 strains for possible biofuel applications (Johnson, 1988). The attempts at selecting for increased hydrogen production (Ward *et al.*, 1985; Spiller and Shanmugam, 1986) have already been mentioned.

The methodology needed for genetic engineering is also being developed for the algae. Both protoplasts and callus can be made from the seaweeds, and the seaweeds can be regenerated from callus (Cheney, 1986), an essential step in the derivation of new varieties after genome manipulation has been accomplished. Attempts are being made to fuse protoplasts of different species (Saga *et al.*, 1986); this technique allows for the combination of favourable genetic traits from different organisms. Tissue culture is being utilized to select for individuals with greater light and temperature tolerances, increased nutritional efficiency, changes in morphology, and disease resistance (Polne-Fuller *et al.*, 1986).

The actual manipulation of the algal genome has been accomplished with cyanobacteria (Lem and Glick, 1985). Attempts to manipulate the eukaryotic algae are currently centring on the isolation of algal viruses that could act as vectors for DNA recombination (Johnson, 1988; Meints *et al.*, 1987). These vectors, or carriers, are necessary to insert selected DNA sequences into the algal genome. As of yet, none of the isolated viruses will replicate once they have been inserted into high productivity strains cultured by SERI.

Conclusions

It is possible that algal systems may provide an environmentally and economically feasible alternative to fossil fuels, for the technology needed to utilize the algae is available. But certain developmental goals must be reached before this programme can be implemented:

(a) The proper methods of cultivation and harvesting must be determined to maximize yields and minimize energy and monetary inputs.

(b) The efficiency of the conversion of biomass to usable fuel must be maximized.

(c) An understanding of the physiological controls of growth and production of energy-rich compounds must be developed before these components can be produced at the highest possible rates.

(d) Selection and/or modification of algal strains with wider environmental tolerances, increased photosynthetic conversion efficiencies, faster growth rates, and increased production of energy-rich components must be accomplished.

Because of the limitations of available coastal areas, low yields, and lack of investment capital, it is doubtful that algal-derived fuels can totally substitute for either petroleum fuels or natural gas. But even if algal-derived fuels comprise only a few per cent of the total global energy use, this still represents an enormous quantity of energy, and could contribute to the improvement of the global carbon dioxide problems. Therefore, an algal biomass programme should be considered in combination with other energy alternatives, such as land-based biomass and energy derived from artificial photosynthetic systems. Such an integrated system could reduce our dependence on fossil fuels, and all the problems associated with their use.

Acknowledgement

This work was supported by the Office of Basic Energy Sciences, Biological Research Division of the US Department of Energy under Contract No. DE-AC03-76SF00098.

References

Arad S 1987 Integrated field-scale production of oil-rich microalgae under desert conditions. In *Fiscal Year 1986 Aquatic Species Program Annual Report*. Solar Energy Research Institute, Golden, CO, USA, pp 139–68

Barclay B, Nagle N, Terry K 1987 Screening microalgae for biomass production potential: protocol modification and evaluation. In *Fiscal Year 1986 Aquatic Species Program Annual Report*. Solar Energy Research Institute, Golden, CO, USA, pp 23–40

Ben-Amotz A 1980 Glycerol production in the alga *Dunaliella*. In San Pietro A (ed) *Biochemical and Photosynthetic Aspects of Energy Production*, Academic Press, pp 191–208

Ben-Amotz A 1986 β-Carotene enhancement and its role in protecting *Dunaliella bardawil* against injury by high irradiance. In Barclay W R, McIntosh R P (eds) *Algal Biomass Technologies*. J Cramer, Berlin, pp 132–5

Benemann J R, Miyamoto K, Hallenbeck P C 1980 Bioengineering aspects of biophotolysis. *Enz. Microbiol. Technol.* **2**: 103–11

Benemann J R, Weare N M 1974 Hydrogen evolution by nitrogen-fixing *Anabaena cylindrica* cultures. *Science* **184**: 174–5

Benemann J R, Weissman J C, Goebel R P, Augenstein D C 1986 Microalgae fuel economics. In Barclay W R, McIntosh R P (eds) *Algal Biomass Technologies*, J Cramer, Berlin, pp 176–91

Bird K T 1986 An historical perspective on economic and system costs of kelp cultivation for bioconversion to methane. In Barclay W R, McIntosh R P (eds) *Algal Biomass Technologies*, J Cramer, Berlin, pp 150–4

Calvin M 1984 Renewable fuels for the future. *J. Appl. Biochem.* **6**: 3–18

Cheesbrough T M, Kolattukudy P E 1984 Alkane biosynthesis by decarbonylation of aldehydes catalyzed by a particulate preparation from *Pisum sativum*. *Proc. Nat. Acad. Sci. USA* **81**: 6613–17

Cheney D 1986 Genetic engineering in seaweeds: applications and current status. In Barclay W R, McIntosh R P (eds) *Algal Biomass Technologies*, J Cramer, Berlin, pp 22–9

Chin L-Y, Engel A J 1981 Hydrocarbon feedstocks from algae hydrogenation. *Biotechnol. Bioeng. Symp.* **11**: 171–86

Chynoweth D P 1980 Anaerobic digestion of marine biomass. In *Biogas and Alcohol Fuels Production*, The JG Press, Emmaus, PA, USA, pp 185–201

Costa E, Uguina A, Aguado J, Hernandez P J 1985 Ethanol to gasoline: effect of variables, mechanism and kinetics. *Ind. Eng. Chem. Process Des. Dev.* **24**: 239–44

Dao L H, Mao R L 1988 Conversion of ethanol to ethylene over zeolite catalysts in the presence of water and other light alcohols. In Klass D L (ed) *Energy from Biomass and Wastes XI*, Institute of Gas Technology, Chicago, pp 877–91

Dugan G L, Cheng E D H, Takahashi P K 1988 Induced mixing characteristics for algal biomass enhancement. In Klass D L (ed) *Energy from Biomass and Wastes XI*, Institute of Gas Technology, Chicago, pp 787–806

Fenical W 1982 Terpenoids of algae. In Zaborsky O R, Mitsui A, Black C C (eds) *CRC Handbook of Biosolar Resources Vol I Part 1 Basic Principles*, CRC Press, Boca Raton, FL, USA, pp 467–77

Gellenbeck K, Chapman D 1986 Feasibility of mariculture of the brown seaweed, *Sargassum muticum* (Phaeophyta): growth and culture conditions, culture methods, alginic acid content and conversion to methane. In Barclay W R, McIntosh R P (eds) *Algal Biomass Technologies*, J Cramer, Berlin, pp 107–15

158   Algal and cyanobacterial biotechnology

Golueke C G 1980 Basic principles of anaerobic digestion. In *Biogas and Alcohol Fuels Production*, The JG Press, Emmaus, PA, USA, pp 7–14
Greenbaum E 1977 The photosynthetic unit of hydrogen evolution. *Science* **196**: 879–80
Greenbaum E, Ramus J 1983 Survey of selected seaweeds for simultaneous photoproduction of hydrogen and oxygen. *J. Phycol.* **19**: 53–7

Han J, Chan H W-S, Calvin M 1969 Biosynthesis of alkanes in *Nostoc muscorum*. *J. Amer. Chem. Soc.* **91** 5156–9
Han J, McCarthy E D, Calvin M, Benn M H 1968 Hydrocarbon constituents of the blue–green algae *Nostoc muscorum, Anacystis nidulans, Phorimidium luridum* and *Chlorogloea fritschii. J. Chem. Soc.* **1968C**: 2785–91
Hanisak M D, Ryther J H 1986 The experimental cultivation of the seaweed *Gracilaria tikvahaiae* as an 'energy crop': an overview. In Barclay W R, McIntosh R P (eds) *Algal Biomass Technologies*, J Cramer, Berlin, pp 212–17

Indergaard M 1983 The aquatic resource. In Cote W A (ed) *Biomass Utilization*, Plenum Press, pp 137–68

Jackson G A 1980 Marine biomass production through seaweed aquaculture. In San Pietro A (ed) *Biochemical and Photosynthetic Aspects of Energy Production*, Academic Press, pp 31–58
Johnson D A 1988 The technology and cost of producing triglyceride liquids from microalgae for use as fuels. In Klass D L (ed) *Energy from Biomass and Wastes XI*, Institute of Gas Technology, Chicago, pp 755–69

Kaufman K R 1984 Testing of vegetable oils in diesel engines. In Schultz E B Jr, Morgan R P (eds) *Fuels and Chemicals from Oilseeds*, Westview Press, Boulder, CO, USA, pp 143–76
King G M, Guist G G, Lauterbech, G E 1986 Stability of carrageenan during anaerobic digestion. In Barclay W R, McIntosh R P (eds) *Algal Biomass Technologies*, J Cramer, Berlin, pp 116–23
Kokotailo G T, Lawton S L, Olson D H, Meier W M 1978 Structure of synthetic zeolite ZSM-5, *Nature (Lond.)* **272**: 437–8

Laws E A, Taguchi S, Hirata J, Pang L 1986 High algal production rates achieved in a shallow outdoor flume. *Biotechnol. Bioeng.* **37**: 191–7
Lem N W, Glick B R 1985 Biotechnological use of cyanobacteria. *Biotechnol. Adv.* **3**: 195–208
Lincoln E P, Koopman B 1986 Bioflocculation of microalgae in mass culture. In Barclay W R, McIntosh R P (eds) *Algal Biomass Technologies*, J Cramer, Berlin, pp 207–11

Mahro B, Grimme L H 1983 Hydrogen from water – the potential role of green algae as a solar energy conversion system. In Cote W A (ed) *Biomass Utilization*, Plenum Press, pp 227–32
Matsunaga T, Izumida H 1984 Seawater-based methane production from blue–green algae biomass by marine bacteria coculture. *Biotechnol. Bioeng. Symp.* **14**: 407–18
Maxwell J R, Douglas A G, Eglinton G, McCormick A 1968 The botryococcenes – hydrocarbons of novel structure from the alga *Botryococcus braunii* Kutzing. *Phytochemistry* **7**: 2157–71
Meints R H, Schuster A M, van Etten J L 1987 Characterization of viruses infecting *Chlorella*-like alga. In *Fiscal Year 1986 Aquatic Species Program Annual Report*. Solar Energy Research Institute, Golden, CO, USA, p 338

Nakas J P, Schaedle M, Tanenbaum S W 1986 Bioconversion of algal biomass to neutral solvents. In Barclay W R, McIntosh R P (eds) *Algal Biomass Technologies*, J Cramer, Berlin, pp 171–5

Pohl P 1982 Lipids and fatty acids of microalgae. In Zaborsky O R, Mitsui A, Black C C (eds) *CRC Handbook of Biosolar Resources Vol I Part 1 Basic Principles*, CRC Press, Boca Raton, FL, USA, pp 383–404

Polne-Fuller M, Saga N, Gibor A 1986 Algal cells, callus, and tissue culture and selection of algal strains. In Barclay W R, McIntosh R P (eds) *Algal Biomass Technologies*, J Cramer, Berlin, pp 30–6

Pryde E H 1984 Chemicals and fuels from commercial oilseed crops. In Schultz E B Jr, Morgan R P (eds) *Fuels and Chemicals from Oilseeds*, Westview Press, Boulder, CO, USA, pp 51–70

Roessler P 1987 Biochemical aspects of lipid accumulation in silicon-deficient diatoms. In *Fiscal Year 1986 Aquatic Species Program Annual Report*, Solar Energy Research Institute, Golden, CO, USA, pp 257–71

Roos J W, Ristroph D L 1988 Storage and utilization of anaerobic digestion intermediates for continuous methane production. In Klass K L (ed) *Energy from Biomass and Wastes XI*, Institute of Gas Technology, Chicago, pp 665–78

Rosenkrans A M, Krasna A I 1984 Stimulation of hydrogen photoproduction in algae by removal of oxygen by reagents that combine reversibly with oxygen. *Biotechnol. Bioeng.* **26**: 1334–42

Ryther J H 1959 Potential productivity of the sea. *Science* **130**: 602–8

Ryther J H, Hanisak M D 1982 Anaerobic digestion and nutrient recycling of small benthic or floating seaweeds. In Klass D L (ed) *Energy from Biomass and Wastes V*, Institute of Gas Technology, Chicago, pp 384–410

Saga N, Polne-Fuller M, Gibor A 1986 Protoplasts from seaweeds: production and fusion. In Barclay W R, McIntosh R P (eds) *Algal Biomass Technologies*, J Cramer, Berlin, pp 37–43

Samson R, LeDuy A 1982 Biogas production from anaerobic digestion of *Spirulina maxima* algal biomass. *Biotechnol. Bioeng.* **24**: 1919–24

Shelef G, Sukenik A, Green M 1986 Separation and harvesting of marine microalgal biomass. In Barclay W R, McIntosh R P (eds) *Algal Biomass Technologies*, J Cramer, Berlin, pp 245–51

Shifrin N S, Chisholm S W 1981 Phytoplankton lipids: interspecific differences and effects of nitrate, silicate and light-dark cycles. *J. Phycol.* **17**: 374–84

Show Jr I T 1981 Marine plants. In Zaborsky O R, McClure T A, Lipinsky E S (eds) *CRC Handbook of Biosolar Resources Vol II Resource Materials*, CRC Press, Boca Raton, FL, USA, pp 471–98

Synder N W, Brehany J J 1988 Kelp to energy utilizing anaerobic digestion. In Klass D L (ed) *Energy from Biomass and Wastes XI*, Institute of Gas Technology, Chicago, pp 1099–121

Spiller H, Shanmugam K R 1986 Genetic modification of *Anabaena variabilis* for the enhancement of H_2 evolution. In Barclay W R, McIntosh R P (eds) *Algal Biomass Technologies*, J Cramer, Berlin, pp 3–5

Spoehr H A, Milner H W 1949 The chemical composition of *Chlorella*; effect of environmental conditions. *Pl. Physiol.* **24**: 120–49

Stewart W D P 1980 Some aspects of structure and function in N_2-fixing cyanobacteria. *A. Rev. Microbiol.* **34**: 497–536

Tadros M G 1987 Screening and characterizing oleaginous microalgal species in the southeastern United States. In *Fiscal Year 1986 Aquatic Species Program Annual Report*, Solar Energy Research Institute, Golden, CO, USA, pp 67–89

Tillett D M, Benemann J R 1988 Techniques for maximizing lipid formation in microalgae production. In Klass D L (ed) *Energy from Biomass and Wastes XI*, Institute of Gas Technology, Chicago, pp 771–87

Waaland J R, Dickson L G, Duffield E C S, Burzycki G M 1986 Research on *Porphyra* acquaculture. In Barclay W R, McIntosh R P (eds) *Algal Biomass Technologies*, J Cramer, Berlin, pp 124–31

Ward B, Reeves M E, Greenbaum E 1985 Stress-selected *Chlamydomonas reinhardtii* for photoproduction of hydrogen. *Biotechnol. Bioeng. Symp.* **15**: 501–10

Weissman J C, Goebel R P 1987 Factors affecting the photosynthetic yield of microalgae. In *Fiscal Year 1986 Aquatic Species Program Annual Report*, Solar Energy Research Insitute, Golden, CO, USA, pp 139–68

Weisz P B, Haag W O, Rodewald P G 1979 Catalytic production of high-grade fuel (gasoline) from biomass compounds by shape-selective catalysis. *Science* **206**: 57–8

Whitesides G, Elliott J 1984 Organic chemicals from marine sources. In Colwell R R, Sinskey A J, Pariser E R (eds) *Biotechnology in the Marine Sciences*, John Wiley and Sons, pp 135–52

Wolf F R 1986 Physiological and metabolic studies on a branched hydrocarbon-producing strain of *Botryococcus braunii*. In Barclay W R, McIntosh R P (eds) *Algal Biomass Technologies*, J Cramer, Berlin, pp 160–70

Wolf F R Nonomura A M, Bassham J A 1985a Growth and branched hydrocarbon production in a strain of *Botryococcus braunii* (Chlorophyta). *J. Phycol.* **21**: 388–96

Wolf F R, Nemethy E K, Blanding J H, Bassham J A 1985b Biosynthesis of unusual acyclic isoprenoids in the alga *Botryococcus braunii*. Phytochemistry **24**: 733–7

Yang P Y 1982 Methane fermentation of Hawaiian seaweeds. In Klass D L (ed) *Energy from Biomass and Wastes V*, Institute of Gas Technology, Chicago, pp 307–27

7

Secondary metabolites of pharmaceutical potential

K.-W. Glombitza and M. Koch

Introduction

Mankind learned very early to use algae for therapeutical purposes by making use of their bioactive substances. The medicinal use of algae can be traced from K'un-pu (*Laminaria japonica*) of the Chinese Pen-ts'ao literature, which supposedly goes back to the legendary Shen-nung (27th–29th century BC), the Indian Ayurveda medicine and the Materia Medica of Dioscurides and the pharmacopoeia of the modern age up to 'Tang' (*Fucus vesiculosus, Ascophyllum nodosum*) of the German Pharmacopoeia DAB 9. In contrast to the worldwide use of algae in folk medicine (Hoppe, 1979), there are few appreciated preparations and only a low number of defined active substances produced from algae in modern medicine. During the last hundred years, natural chemistry was synonymous with terrestrial natural product chemistry (Scheuer in Okuda *et al.*, 1982) and not with the chemistry of marine natural products. Only the polymeric carbohydrates (alginates, agars, carrageenans) have been of scientific interest and are economically important not as active substances but as inactive ingredients (Bonotto, 1979; McLachlan, 1985).

In the 1950s screening of thallus preparations or extracts of algae for various effects began; tests for antibiotic activity became important due to the importance which antibiotic therapy had attained and the availability of simple assay methods. Soon other screening systems were devised with many algae exhibiting antimicrobial, i.e. antibacterial, antifungal or antialgal as well as anthelmintic, antiviral, antimitotic, anticoagulating or haemagglutinating activities. Convulsive or spasmolytic, paralytic, and hypocholesterolaemic effects were demonstrated (review: Ragan, 1984). Ichthyotoxic or antifeedant properties of marine, limnetic or terrestrial algae have frequently initiated intensive research work. Thereby, interesting insights into the biological complexity of the marine environment and the dependence of the individual organisms on each other have often been demonstrated.

In the search for the principles responsible for the effects, a great number of substances have been identified and new, often surprising, structures found. The development began in the 1970s and is still continuing today. It reached a culminating point in the work of the Roche Research Institute of Marine Pharmacology, located in Australian Dee Why, which was only in existence seven years and which investigated many marine organisms including algae with systematic screening philosophy (Baker, 1984).

The quickly growing knowledge about the new compounds soon made review articles necessary. Of particular importance are the papers published by Faulkner in which many newly discovered structures of marine products were listed (Faulkner, 1977 (ref. up to 1976), 1978, 1984 (ref. 1977–Oct. 83), 1986 (ref. Oct. 1983–Jul. 85).

But many other reviews have also appeared, including general reviews: Bhakuni and Silva (1974); Aubert *et al.* (1979); Chapman (1979); Diaz-Pifferer (1979); Glombitza (1979); Hashimoto (1979); der Marderosian (1979); Hoppe (1982); Ragan (1984); Stein and Borden (1984); Wright (1984); Berdy *et al.* (1985); Lindequist and Teuscher (1987); and special reviews: Fenical (1978); Martin and Darias (1978); Moore (1978); Higa (1981); Moore (1981); Erickson (1983); Naylor *et al.* (1983); Shimizu and Kamiya (1983); Shimizu (1984); Blunden and Gordon (1986) and Ragan and Glombitza (1986) (for other references see later sections).

The authors realize that these lists are incomplete, moreover, quotations in the following sections do not make them complete. The list in general is intentionally restricted to reports in which biological activities are attributed to certain well defined substances. In numerous cases, the discovery of a biological activity in an extract initiated the search for the substances responsible. However, only the structure of the substance was of interest afterwards and either the interest or the necessary amount of pure substance was lacking for correlation with the biological activity. More or less purified fractions of substances with attractive names like chonalgin, sarganin (Nadal *et al.*, 1966), aponin (Kutt and Martin, 1975) but which are not exactly defined are not discussed in this article.

Rhodophyceae

Bioactive metabolites from red algae

Numerous red algae have been used in folk medicine throughout history. In many cases the activity can be traced to carrageenans or other polymeric carbohydrates. The unidentified highly antibacterial fraction from *Chondrus crispus* may be such a substance (Natsumo *et al.*, 1986). In some cases certain secondary plant substances could be held responsible for the activity.

Unusual nitrogen-containing compounds from red algae

Various species of the genera *Chondria*, *Digenea* and *Alsidium* (Rhodomelaceae) have been used traditionally because of their anthelmintic activity against ascariasis: *C. armata* in south Japan under the name of 'domoi', *C. sanguinea* and *C. vermicularis* in Brazil, *Digenea simplex* as basic stock of 'tse ko-tsoi' in south China and *Alsidium helminthocorton* as 'Mousse de Corse'. The effective substance, α-kainic acid (**1**) and the less effective allo-kainic acid (**2**) were found in *D. simplex* (Ueno *et al.*, 1955) and in *Alsidium helminthocorton* (Balansard *et al.*, 1983) together with domoic acid (**3**) (Takemoto and Daigo, 1960). A series of less effective isomeric compounds (**4–6**, Maeda *et al.*, 1985, 1986) were also isolated from *C. armata*.

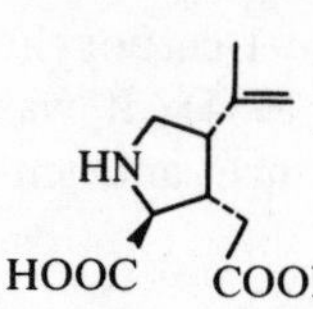

1 α-kainic acid 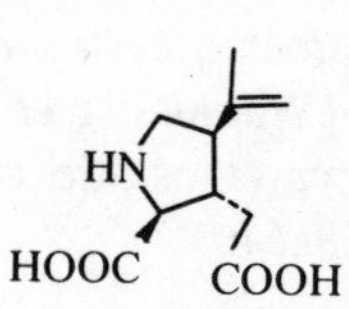   2 allo-kainic acid

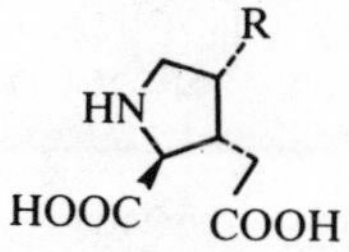

R=

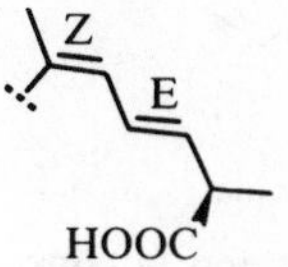

3 domoic acid 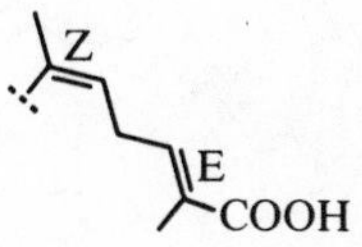   4 isodomoic acid A 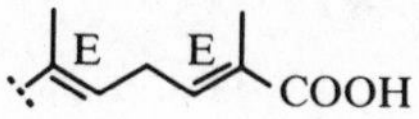   5 isodomoic acid B 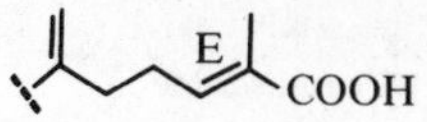   6 isodomoic acid C

Fig. 7.1 Structure of kainic and domoic acids.

Domoic acid has an intraperitoneal activity against American cockroaches (*Periplaneta americana*), is 100-fold stronger insecticide than pyrethrin (Maeda *et al.*, 1986) and is also more effective than kainic acid (Maeda *et al.*, 1984). When applied topically, it is highly effective against houseflies and German cockroaches (*Blattela germanica*) (Suntory Ltd, 1982; Maeda *et al.*, 1984), but the mechanism of the insecticidal activity is not fully understood. α-Kainic acid and domoic acid are potent neurotoxic, glutamate-like excitants of rat cerebral and cat, frog and rat spinal neurons (review: McGeer *et al.*, 1978). Both have similar action on the mammalian central nervous system and the crayfish muscular junction (Shinozaki, 1978). α-Kainic acid causes an acute and prolonged seizure disorder, that appears to involve the limbic system selectively (Nadler *et al.*, 1978) and, additionally, a highly irregular pattern of neuronal degeneration occurs in which some neurons are relatively resistant whereas others are remarkably sensitive (Biziere et al., 1981). It damages most of the neostriatum and lesion sites are characterized by gliosis and the absence of neurons (Walker and McAllister, 1986). It causes dose-correlated neurochemical changes, irreversible losses of the enzyme markers glutamic acid decarboxylase and choline acetyltransferase and incomplete parenchymal necrosis and haemorrhages (Sperk et al., 1985). α-Kainic acid also causes large stationary depolarization of cell membranes with profound ionic imbalance (Garthwaite et al., 1986) which may explain, at least partly, its stimulatory and neurotoxic effects. They have become a potent tool in neurobiology.

Carnosadine (14) is 1-amino-2-guanidinomethylcyclopropane-1-carboxylic acid and is found in Grateloupia armata (Wakamiya et al., 1984). It was patented as an anti-inflammatory drug, carcinostatic substance and immunity enhancer (Shiba and Wakamiya, 1985).

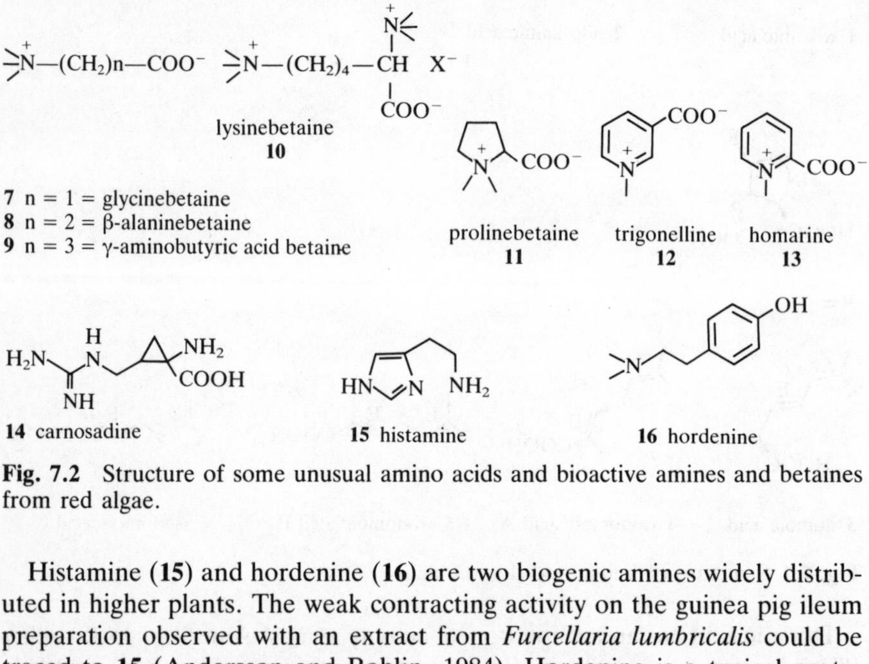

$\overset{+}{\geq}$N—(CH$_2$)n—COO$^-$ $\overset{+}{\geq}$N—(CH$_2$)$_4$—CH   X$^-$

lysinebetaine
10

7 n = 1 = glycinebetaine
8 n = 2 = β-alaninebetaine
9 n = 3 = γ-aminobutyric acid betaine

prolinebetaine
11

trigonelline
12

homarine
13

14 carnosadine   **15** histamine   **16** hordenine

Fig. 7.2 Structure of some unusual amino acids and bioactive amines and betaines from red algae.

Histamine (**15**) and hordenine (**16**) are two biogenic amines widely distributed in higher plants. The weak contracting activity on the guinea pig ileum preparation observed with an extract from Furcellaria lumbricalis could be traced to **15** (Andersson and Bohlin, 1984), Hordenine is a typical protoalkaloid which was first found in Phyllophora nervosa (Güven et al., 1970), Ahnfeltia paradoxa (Kawauchi and Sasaki, 1978) and later in other red algae as well. The betaines β-alaninebetaine (**8**), lysinebetaine (**10**), trigonelline (**12**), and homarine (**13**), which occur in red algae, reduce both total and free cholesterol levels in blood plasma of rats and cause autonomic excitation which may culminate in death. Glycinebetaine (**7**) has an anticonvulsant effect and decreases seizures induced by pentylenetetrazol and electroconvulsive shock. Prolinebetaine enhances the coagulability of blood (review: Blunden and Gordon, 1986).

Four brominated indoles (**17–20**) were isolated from antimicrobially

	R¹	R²	R³	R⁴	R⁵	R⁶	R⁷
17	CH₃	Br	Br	H	H	Br	H
18	CH₃	Br	Br	H	Br	H	H
19	H	Br	Br	H	Br	Br	H
20	CH₃	Br	Br	H	Br	Br	H
21	H	Cl	Cl	H	H	H	Cl
22	H	Cl	Cl	H	H	H	Br
23	H	Br	Cl	H	H	H	Br
24	H	Br	Br	H	H	H	Br
25	H	Cl	Cl	Cl	H	H	Cl
26	H	Cl	Cl	Br (Cl)	H	H	Cl (Br)
27	H	Cl	Cl	Br	H	H	Br
28	H	Br (Cl)	Cl (Br)	Br	H	H	Br
29	H	Br	Br	Br	H	H	Br
30	H	Cl	Cl	Cl	H	H	H

Fig. 7.3 Halogenated indoles from red algae.

active extracts of *Laurencia brongniartii*, from which, however, only **19** is effective against *B. subtilis* and *Saccharomyces cerevisiae* (Carter *et al.*, 1978). Which of the numerous halogenated indoles (**21–30**) found in *Rhodophyllis membranacea* is responsible for the antifungal activity of the raw extract was not investigated (Brennan and Erickson, 1978). Three basic indolalkaloids were isolated from the red alga *Martensia fragilis*: fragilamide, martensine A (**31**) and B. An antibacterial activity was detected for **31** (Kirkup and Moore, 1983). Only a few antimetabolites with a purine or pyrimidine skeleton have been isolated from algae so far. An especially interesting compound is the 5-deoxy-5-iodotubercidin (**32**), from *Hypnea valentiae* (Kazlauskas *et al.*, 1983), which causes potent muscle relaxation and hypothermia when injected into mice. It is a very potent inhibitor of adenosine uptake in brain slices and the most potent adenosine kinase inhibitor yet described and may prove to be the most specific (Davies *et al.*, 1984).

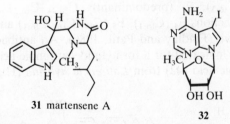

31 martensene A

32

Fig. 7.4 Structure of red algal heterocyclic alkaloids.

Sulphur and sulphur-containing heterocyclic compounds

The antibiotic activity of *Ceramium rubrum* against *B. subtilis* is traced back to the occurrence of elemental sulphur (0.008% dry weight, Ikawa *et al.*, 1973). In *Chondria californica*, a species belonging to the Rhodomelaceae, various sulphur-containing heterocyclic compounds were found: cyclic polysulphides and their oxidation products, several thiepanes (**33–36**) and thiolanes (**37–40**) some of which (**33, 34** and especially **37**) are antibiotically active (Wratten and Faulkner, 1976). Substance **37** has a weak activity in both the sea urchin egg cell division (SUED) assay and a microtubule assembly assay (Jacobs *et al.*, 1981). Similar cyclic polysulphides are also present in the brown alga *Dictyopteris* sp. (Roller *et al.*, 1971).

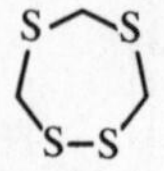

33 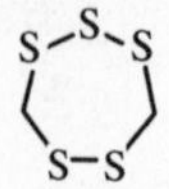   **34**   lenthionin 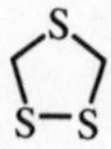   **35** 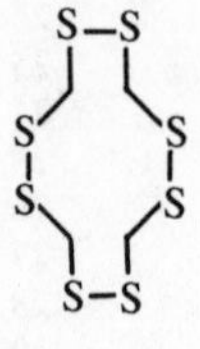   **36**

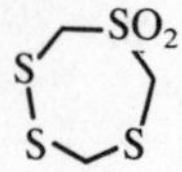

37    **38** 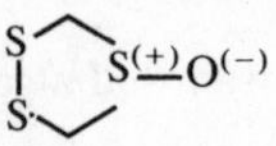   **39** 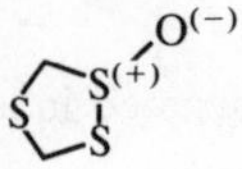   **40**

Fig. 7.5   Sulphur-containing compounds from red algae.

Unsaturated aliphatic carboxylic acids

Several red algae, in particular from the order Ceramiales (e.g. *Polysiphonia, Rhodomela*) contain a relatively large amount of dimethylpropiothetine (**184**), which releases acrylic acid after cleavage (Glombitza, 1979). This contributes considerably, at least with representatives of the genus *Polysiphonia*, to antimicrobial inhibition zones obtained with thallus preparations and extracts in the agar-diffusion test. In many red algae, longer chained, polyunsaturated fatty acids and their derivatives were detected (Pohl and Zurheide, 1979). Such fatty acid fractions from *Cystoclonium purpureum* proved to be partly antibiotically active. An inhibiting effect against various Gram-positive and Gram-negative bacteria was detected for the fractions F_1 (predominantly $C_{16:1}$, $C_{16:3}$), F_2 (predominantly $C_{20:4}$, $C_{16:2}$), F_3 ($C_{16:2}$, $C_{20:4}$, $C_{18:3}$), while the fractions F_4 ($C_{16:1}$), F_5 (C_{14}, C_{16}, $C_{16:1}$) and F_6 ($C_{16:1}$ and $C_{20:5}$) were ineffective (Findlay and Patil, 1986). An antibacterial and antifungal activity is also indicated for a formylundecatrienoic acid (**41**) and a hydroxyeicosapentaenoic acid (**42**) from *Laurencia hybrida* (Higgs, 1981).

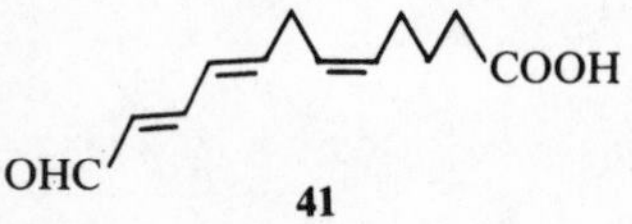

41 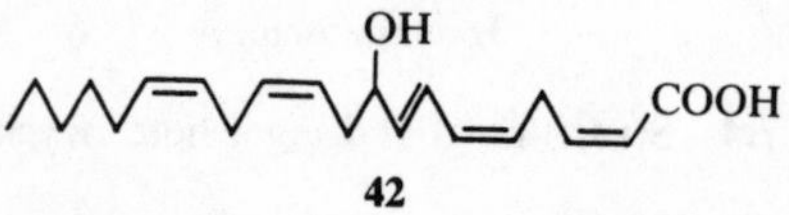   **42**

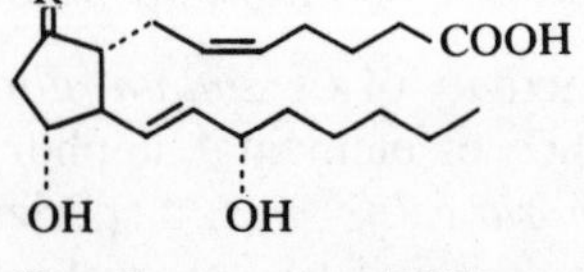

43   R = O; PGE_2
44   R = α OH, β H: $PGF_{2\alpha}$

Fig. 7.6   Unsaturated carboxylic acids and prostaglandins from red algae.

In animals, prostaglandins are widely distributed derivatives of higher, unsaturated fatty acids. They occur in soft corals (*Plexaura homomalla*) in substantial quantities. Surprisingly, the prostaglandins PGE_2 (**43**) and $PGF_{2\alpha}$ (**44**) were also conclusively identified in an antihypertensive extract from *Gracilaria lichenoides* (0.07–0.1% dry weight alga) and identified with some certainty for *G. confervoides* (Gregson *et al.*, 1979).

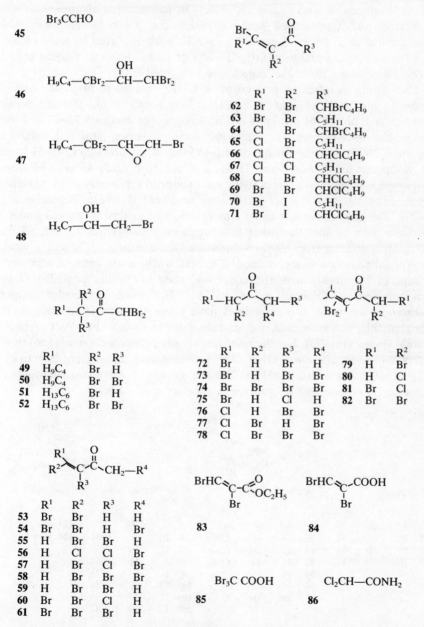

Fig. 7.7   Antibiotically active, halogenated compounds from Bonnemaisoniaceae.

Halogenated low molecular weight aliphatic compounds

Extracts of *Ptilonia, Asparagopsis, Bonnemaisonia hamifera* and its tetra-sporophyte *Falkenbergia* turned out to be exceptionally antibiotically active. In solutions of raw substance mixtures, free iodine is developed very quickly (McConnell and Fenical, 1977a). Upon addition of sodium thiosulphate to the extracts the activity sometimes disappears. In the highly volatile fraction of the Bonnemaisoniaceae a great variety of halogenated alkanes, alkenes, saturated and unsaturated ketones, aldehydes, alcohols, epoxides and halogenated derivatives of acetic and acrylic acids (isolated by their esters) have been found usually by GC, GC-MS or LPLC (review: Fenical *et al.*, 1979; Glombitza, 1979; McConnell and Fenical, 1980).

The strong disinfecting properties of $CHBr_3$ (bromoform) from *Aspar-agopsis* and *Falkenbergia* and CHI_3 (iodoform) from *A. taxiformis* (Burreson *et al.*, 1976) are known. The halogenated acetones **72–78** and the butenones **79–82** from *A. taxiformis* have a strong antibiotic activity (Fenical, 1974). For the substances **62–68** from *B. asparagoides*, **46–51** from *B. nootkana* or *B. hamifera*, **53–61**, and the free acids **45, 46, 84** from *Asparagopsis*, the antibiotic activity was examined separately or in mixtures of closely related substances (McConnell and Fenical, 1977b; Fenical *et al.*, 1979). The substances are usually more effective against *S. aureus* and *C. albicans* than against the other test organisms. Substances **62, 63, 69–71** were also isolated from *Delisea fimbriata* (Rose *et al.*, 1977) and a weak antifungal activity was ascertained. The MIC of the more active compounds compares favourably with the commonly used antibiotics penicillin G or amphotericin B (Fenical *et al.*, 1979). The lower molecular weight substances are toxic and sometimes have properties similar to teargas. In spite of this, the Hawaians use *A. taxiformis* to savour their diet (Abbott and Williamson, 1974). For the relatively simple dichloroacetamide (**86**) from *Marginisporum aberrans*, Ohta (1979) determined an antifungal activity. The tribromoacetic acid (**85**) inhibits the growth of epiphytic microorgan-isms on *A. armata* and in cultures (Codomier *et al.*, 1981).

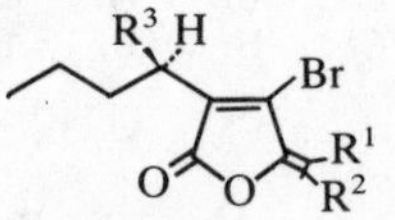

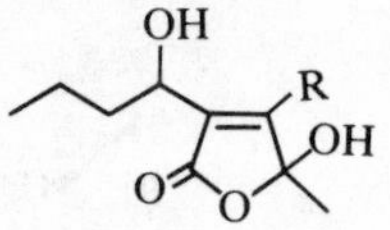

fimbrolides   acetoxyfimbrolides   hydroxyfimbrolides   beckerelides

	R^1	R^2	R^3		R^1	R^2	R^3		R^1	R^2	R^3		
87	Br	H	H	**90**	Br	H	OAc	**96**	Br	H	OH	**103**	R = Br
88	H	Br	H	**91**	H	Br	OAc	**97**	H	Br	OH	**104**	R = Cl
89	Br	Br	H	**92**	I	H	OAc	**98**	I	H	OH		
				93	H	I	OAc	**99**	H	I	OH		
				94	H	Cl	OAc	**100**	Cl	H	OH		
				95	Br	Br	OAc	**101**	H	Cl	OH		
								102	Br	Br	OH		

Fig. 7.8 Structure of the fimbrolides from Bonnemaisoniaceae.

The thallus surface of *Delisea fimbriata* is nearly free of bacterial and other epiphytes. The strong antibiotic activity of dichloromethane extracts of this alga (Kazlauskas *et al.*, 1977) is due to the fimbrolides (**87–89**), hydroxyfimbrolides (**96–102**) and acetoxyfimbrolides (**90–95**). The beckerelides (**103, 104**) from *Marginisporum aberrans* show an intensive inhibition against *B. subtilis* (Ohta, 1977).

Various species of the genus *Laurencia*, as a rule contain, over an ether bridge, cyclized halogenated C_{15}-compounds with a terminal acetylene group (lauristane-skeleton) which are possibly derived from the aliphatic acetylene laurediol (**105**), in *Laurencia nipponica*. For chondriol (**106**) from *Laurencia yamada* (Fenical and Norris, 1975), an antibiotic activity comparable with prepacifenol (**156**) and debromolaurinterol (**159**) has been ascertained (Sims *et al.*, 1975). For laureatin (**108**), isolaureatin (**109**) and laurefucin (**110**) from *Laurencia nipponica* (Kurosawa *et al.*, 1973), *trans*-olefine, and mixtures of *cis*- and *trans*-isomers (Faulkner, 1977) from *L. subopposita* and laurencin (**107**) from *L. glandulifera* (Irie *et al.*, 1968) it was shown that they prolong the phenobarbitone induced sleep-time. This group of substances appears to act by inhibiting the metabolism of pentobarbitone. The substances, with the applied doses, do not have a pharmacological effect of their own but they constitute a novel group of drug metabolism inhibitors (Kaul *et al.*, 1978).

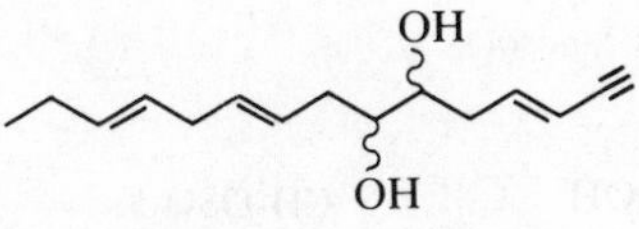

105 laurediol 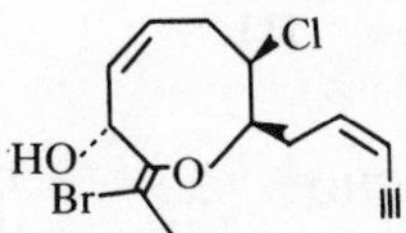   **106** chondriol 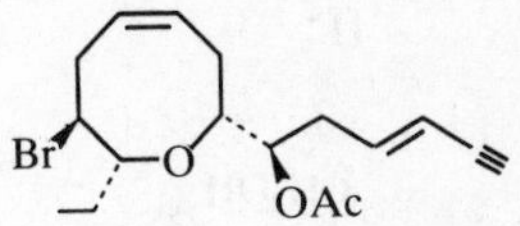   **107** laurencin

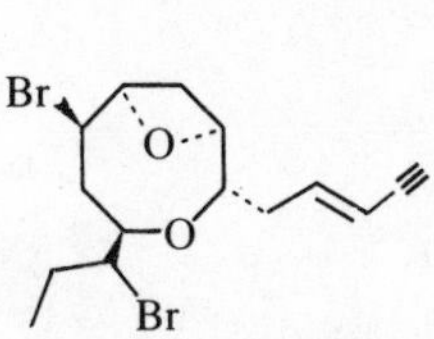

108 laureatin 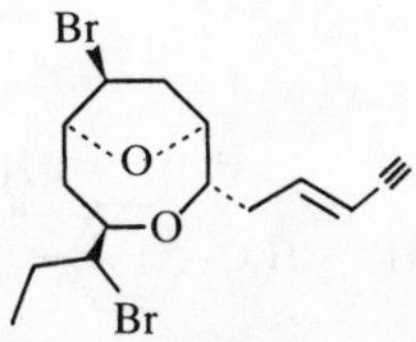   **109** isolaureatin 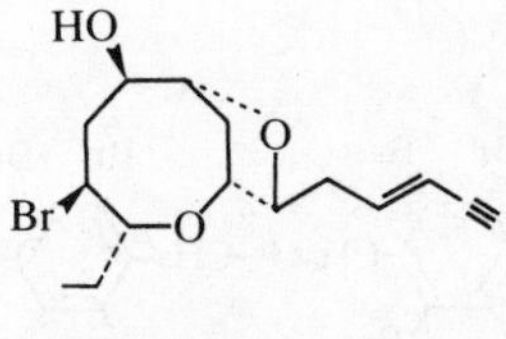   **110** laurefucin

Fig. 7.9 Biologically active C_{15}-enines (laurestanes) from *Laurencia*.

Kazlauskas *et al.* (1982) isolated a series of γ-pyrones from the alga *Phacelocarpus labillardieri* which belongs to the Gigartinales and is found in South Australian waters. The crude dichloromethane extract gave fractions which were active in mitochondrial uncoupling and polysynaptic reflex blocking tests. The substance responsible for these effects was not found, but **111** proved to be very toxic (Baker, 1984).

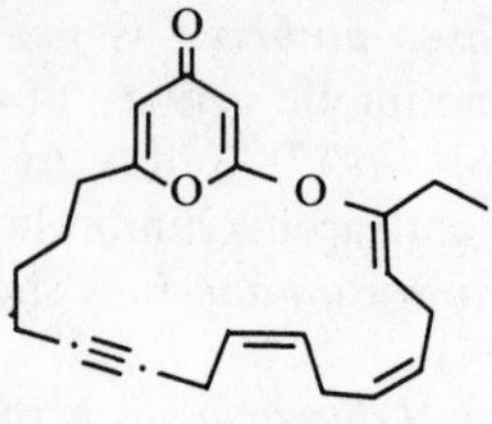

111

Fig. 7.10 Toxic γ-pyrone from *Phacelocarpus labillardieri.*

Antibiotically active, non-terpenoid phenols from red algae

The simplest antibiotically active phenol found in red algae, *p*-hydroxybenzaldehyde (**112**), was isolated from *Dasya pedicellata* var. *stanfordiana* (Fenical and McConnell, 1976) and *Marginisporum aberrans* (Ohta and Takagi, 1977). Substance **112** is one of the biogenetic precursors from

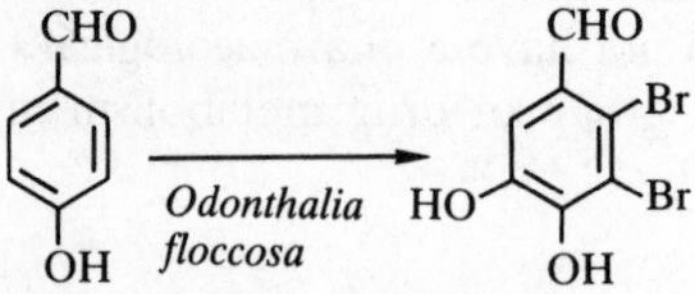

112 **113**

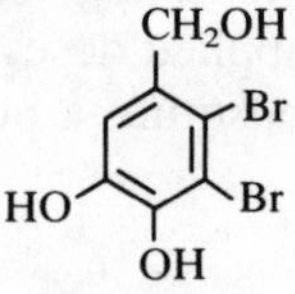

114 lanosol

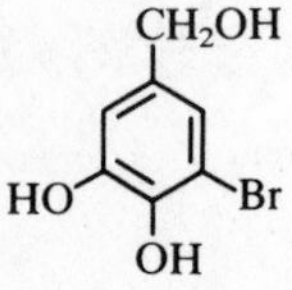

115

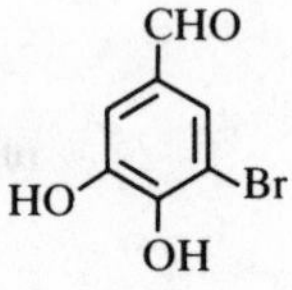

116

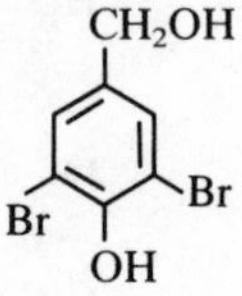

117

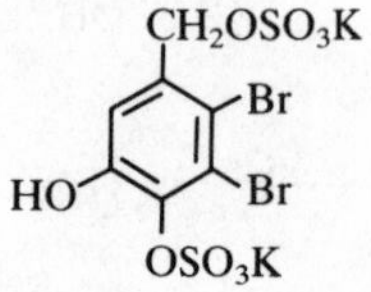

118 lanosoldisulphate

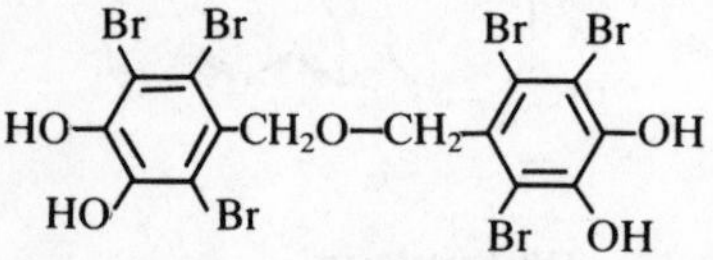

119

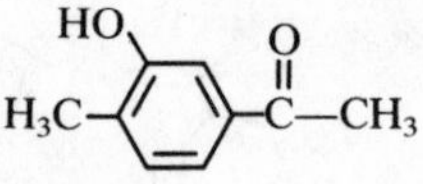

120

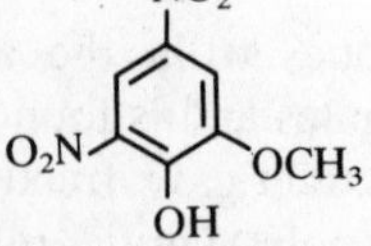

121

Fig. 7.11 Bioactive phenols from Rhodophyceae.

Table 7.1 Genera containing brominated phenols

Cyanophyceae	*Dasya*
Calothrix	*Halopitys*
	Lenormandia
Phaeophyceae	*Odonthalia*
Fucus	*Phycodrys*
	Polysiphonia
Rhodophyceae	*Pterosiphonia*
Antithamnion	*Rhodomela*
Ceramium	*Rhytiphlaea*
Corallina	*Symphocladia*
Cystoclonium	*Vidalia*

which, for example, a 3,4-dibromodihydroxybenzaldehyde (**113**) in *Odonthalia floccosa* can be formed (Manley and Chapman, 1980). The aldehyde can probably be reduced to the alcohol lanosol (**114**), a compound which sometimes occurs in substantial amounts in many Rhodomelaceae (Table 7.1). Compounds **114–117** inhibit growth of bacteria. Some of them (**113, 117**) have also been found in the brown alga *Fucus vesiculosus* (Pedersén and Fries, 1975) and the cyanobacterium *Calothrix brevissima* (Pedersén and Da Silva, 1973). Lanosoldisulphate (**118**), which was discovered in 1966 (Hodgkin *et al.*, 1966) and for which a changed structure was later presented (Glombitza and Stoffelen, 1972) is widely distributed but antibiotically inactive. The diphenylether **119** (isolation artefact?) from *Symphocladia latiuscula* (Kurata and Amiya, 1980) inhibits fungi and 13 different bacterial strains. The substances mentioned are just a few examples of a larger group of widely distributed compounds in the Rhodomelaceae (review: Fenical, 1975; Faulkner, 1978; Glombitza, 1979). The weakly antibiotically active acetophenone (**120**) was isolated from *Laurencia chilensis* (Valdebenito *et al.*, 1982) and the 3,5-dinitroguaiacol (**121**), which is more likely to be an industrial waste than a natural product, was isolated from *Marginisporum aberrans* (Ohta and Takagi, 1977).

Aliphatic and alicyclic monoterpenes from red algae

Aliphatic and three different groups of monocyclic monoterpenes, which can be substituted once or several (up to six) times with bromine or chlorine, were isolated from various representatives of the Plocamiaceae (*Plocamium*, seven species), Rhizophyllidaceae (*Chondrococcus*, *Ochtodes*) and Ceramiaceae (*Microcladia*) (review: Naylor *et al.*, 1983) (Fig. 7.12). The variety of isolated compounds is caused by the occurrence of several morphologically indistinguishable 'chemotypes' which were examined more closely with *Plocamium violaceum* (Naylor *et al.*, 1983) and *P. cartilagineum* (San-Martin and Rovirosa, 1986). The occurrence of comparable quantity ratios of similar compounds in species that are not related to each other but grow

type A B C D

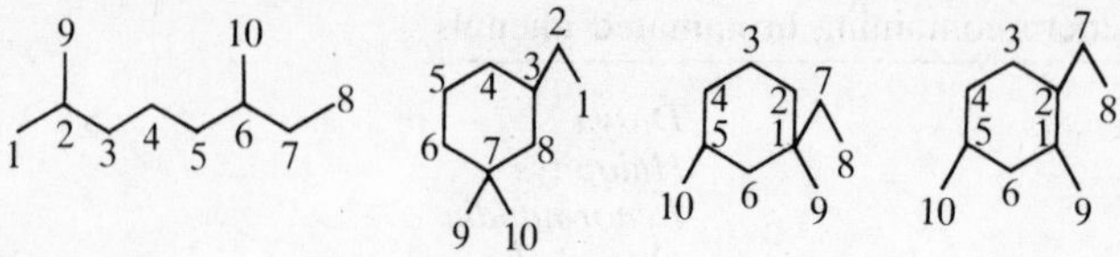

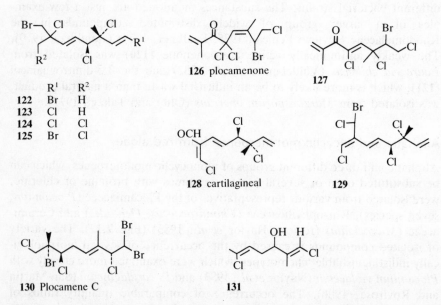

Fig. 7.12 Structural types of red algal monoterpenes.

together, e.g. *P. cartilagineum*/*Microcladia coulteri*, *P. violaceum*/*M. borealis*, *P. cartilagineum*/*Schottera nicaeensis*, supports the supposition that the substances found in each second species were not synthesized from this but only absorbed and stored (Crews *et al.*, 1976; Rivera *et al.*, 1987). The same or at least very similar compounds also occur in opisthobranch molluscs (*Aplysia*, sea hare) which feed on algae. They store these substances in their digestive glands as chemical deterrents against predators (Stallard and Faulkner, 1974).

Of the approximately 50 halogenated aliphatic monoterpenes isolated from red algae, fewer than 20% (Fig. 7.13) have been examined for their biological activity (Table 7.2). They have mainly a mutagenic effect. The activity spectrum of the examined cyclic compounds (Fig. 7.14) is broader (Table 7.2). Aside from a relatively weak activity against bacteria, insecticidal properties have been found. Interestingly, the raw extract from *Plocamium cartilagineum* is tenfold more active against *Aedes* mosquito larvae (Crews *et al.*, 1984b) than the most active pure substance plocamene B (**141**). Certain algal species have often been examined because it was observed that herbivores avoid them. They inhibit feeding and are toxic to

	R^1	R^2
122	Br	H
123	Cl	H
124	Cl	Cl
125	Br	Cl

126 plocamenone

127

128 cartilagineal

129

130 Plocamene C

131

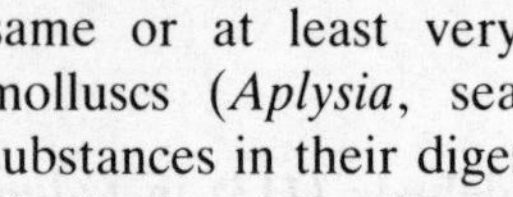

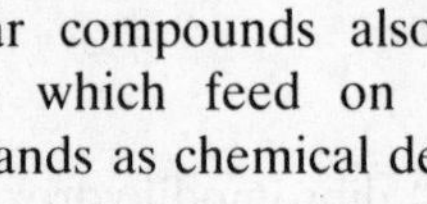

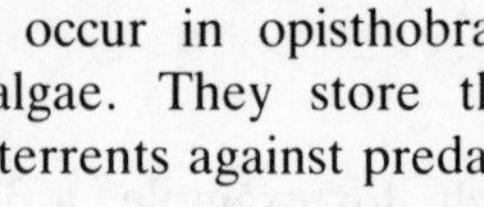

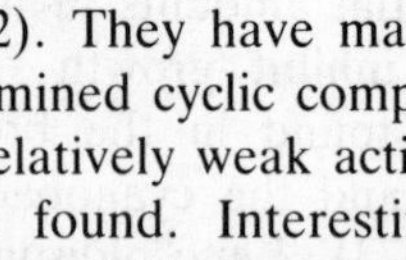

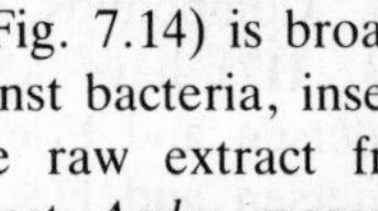

Fig. 7.13 Bioactive halogenated aliphatic monoterpenes from red algae.

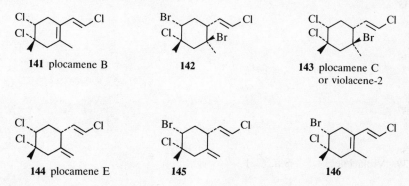

type B   ochtodane

132 ochtodene   **133**   **134**   **135**

136   **137**   **138**

type C   1-ethyl-1,3-dimethylcyclohexanes

139 violacene   **140**

type D   1-ethyl-2,4-dimethylcyclohexanes

141 plocamene B   **142**   **143** plocamene C
                                              or violacene-2

144 plocamene E   **145**   **146**

Fig. 7.14 Bioactive halogenated cyclic monoterpenes from red algae.

fish. Other pharmacological effects, such as influences on brain activity and
the guinea pig ileum model, do not give comparable results. Slight changes
in the structure (e.g. **141**, **144**) can make an acetylcholine antagonist out of
a substance increasing the acetylcholine effect (Crews *et al.*, 1984b). A therapeutical use is not apparent for any of these compounds.

Table 7.2   Bioactivity of halogenated red algal monoterpenes

Substance		Antimicrobial activity
122		Weak antifungal, C.c.* (34)[†]
123		Weak antifungal, C.c. (34)
124		Weak antifungal, C.c. (34)
125		Weak antifungal, C.c. (34)
126	Plocamenone	
127		
128	Cartilagineal	
129		
130		Weak antibacterial, B.s. (1)
131		Weak antibacterial, B.s. (1)
132	Ochtodene	S.a. (26)
133		
134		
135		
136		
137		
138		
139	Violacene	S.a. (27)
140		B.s., S.a., St.f., P.a. (18)

Cytotoxic and other activities	Found in species
Mutagenic Ames reversion assay (25)	*Plocamium* sp. (34)
Mutagenic Ames reversion assay (25)	*Plocamium* sp. (34)
Mutagenic Ames reversion assay (25)	*Plocamium* sp. (34)
Mutagenic Ames reversion assay (25)	*Plocamium* sp. (34)
Mutagenic Ames reversion assay (35)	*Plocamium* sp. (35)
Mutagenic Ames reversion assay (25)	*Plocamium* sp. (33)
Guinea pig ileum histamine antagonism (9)	*Plocamium cartilagineum* (5)
Inhibition of SUED, inhibition of fly larvae (*Musca*), guinea pig ileum, histamine antagonism, acetylcholine antagonism (9)	*Plocamium cartilagineum* (8)
	Plocamium cruciferum (1)
	Plocamium cruciferum (1)
	Ochtodes secundiramea (26)
Sedation for fish P.c., 2.5 μg ml⁻, toxic (P.c. 10 μg ml⁻¹), feeding inhibition 300 ppm (30)	*Ochtodes crockeri* (30)
Sedation for fish P.c. 5 μg ml⁻¹, toxic (P.c. 10 μg ml⁻), feeding inhibition 300 ppm (30)	Ochtodes crockeri (30)
Sedation for fish P.c. 2 μg ml⁻¹, toxic (P.c. 10 μg ml⁻¹), feeding inhibition 300 ppm (30)	*Ochtodes crockeri* (30)
Sedation for fish P.c. 2 μg ml⁻, toxic (P.c. 10 μg ml⁻¹), feeding inhibition 100 ppm (30)	*Ochtodes crockeri* (30)
Sedation for fish P.c. 10 μg ml⁻¹, not toxic for P.c., feeding inhibition 100 ppm (30)	*Ochtodes crockeri* (30)
Inhibition of SUED (9)	*Chondrococcus hornemanni* (4, 9)
Inhibition of mosquito larvae (7, 9), tobacco hornworm larva-III (*Manduca*), fly larvae (*Musca*), budworm larva-II (*Heliothis*) (9), toxic for goldfish (7), forces guinea pig heart auricle contraction (9)	*Plocamium violaceum, P. cartilagineum* (7)
Cytostatic on HeLa cells (18)	*Plocamium cartilagineum* (16)

Table 7.2 Cont.

Substance	Antimicrobial activity
141 Plocamene B	
142	B.sp., S.a., St.f., P.a. (18)
143 Plocamene C	inactive against S.a. (27)
144 Plocamene E	
145	S.a. (18)
146	S.a. (18)

* For list of abbreviations used in Tables, see p 218.
† For reference list, see Table 7.3.

Table 7.3 Biological activity of sesquiterpenoids from red algae

Substance	Antimicrobial activity
147 Caespitol	B.sph., S.a., P.a. (18)
148 Isocaespitol	B.sph., S.a., P.a. (18)
149 Isocaespitolacetate	B.sph., S.a., St.f., P.a. (18)
150 Obtusol	B.sph., S.a., St.f., P.a. (18)
151 Isoobtusol	B.sph., S.a., St.f., P.a. (18)
152 Isoobtusol-acetate	B.sph., S.a., St.f., P.a. (18)
153 Elatol	C.c. (3)
154 Elatone	
156 Prepacifenol	S.a., M.s. (32)
157 'Chamigrene alcohol 1'	
158 Laurinterol	S.a., M.s., C.a. (32), St.h. (24), S.e., B.s. (29)
159 Debromolaurinterol	S.a., M.s., C.a. (32), St.h., S.e., P.v. (24)
160 Neolaurinterol	St.h., S.a. (both weak) (24)
161	St.h., S.a. (both weak) (24)
162 Isolaurinterol	St.h., S.a., S.e. (24), B.s. (29)
163 Debromoisolaurinterol	St.h., S.a., S.e. (all weak) (24)
164 7-Hydroxylaurene	antimicrobial (11)
165 allo-Laurinterol	antimicrobial (11)
166 Cycloeudesmol	S.a., S.c., M.s., C.a. (32, 12)
167 3-Bromo-8-epica-parapioxide	S.a. (10)

Reference list for Tables 7.2 and 7.3: (1) Bates *et al.* (1979); (2) Bittner *et al.* (1985); (3) Brennan *et al.* (1987); (4) Burreson *et al.* (1975); (5) Crews and Kho (1974); (6) Crews and Kho (1975); (7) Crews *et al.* (1978); (8) Crews *et al.* (1984a); (9) Crews *et al.* (1984b); (10) Faulkner (1976); (11) Faulkner (1978); (12) Fenical and Sims (1974); (13) González *et al.* (1973); (14) González *et al.* (1974b); (15) González *et al.* (1976); (16) González *et al.* (1978); (17) González *et al.* (1979); (18) González *et al.* (1982); (19) Higgs *et al.* (1977); (20) Irie *et al.*

Cytotoxic and other activities	Found in species
Inhibition of mosquito larvae (*Aedes*) (7, 9), toxic for goldfish (6), guinea pig ileum acetylcholine and histamine potentiation (9)	*Plocamium violaceum* (30)
Anticonvulsant, uncoupling of oxidative phosphorylation (22)	*Plocamium cartilagineum* (22)
	Plocamium violaceum (28)
Guinea pig ileum acetylcholine antagonism, forces heart auricle contractions (9)	*Plocamium cartilagineum* (7)
Cytostatic on HeLa cells (18)	*Plocamium cartilagineum* (19)
Cytostatic on HeLa cells (18)	*Plocamiun cartilagineum* (19)

Cytotoxic and other activities	Found in species
Weak cytostatic on HeLa cells (18)	*Laurencia caespitosa* (11, 13, 14)
Weak cytostatic on HeLa cells (18)	*Laurencia caespitosa* (11, 14)
Weak cytostatic on HeLa cells (18)	*Laurencia caespitosa* (11, 17)
Weak cytostatic on HeLa cells (18)	*Laurencia obtusa* (15)
Weak cytostatic on HeLa cells (18)	*Laurencia obtusa* (15)
Weak cytostatic on HeLa cells (18)	*Aplysia dactylomela* (18)
SUED (37, 39), toxic to brine shrimp hatchlings (3)	*Laurentia elata* (31)
SUED through inhibition of mitoses (38, 37)	From elatol by oxidation (31)
Ichthyotoxic (P.c. 15 μg ml^{-1}) (2)	*Laurencia* sp. (2)
Ichthyotoxic (P.c. 15 μg ml^{-1}) (2)	*Laurencia* sp. (2)
	Laurencia okamurai (20)
	Laurencia okamurai (20)
	Laurencia okamurai (36)
	Laurencia okamurai (36)
	Laurencia intermedia (21)
	Laurencia intermedia (21)
	Laurencia subopposita (40)
	Laurencia filiformis (23)
	Chondria oppositiclada (12)
	Laurencia obtusa (10)

(1969); (21) Irie *et al.* (1970); (22) Jamieson and Taylor (1979); (23) Kazlauskas *et al.* (1976); (24) Kurosawa and Suzuki (1983); (25) Leary *et al.* (1979); (26) McConnell and Fenical (1978); (27) Mynderse (1975); (28) Mynderse *et al.* (1975); (29) Ohta (1979); (30) Paul *et al.* (1980); (31) Sims *et al.* (1974); (32) Sims *et al.* (1975); (33) Stierle (1977); (34) Stierle *et al.* (1979); (35) Stierle and Sims (1984); (36) Suzuki and Kurosawa (1978); (37) White and Jacobs (1979); (38) White and Jacobs (1981); (39) White *et al.* (1978); (40) Wratten and Faulkner (1977).

(a) bisabolanes

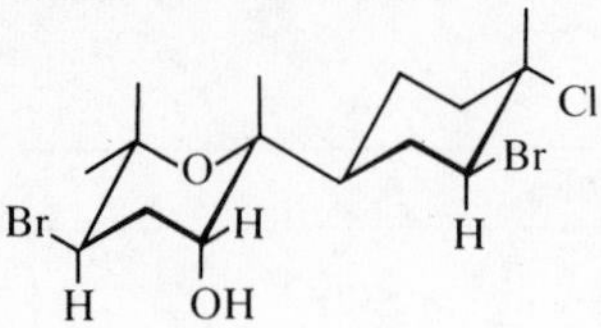

147 caespitol

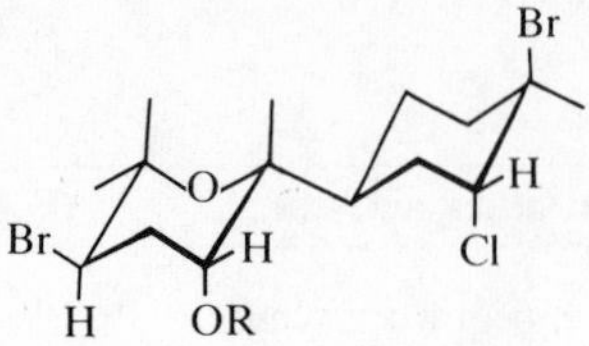

148 R = H = isocaespitol
149 R = Ac = isocaespitolacetate

(b) chamigranes

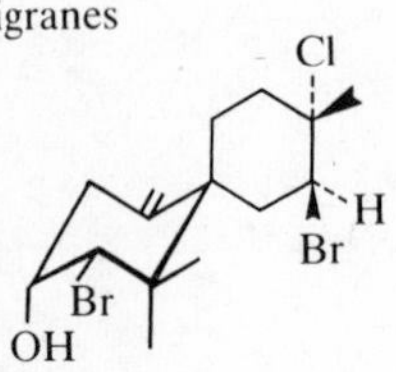

150 obtusol

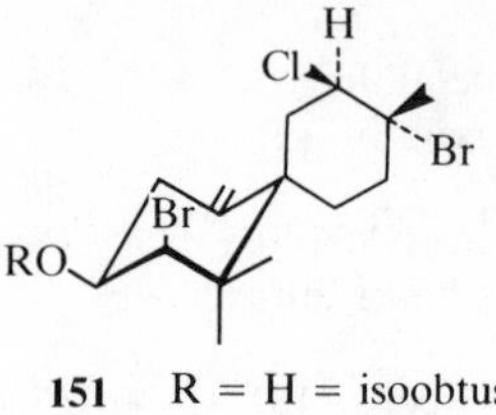

151 R = H = isoobtusol
152 R = Ac = isoobtusolacetate

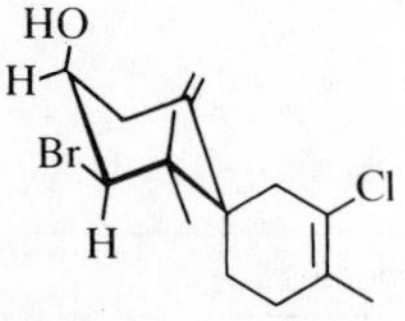

153 elatol

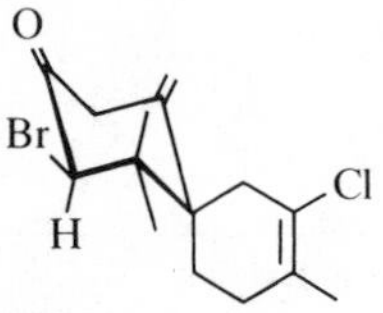

154 elatone

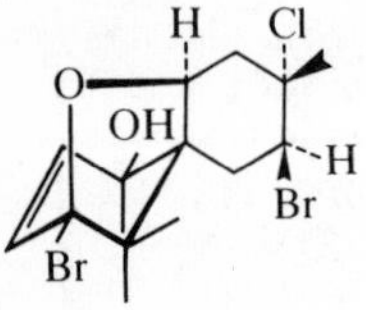

155 pacifenol

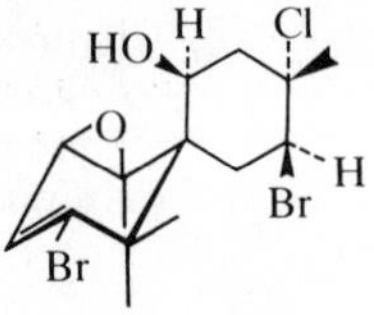

156 prepacifenol

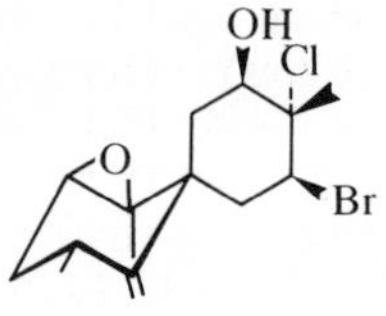

157

(c) cuparanes

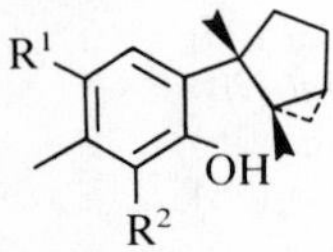

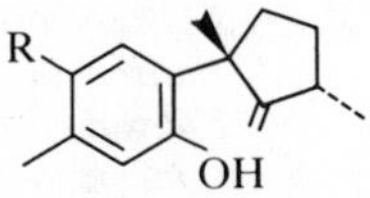

162 R = Br isolaurinterol
163 R = H debromoisolaurinterol

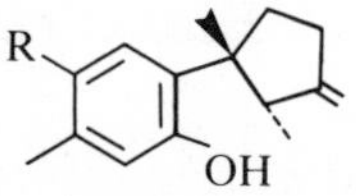

164 R = H   7-hydroxylaurene
165 R = Br   allo-laurinterol

	R¹	R²	
158	Br	H	laurinterol
159	H	H	debromolaurinterol
160	H	Br	neolaurinterol
161	Br	Br	bromolaurinterol

(d) selinanes

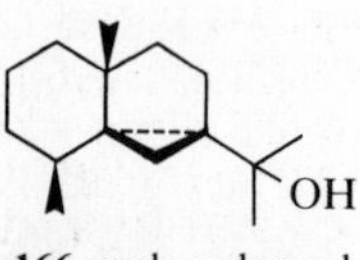

(e) other type

166 cycloeudesmol

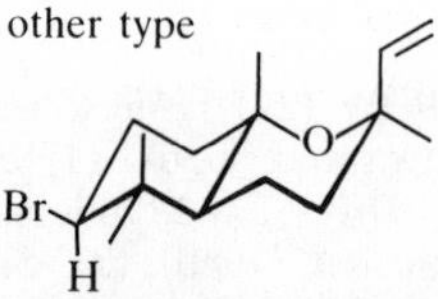

167 3-bromo-8-epicaparrapioxide

Fig. 7.15 Bioactive sesquiterpenes from red algae.

Sesquiterpenes from red algae

Many sesquiterpenes, with different skeletons, were found in *Laurencia* species. For several bisabolanes (**147, 148**), chamigranes (**150–157**) cuparanes (**158–165**) and selinanes (**166**), antimicrobial effects against several Gram-positive bacteria, but fewer against Gram-negative strains were ascertained. The substances are often found in the same or chemically changed form in sea hares of the genus *Aplysia* (e.g. Irie *et al.*, 1969; Stallard and Faulkner, 1974). It was found that some of the isolated chamigranes from *Aplysia dactylomela* have such a good effect against Herpes simplex virus type I and vesicular stomatitis virus *in vitro* that a patent was applied for (Snader and Higa, 1986). In other cases, a cytotoxic and antimycotic effect was determined (Table 7.3).

Diterpenes from red algae

Komura and Nagayama (1983) observed a stimulation of *in vitro* pancreatic carboxylesterase activity towards triacetine with certain fractions of unsaponifiable matter from *Porphyra tenera*. Komura *et al.* (1974) identified similar effects with the brown alga *Hizikia fusiformis*. In both cases the ubiquitous diterpene phytol (**168**) seemed to be responsible for the effect. Only a few of the numerous other, usually brominated, di- and tricyclic diterpenes isolated from various species of the genus *Laurencia* and from *Sphaerococcus coronopifolius* (review: Faulkner 1977, 1984, 1986, Fenical, 1978) have been investigated for biological activity. For example, laurencianol (**169**) from *Laurencia obtusa* (Caccamese *et al.*, 1982) shows activity against *B. subtilis* and *E. coli*.

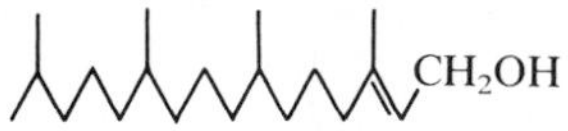

168 phytol

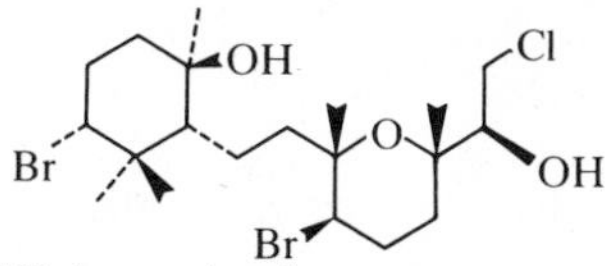

169 laurencianol

Fig. 7.16 Diterpenes from red algae.

Triterpenes from red algae

Marine organisms have a variety of sterols which far exceeds that of terrestrial organisms. Information is lacking about the biological meaning of this variety. An ecdysone-like moulting hormone activity was detected for several pinnasterols from *Laurencia pinnata* (Fukuzawa *et al.*, 1986). In the *Sarcophaga* method, the compounds showed an ED_{50} of 0.25 μg (**173**), 0.54 μg (**170**), 2 μg (**172**), 6 μg (**171**), (ecdysone ED_{50} 0.018 μg).

Squalene derivatives, which were cyclized over ether bridges to several tetrahydropyranoids and also partly to a further tetrahydrofuranoid ring, have been isolated from several *Laurencia* species. The thyrsiferol (**174**) from *Laurencia thyrsifera* (Blunt *et al.*, 1978) and especially the thyrsiferol-acetate (**175**) from *L. obtusa* have a cytotoxic activity against leukemia

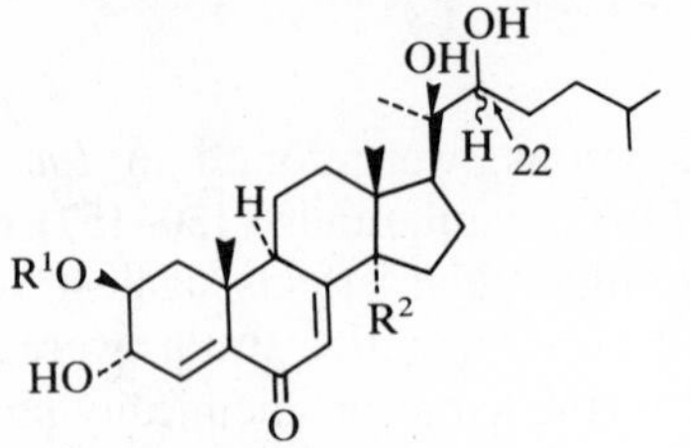

	R^1	R^2	C-22
170	Ac	H	β-H
	Acetylpinnasterol		
171	Ac	OH	β-H
172	Ac	OH	β-H
173	H	OH	β-H

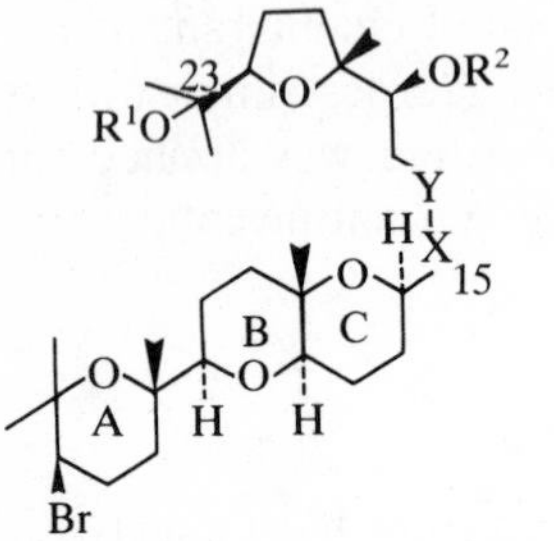

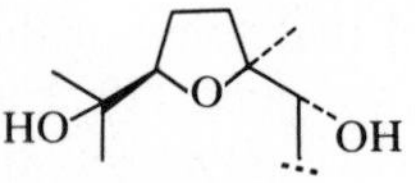

stereochemistry of rings A, B, C and C15 the same as in venustatriol

180

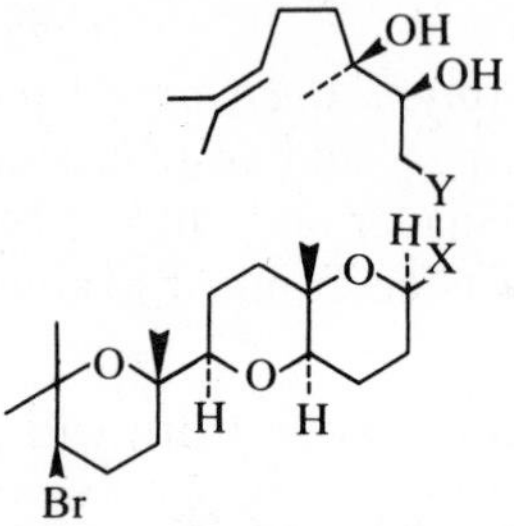

	R^1	R^2	XY	
174	H	H	I	thyrsiferol
175	Ac	H	I	thyrsiferol-23-acetate
176	Ac	Ac	I	thyrsiferol-23, 18-diacetate
177	H	Ac	I	thyrsiferol-18-acetate
178	Ac	Ac	II	15(28)-anhydrothyrsiferol-diacetate
179	Ac	Ac	III	15-anhydrothyrsiferol-diacetate

181	I	magireol A
182	II	magireol B
183	III	magireol C

elements XY

I =
$$\overset{Me}{\underset{\underset{OH}{|}}{\overset{15\quad16}{-CHCH_2-}}}$$

II =
$$\overset{CH_2}{\overset{||}{\underset{15\ \ 16}{-C-CH_2-}}}$$

III =
$$\overset{Me}{\overset{|}{\underset{15\ \ 16}{-C=CH-}}}$$

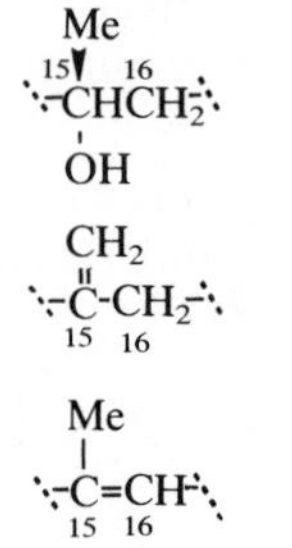

Fig. 7.17   Triterpenes from red algae.

P388 cells and were patented as antitumour agents (Imanaka and Kato, 1986). A deviating acetylation (**176, 177**) weakens the activity in relation to **175**, more than 10^4-fold. Sakemi *et al.* (1986) revised the absolute configuration for **174** and showed that **174, 175** and the isolated enantiomeric venustatriol (**180**) from *L. venusta* have an antiviral activity. The compounds **175, 178, 179** isolated from *L. obtusa* and the non-furanoids, magireols, **181–183** have a cytotoxic activity (Suzuki *et al.*, 1987).

Chrysophyceae

Bioactive metabolites

The few Chrysophyceae examined so far have been found to produce antimicrobial and ichthyotoxic effects (Ragan, 1984). In *Stichochrysis immobilis*, the antibiotic activity is possibly caused by a peptide (Berland *et al.*, 1972); in *Ochromonas malhamensis* it is associated with unidentified chlorophyllides (Hansen, 1973). The ichthyotoxicity of the chrysophyte *Prymnesium parvum*, which is widespread in brackish and marine waters has been studied for several years. The toxic principle was found to be a family of closely resembling substances causing the cytotoxic, haemolytic, neurotoxic and lethal effects on isolated cells and organisms (reviews: Shilo, 1981; Carmichael, 1986). Though the chemical properties of the toxins have been studied by several workers, the conclusions are contradictory and confusing, but Kozakai *et al.* (1982) succeeded in elucidating the structure of two hemolysins. Hemolysin I is identical with **297** from the dinoflagellate *Amphidinium carteri* and hemolysin II is similar to **298** but the second galactosyl moiety is β-C-6-bound.

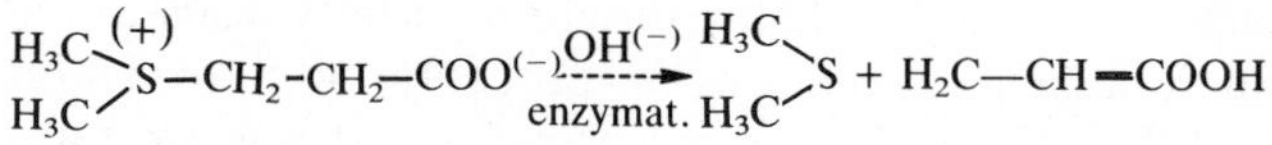

$$\underset{H_3C}{\overset{H_3C}{>}}\!\!\overset{(+)}{S}\!-\!CH_2\!-\!CH_2\!-\!COO^{(-)} \underset{\text{enzymat.}}{\overset{OH^{(-)}}{\xrightarrow{\hspace{1cm}}}} \underset{H_3C}{\overset{H_3C}{>}}\!S + H_2C\!-\!CH\!=\!COOH$$

184   **185**

Fig. 7.18   Formation of acrylic acid from dimethylpropiothetine.

Acrylic acid (**185**) was the first antibiotically active compound from algae (*Phaeocystis pouchetii*) to be unambiguously identified and correlated with a biological effect (Sieburth, 1961). It occurs partly free, and partly bound as the inactive thetine (**184**) which may be split enzymatically or chemically to dimethylsulphide and **185**. Subsequently, both were found in many algae (Glombitza, 1979).

Bacillariophyceae (diatoms)

Bioactive metabolites

Gauthier *et al.* (1978a) showed that the antibiotic activity of *Asterionella japonica* and *Chaetoceros lauderi* is generated by lipophilic substances, can be photoactivated, and in *C. lauderi*, could involve fatty acids (Gauthier *et al.*, 1978b). Pesando (1972) isolated a *cis*-eicosapentaenoic acid (**186**) from *Asterionella japonica* and Findlay and Patil (1984) found the corresponding phytylester (**187**) in *Navicula delognei*. It only concerns one (often detected) of many multiple unsaturated fatty acids which have been isolated from diatoms (Pohl and Zurheide, 1979).

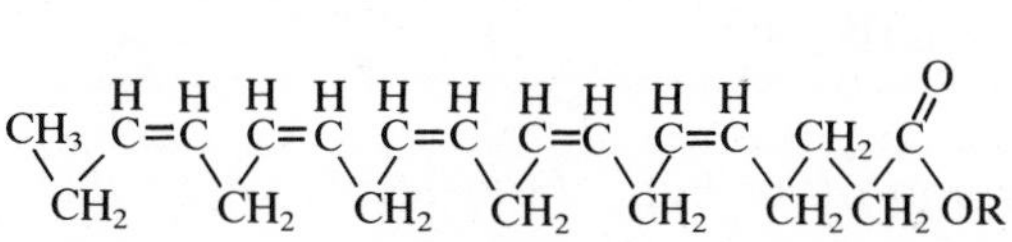

186 R = H
187 R = phytyl

Fig. 7.19   Antibiotically active fatty acid derivatives from Bacillariophyceae.

Phaeophyceae

Low molecular weight substances and other compounds

Unsaturated compounds

The antibiotically effective acrylic acid (185) which is found in large quantities in several green and red algae and its biological precursor dimethylpropiothetine could be detected only in small amounts in brown algae (Glombitza, 1979). It has no significance for the antimicrobial activity of these algae. An *in vitro* mixture (ED_{50}, 15 μg ml^{-1}) of an ethyl ester of an unsaturated fatty acid (188) and the corresponding aldehyde (189) from lipophilic extracts of *Cystoseira barbata* affects P388 lymphatic leukaemia in mice (Banaigs *et al.*, 1984). Highly unsaturated fatty acids can act antibacterially under suitable conditions – probably over oxidation products (Glombitza, 1979). With *Colpomenia peregrina* extracts a cytotoxic activity was found in a mixture of saturated and unsaturated fatty acids and even with the carotenoid fucoxanthin (Biard and Verbist, 1981). The distribution pattern of unsaturated fatty acids varies within the individual algal classes. Chlorophyceae possess preferably types $C_{16:3(n-3)}$, $C_{16:4(n-3)}$, Rhodophyceae, $C_{20:4(n-6)}$, $C_{20:5(n-3)}$, and Phaeophyceae $C_{18:3(n-3)}$, $C_{18:4(n-3)}$, $C_{20:4(n-6)}$, $C_{20:5(n-3)}$, (Takagi *et al.*, 1985).

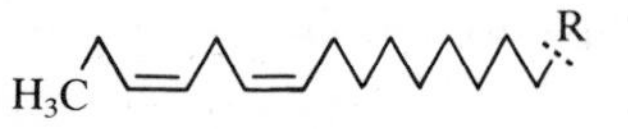

R=COOC$_2$H$_5$	188
R=CHO	189

Fig. 7.20 Unsaturated lipids from *Cystoseira barbata*.

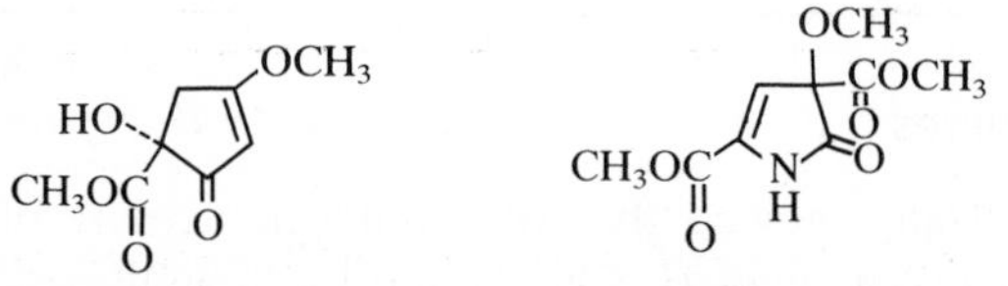

190 kjellmanianone   191 sargassumlactam

Fig. 7.21 Antibiotically active substances from *Sargassum kjellmanianum*.

The antibiotically active, unsaturated cyclopentenone (+)-kjellmanianone (190; Nakayama *et al.*, 1980) and the unsaturated sargassumlactam (191: Nozaki *et al.*, 1980) were identified in *Sargassum kjellmanianum* and both have a weak antibacterial activity.

Betaines, sterols and other hypocholesterolaemic substances

Laminine is a betaine of lysine and is formed by the trimethylization of the ω-amino group. It was first found in *Laminaria angustata* and subsequently in other Laminariales (0.01–0.18–0.34 μmol g^{-1}; Takemoto *et al.*, 1965).

It transitorily depresses blood pressure, suppresses heart motion and the contraction of excised smooth muscles in urethane anaesthetized rabbits (Ozawa *et al.*, 1967). A relatively large number of brown algae contain the following betaines: glycinebetaine (**7**) which acts as a methyl donor and has an anticonvulsant effect, aminobutyric acid betaine (**9**) which has an excitatory acetylcholine-like effect, prolinebetaine (**11**) which has a positive chronotropic and a negative inotropic effect, and β-alaninebetaine (**8**) which lowers plasma cholesterol levels (review: Blunden and Gordon, 1986).

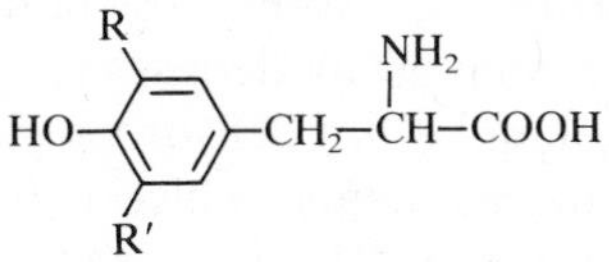

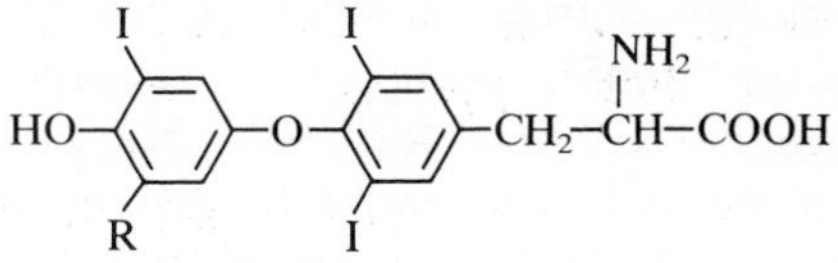

| 192 | R=I R'=H | 194 | R=H |
| 193 | R=R'=I | 195 | R=I |

Fig. 7.22   Iodinated amino acids from brown algae.

Ito *et al.*, (1976) isolated monoiodotyrosine (**192**), diiodotyrosine (**193**), triiodothyronine (**194**) and thyroxine (**195**) from the same alga after hydrolysis of the algal extract and also demonstrated for **194** and **195** a lowering of blood cholesterol concentration in rats. The occurrence of these iodinated amino acids in brown algae has been known for a long time (Roche *et al.*, 1951) and has been confirmed repeatedly (e.g. for *Sargassum*: Ito *et al.*, 1976). They are thought to be responsible for the activity of the algae on iodine deficiency diseases of the thyroid gland. The plasma cholesterol level can also be reduced by the addition of algal sterols to the diet. One example is cited here from a large number of papers. Cholesterol and algal sterols were added to the diet of young leghorn cockerels. An admixture of 1% fucosterol (**196**) from *Fucus gardneri*, to a diet containing 1% cholesterol, lowered the cholesterol plasma level by 83% and the addition of a fucosterol/sargasterol (**197**) mixture from *Sargassum muticum* lowered it by 59% (Reiner et al., 1967). An antihyperlipidaemic activity is attributed to fatty acids with a large number of double bonds arranged in the correct positions (review: Pohl and Zurheide, 1979).

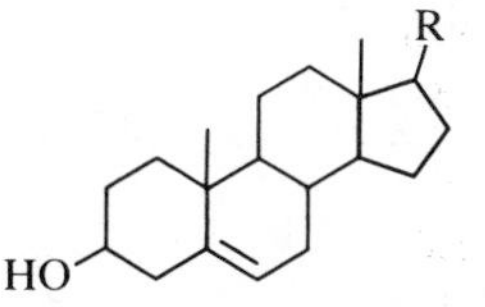

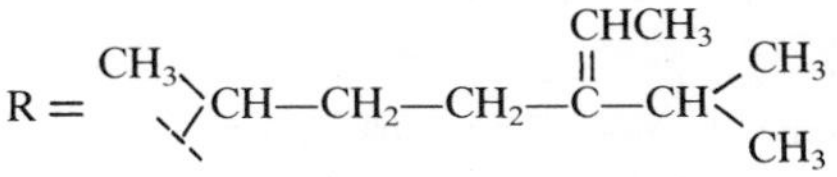

196 fucosterol

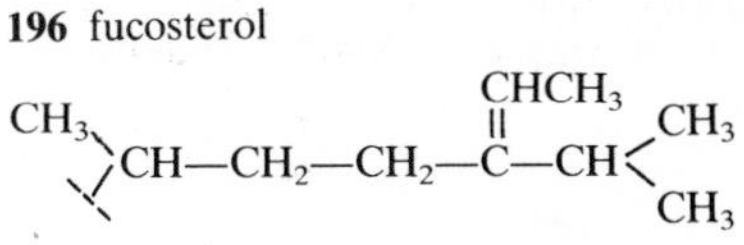

197 sargasterol

Fig. 7.23   Sterols from brown algae.

Polyhydroxyphenols

When thallus fragments or extracts of many brown algae are examined for antibiotic activities with the agar diffusion test, comparatively small inhibition zones are measured in most cases (Glombitza, 1979). Frequently, these zones are caused by phenolic tannin-like substances (phlorotannins). As with phenols and tannins from terrestrial plants, they are non-specific enzyme inhibitors (Esping, 1957a, b) active against bacteria (Glombitza, 1979; Ragan, 1985), yeasts (Seshadri and Sieburth, 1971), algae and a variety of marine animals including hydroids, worms (Sieburth and Conover, 1965), pycnogonids, copepods, and echinoderms. They have antifeedant activities for snails. The phlorotannins are dehydropolymers of phloroglucinol. The antibiotic activity of the monomer, which is found in many algae, has been unambiguously proven (McLachlan and Craigie, 1966), but the oligomers have been tested only in mixtures. Up to now five groups have been detected. Fucols (**198–201**) consist of phloroglucinol units linked through aryl–aryl bonds, phlorethols (**202–208**) consist of units which are linked through diaryl ether bonds, and fucophlorethols (**209–215**) contain both direct carbon–carbon and diaryl ether bonds. Due to the great numbers of different potential linkage sites, the number of isomers is very large. Fuhalols (**216–218**) and isofuhalols (**219–221**) are ether linked phloroglucinol units which contain at least one additional hydroxyl group. More than 80 different substances have been identified up to now from the Fucales, Laminariales and other orders (Table 7.4, review: Ragan and Glombitza, 1986).

Eckols are based on a dehydrooligomerization of three phloroglucinol units, two of these being further cyclized to a dibenzo [1,4] dioxin. The basic unit, eckol (Fukuyama *et al.*, 1985) can be dehydrated to form a second

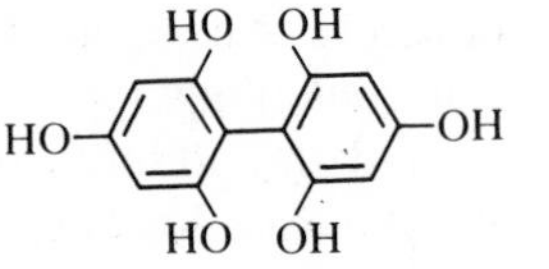

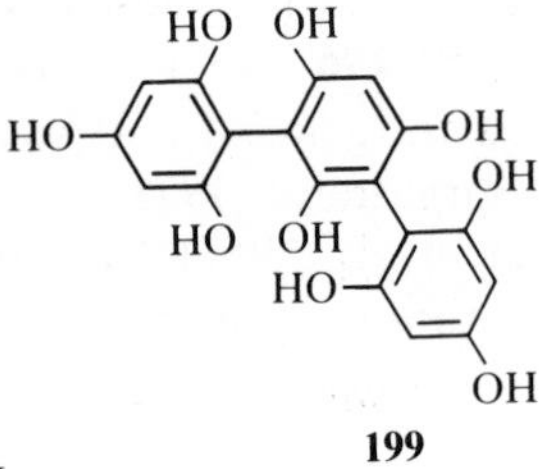

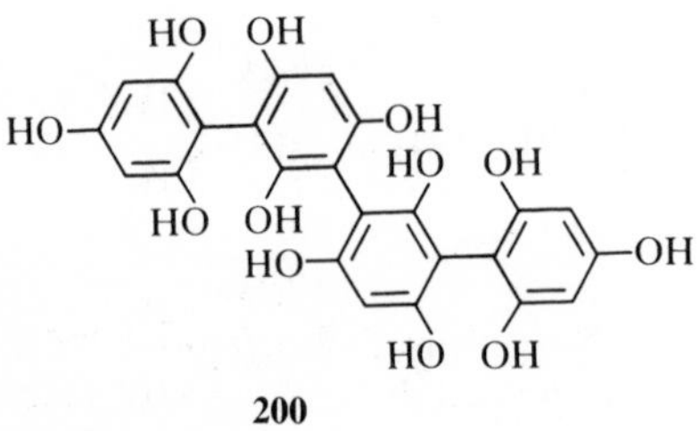

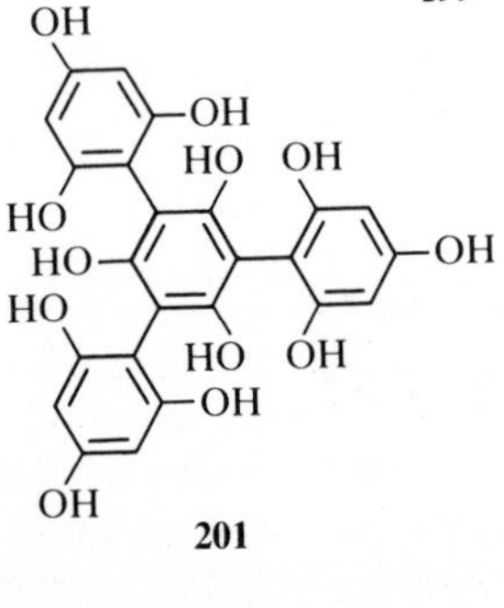

Fig. 7.24 Fucols from brown algae.

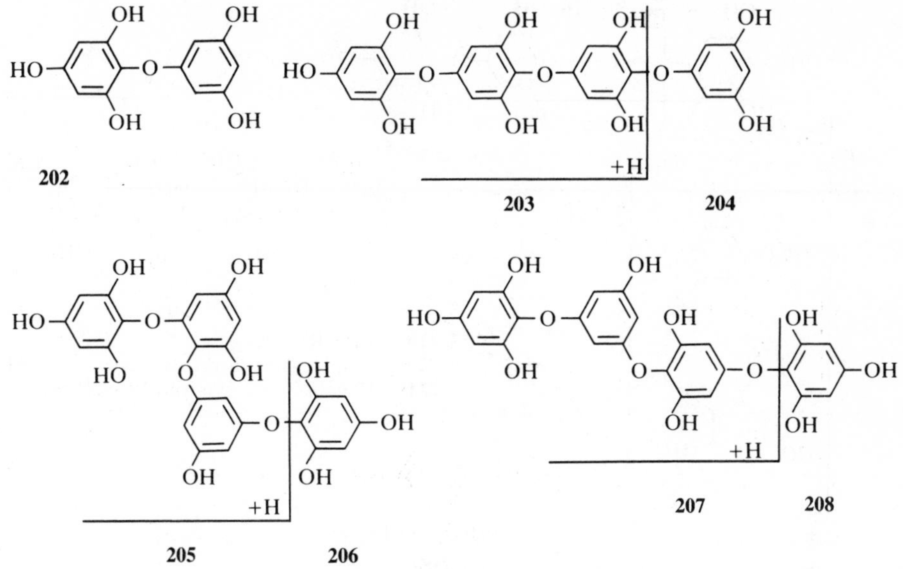

Fig. 7.25 Phlorethols from brown algae.

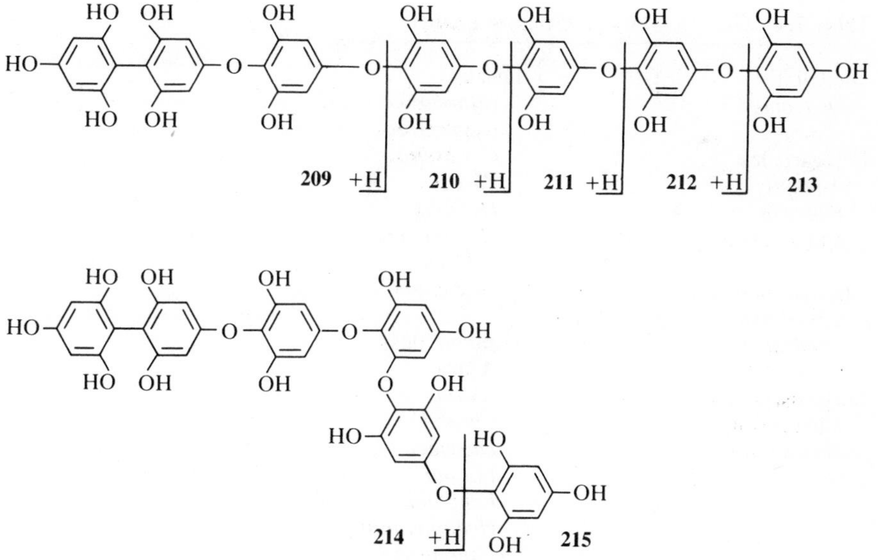

Fig. 7.26 Fucophlorethols from brown algae.

dibenzo [1,4] dioxin system and can be hydroxylated or phenoxylated with a fourth phloroglucinol unit. The fourth phloroglucinol unit can also be attached to the eckol moiety through a direct carbon–carbon bond, and, by elimination of water, form a dibenzofuran. These compounds can then be joined through an ether bridge to a fifth phloroglucinol ring. Dehydrodimers

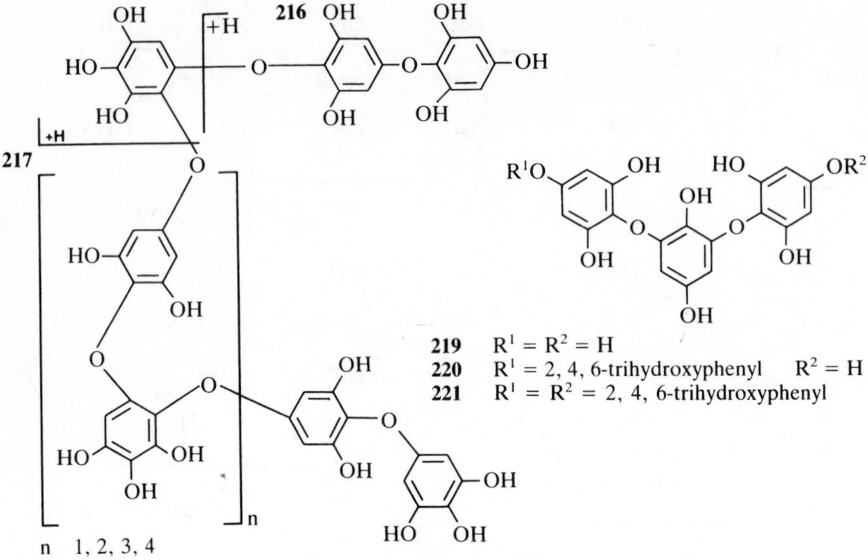

219 $R^1 = R^2 = H$
220 R^1 = 2, 4, 6-trihydroxyphenyl $R^2 = H$
221 $R^1 = R^2$ = 2, 4, 6-trihydroxyphenyl

Fig. 7.27 Fuhalols and isofuhalols from brown algae.

Table 7.4 Genera containing phlorotannins

Dictyotales	Fucales
Dictyota	*Bifurcaria*
	Cystophora
Ectocarpales	*Cystoseira*
Analipus	*Fucus*
Pilayella	*Halidrys*
Spongonema	*Himanthalia*
	Pelvetia
Chordariales	*Sargassum*
Chordaria	
Elachista	Laminariales
	Agarum
Dictyosiphonales	*Alaria*
Dictyosiphon	*Chorda*
Scytosiphon	*Durvillea*
	Eisenia
	Ecklonia
	Hedophyllum
	Laminaria
	Pleurophycus

are called dieckols if the eckols are linked through an ether bond (**222**) and bieckols if they are linked through a direct carbon–carbon bond (**223–227** (Fukuyama *et al.*, 1985; Glombitza and Gerstberger, 1985; Glombitza and Vogels, 1985; for further structures see Ragan and Glombitza, 1986). These compounds were discovered independently in the laboratories of Otsuka

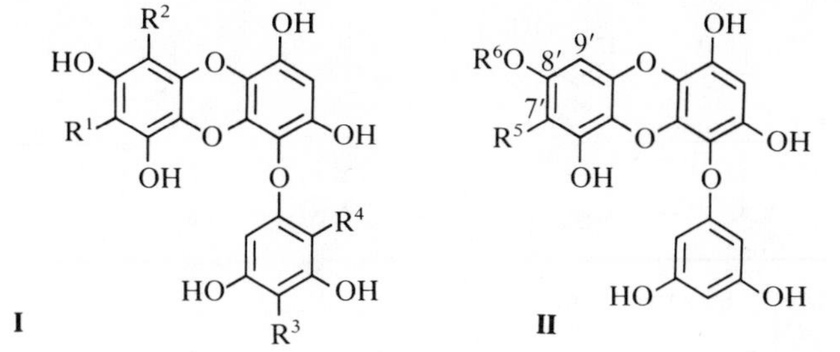

Fig. 7.28   Eckols from brown algae.

	R^1	R^2	R^3	R^4	R^5	R^6
222	H	H	II–(OC–8′)	H	H	(I)
223	II–(C–7′)	H	H	H	(I)	H
224	II–(C–9′)	H	H	H	H	H
225	H	II–(C–9′)	H	H	H	H
226	H	H	H	II–(C–7′)	(I)	H
227	OH	II–(C–9′)	H	H	OH	H

Pharmaceutical Co. from *Ecklonia kurome* and by a German research group from *Eisenia arborea* (Glombitza and Gerstberger, 1985) and *Ecklonia maxima* (Glombitza and Vogels, 1985). The Japanese pharmaceutical industry recognizes the suitability of the free phenols or their alkanoyl derivatives as antiplasmin inhibitors (Otsuka Pharmaceutical Co., 1983a, b, c), i.e. they have strong inhibitory activity on plasma α_2-macroglobulin and α_2-plasmin inhibitor which are important in controlling the fibrinolytic system.

Diterpenoids

Several aliphatic and many cyclic diterpenes have been isolated from certain groups of brown algae, in particular the Dictyotaceae. In the search for new active substances, dictyotaceaen extracts have often shown antimicrobial, cytotoxic, piscicidal and antifeedant activity against herbivores together with other pharmacological effects. The physiological effects observed with the algal extracts could be traced back to certain substances in only a few cases.

The diterpendiol crinitol (**228**) inhibits the growth of insects and Gram-negative bacteria (Kubo *et al.*, 1985). It was isolated from *Cystoseira crinita* (Fattorusso *et al.*, 1976) and *Sargassum tortile* (Kubo *et al.*, 1985). The similarly built éléganolone (**230**) from *Cystoseira elegans* (Francisco *et al.*, 1978) and *Bifurcaria bifurcata* (Biard *et al.*, 1980) inhibits *B. subtilis* slightly and mycobacteria (Biard *et al.*, 1980) a little more strongly. The farnesylacetone epoxide (**229**) has anticonvulsant activity and protects mice against all phases of the convulsive response to electroshock but only after an i.p. dose of 370–550 mg kg^{-1} (Spence *et al.*, 1979). It was isolated from *Cystophora moniliformis* (Kazlauskas *et al.*, 1978).

Many cyclic diterpenes have been found in members of the Dictyotaceae. Often the same or similar structures could be isolated from sea hares (*Aplysia*), soft corals (*Xenia*) or other animals. Faulkner (1984) divides the

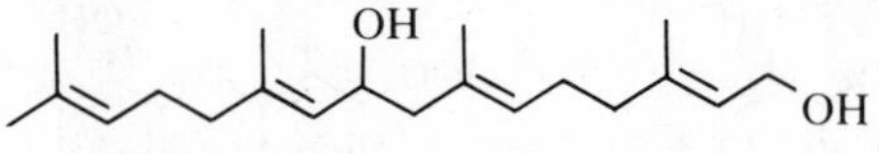

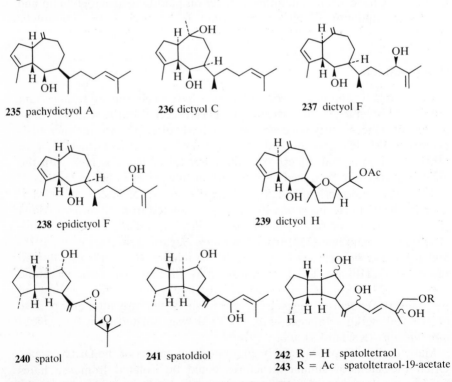

228   crinitol

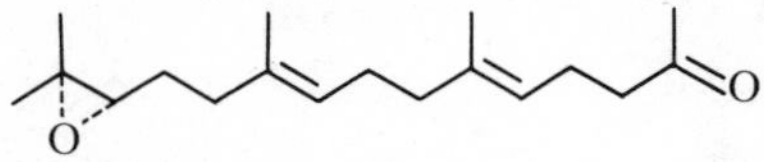

229   farnesylacetone epoxide

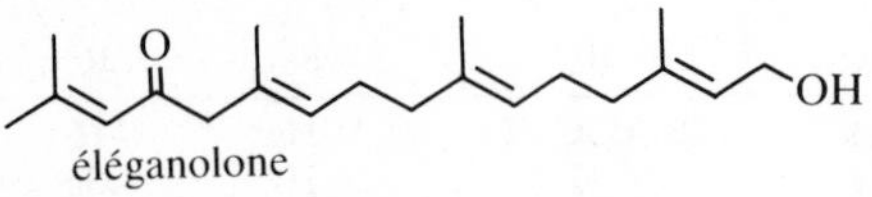

230   éléganolone

Fig. 7.29   Aliphatic diterpenes from brown algae.

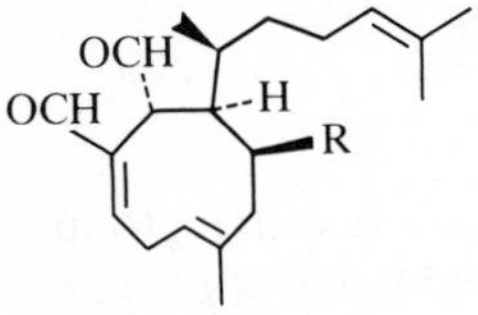

231 R = OH hydroxydictyodial   **233** dilophic acid
232 R = H dictyodial

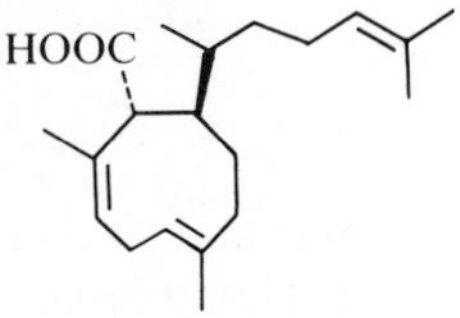

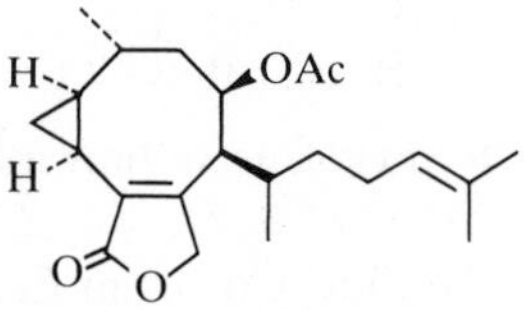

234 acetoxycrenulide

Fig. 7.30   Bioactive xenicanes from Dictyotaceae.

235 pachydictyol A   **236** dictyol C   **237** dictyol F

238 epidictyol F   **239** dictyol H

240 spatol   **241** spatoldiol   **242** R = H   spatoltetraol
                                                243 R = Ac spatoltetraol-19-acetate

Fig. 7.31   Bioactive 'extended sesquiterpenes' from Dictyotaceae.

diterpenes into three groups, based on skeletal features. The first group can be derived from the nine-membered ring 'xenicane' skeleton (Fig. 7.30), first discovered in a soft coral (*Xenia*). The second group has a familiar sesquiterpene ring system with an additional isoprene unit in the side-chain ('extended sesquiterpenes', Fig. 7.31). The third group is based on the 'dolabellane' skeleton (Fig. 7.32) which was first encountered in an opistho-branch mollusc but which is an algal metabolite. Only a few of the numerous substances isolated were examined for their biological effective-ness. Substances have been found with antimicrobial (**231, 232, 235–239, 243–254, 260, 261**), cytotoxic (**239–242, 244, 249, 259**), ichthyotoxic (**233, 234, 255–257**) and molluscicidal (**248, 249, 254**) activity. A possible cyto-toxic effect is often ascertained by the easily detectable inhibition of cell division in fertilized sea urchin eggs (Gerwick *et al.*, 1980). Examples of the results of tests for biological effects are shown in Table 7.5. With an increase

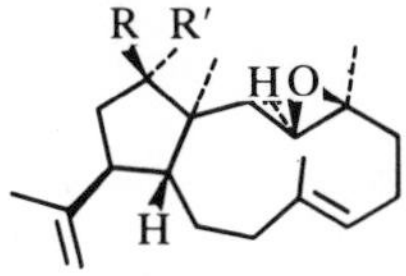

244 R= =O
       R'=

245 R = OH, R' = H
246 R = R' = H

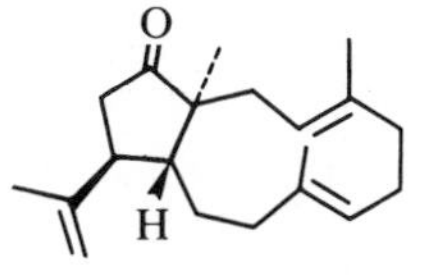

247

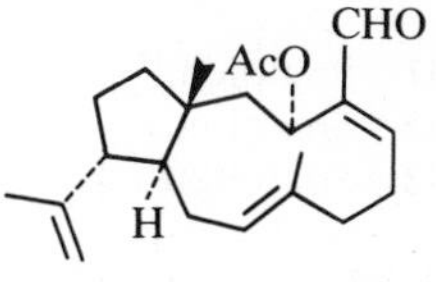

248

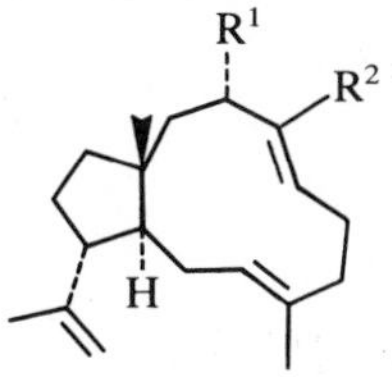

249	R¹ = OAc,	R² = CH₂OH	**255**	R¹ = Ac,	R² = R³ = H
250	R¹ = OH,	R² = CH₃	**256**	R¹ = R² = Ac, R³ = H	
251	R¹ = OH,	R² = CH₂OH	**257**	R¹ = R² = R³ = Ac	
252	R¹ = H,	R² = CH₂OH			
253	R¹ = OH,	R² = CH₂OAc			
254	R¹ = OAc,	R² = CHO			

249 $R^1 = OAc,$ $R^2 = CH_2OH$
250 $R^1 = OH,$ $R^2 = CH_3$
251 $R^1 = OH,$ $R^2 = CH_2OH$
252 $R^1 = H,$ $R^2 = CH_2OH$
253 $R^1 = OH,$ $R^2 = CH_2OAc$
254 $R^1 = OAc,$ $R^2 = CHO$

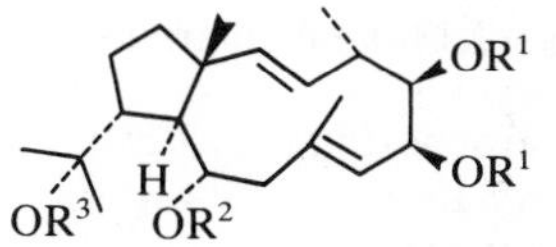

255 $R^1 = Ac,$ $R^2 = R^3 = H$
256 $R^1 = R^2 = Ac, R^3 = H$
257 $R^1 = R^2 = R^3 = Ac$

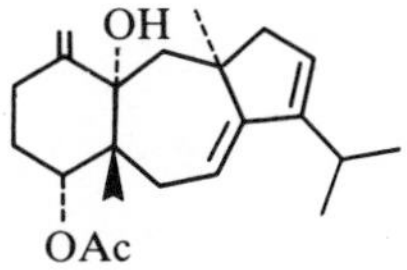

258

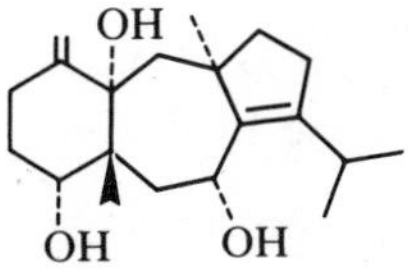

259

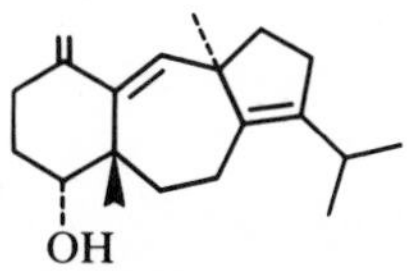

260 amijitrienol

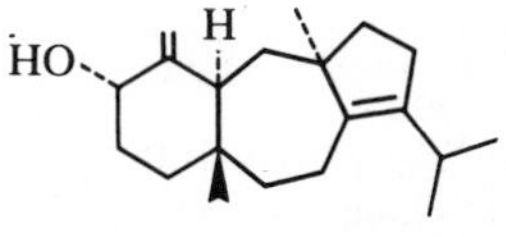

261   14-deoxyisoamijiol

Fig. 7.32 Bioactive dolabellanes from Dictyotaceae.

Table 7.5   Bioactivities of cyclic diterpenes from Dictyotaceae

Substance	Antimicrobial activity
'Xenicane group'	
231 Hydroxydictyodial	S.a., B.s.* (16)
232 Dictyodial	S.a., B.s., C.a. (6)
233 Dilophic acid	B.s. (weak) (13)
234 Acetoxycrenulide	
'Extended sesquiterpenes'	
235 Pachydictyol A	S.a. (weak) (10)
236 Dictyol C	S.a., E.c., B.s., M.l., M.s., C.a. (5)
237 Dictyol F	S.a., E.c., B.s., M.l., M.s., C.a. (5)
238 Epidictyol F	S.a., E.c., B.s., M.l., M.s., C.a. (5)
239 Dictyol H	E.c., P.a. (3)
240 Spatol	
241 Spatoldiol	
242 Spatoltetraol	
243 Spatoltetraol-19-acetate	S.a. (14)
'Dolabellane group'	
244 3,4-Epoxy-14-oxo-7,18-dolabelladiene	S.a., En.c., C.f. (1)
245 3,4-Epoxy-14-hydroxy-7,18-dolabelladiene	S.a., E.c., C.f. (1)
246 3,4-Epoxy-7,18-dolabelladiene	K.p., E.c. (1)
247 14-Oxo-3,7,18-dolabellatriene	K.p., En.c. (1)
248 3-Acetoxy-4,8,18-dolabellatriene-16-al (C-4 *cis*)	C.c. (19)
249 3-Acetoxy-16-hydroxy-4,8,18-dolabellatriene	
250 3-Hydroxy-4,8,18-dolabellatriene	S.a., S.m., A.n., M.r. (17)
251 3,16-Dihydroxy-4,8,18-dolabellatriene	A.n., M.r. (17), C.c. (19)
252 16-Hydroxy-4,8,18-dolabellatriene	S.a., S.m., M.r. (17)
253 3-Hydroxy-16-acetoxy-4,8,18-dolabellatriene	C.c. (19)
254 3-Acetoxy-4,8,18-dolabellatriene-16-al (C-4 *trans*)	C.c. (19)
255 5,6-Diacetoxy-10,18-dihydroxy-2,7-dolabelladiene	
256 5,6,10-Triacetoxy-18-hydroxy-2,7-dolabelladiene	
257 5,6,10,18-Tetraacetoxy-2,7-dolabelladiene	
258 4-Acetoxy-14-hydroxy-1(15),7,9-dolostatriene	

Cytotoxic and other activities	Found in species
Antifeedant (15)	*Dictyota spinulosa* (16)
	Dictyota crenulata, D. flabellata (6)
Ichthyotoxic (*C.* au.) 50 μg ml^{-1} (13)	*Dilophus guineensis* (13)
Ichthyotoxic (E.l.) antifeedant (15)	*Dictyota crenulata* (3)
	Pachydictyon coriaceum (10)
	Dictyota dichotoma (5)
	Dictyota dichotoma (5)
	Dictyota dichotoma (5)
Antitumour (KB 9) (3)	*Dictyota dentata* (3)
Inhibit SUED, ED$_{50}$ 1.2 μg ml^{-1} (7) and	*Spatoglossum schmittii* (7),
other cytotoxicity, piscicidal 100 μg ml^{-1};	*Stoechospermum marginatum* (8)
antialgal (9)	
Like spatol but less active (9)	*Spatoglossum howlei* (9)
Like spatol but less active (9)	*Spatoglossum howlei* (9)
	Stoechospermum marginatum (14)
Cytotoxic (2), antiviral (influenza,	*Stoechospermum marginatum* (14)
adenovirus) (9)	
	Dictyota dichotoma (1)
	Dictyota dichotoma (1)
	Dictyota dichotoma (1)
Molluscicidal on B.g (19)	*Dictyota* sp. (18)
Cytotoxic (20), molluscicidal on B.g. (19)	*Dictyota* sp. (17)
	Dictyota sp. (17)
	Dictyota sp. (17)
	Dictyota sp. (17)
	Dictyota sp. (20)
Molluscicidal on B.g. (19)	*Dictyota* sp. (20)
Ichthyotoxic (G.p.) 50 ppm, phytotoxic	*Dilophus fasciola* (12)
(H.v.) 100 ppm (12)	
Ichthyotoxic (G.p.) 50 ppm, phytotoxic	*Dilophus fasciola* (12)
(H.v.) 100 pm (12)	
Ichthyotoxic (G.p.) 50 ppm, phytotoxic	*Dilophus fasciola* (12)
(H.v.) 100 ppm (12)	
Histamine antagonism (16 μg ml^{-1}) (4)	*Dictyota linearis* (4)

Table 7.5 Cont.

Substance	Antimicrobial activity
259 4,7,14-Trihydroxy-1(15),8-dolostadiene	
260 Amijitrienol	S.a., M.m.(weak) (11)
261 14-Deoxyisoamijiol	S.a., M.m.(weak) (11)

* For abbreviations used, see p 218.

Reference list: (1) Amico *et al.* (1980); (2) Amico *et al.* (1982); (3) Alvarado and Gerwick (1985); (4) Crews *et al.* (1982); (5) Enoki *et al.* (1983); (6) Finer *et al.* (1979); (7) Gerwick *et al.* (1980); (8) Gerwick *et al.* (1981); (9) Gerwick and Fenical (1983); (10) Hirschfeld *et al.* (1973); (11) Ochi *et al.* (1986); (12) Rosa de et al. (1984); (13) Schlenk and Gerwick (1987); (14) Silva de *et al.* (1982); (15) Sun *et al.* (1983a); (16) Tanaka and Higa (1984); (17) Tringali *et al.* (1984a); (18) Tringali *et al.* (1984b); (19) Tringali *et al.* (1986); (20) Tringali *et al.* (1984c).

in comparable test results, it will be possible to consider the structure/activity relationships (Tringali *et al.*, 1986).

Isoprenylated and acylated phenols

Antibiotic effects (**270c, 271a**; Banaigs *et al.*, 1983), a promoting effect on the larval settlement of epiphytic hydrozoa (**271a, 271b, 273b**; Kato *et al.*, 1975) and an antitumour activity have been the guidelines for the isolation of a large number (**267–274**) of prenylated quinones and hydroquinones from numerous brown algae (Faulkner, 1984). Stereochemically, some of these metabolites are very poorly defined and details are absent for a disappointingly large number of compounds. In some cases, one may question whether the molecules might not be artefacts, caused by isolation procedures. Until now the group has been selectively, but not systematically, examined for biological effects. According to the structure, it concerns α-tocopherol, vitamin E (α-tocotrienol, **271a**) its analogous oxygenated substances or those derived from it; thus interesting effects are to be expected.

Bifurcarenone (**262**) from *Bifurcaria galapagensis* (Sun *et al.*, 1980) and mediterraneol A (**263**) and B (**264**) from *Cystoseira mediterranea* (Francisco *et al.*, 1986) belong biochemically to the same family but differ from one another either by cyclization to a cyclopentane ring (**262**) in the middle of the prenyl side-chain or by a condensed ring system (**263, 264**) at the end of the chain which does not, however, correspond to the 'isoprene rule'. Compound **262** has a weak antibiotic effect. All three inhibit mitotic cell division in sea urchin eggs (Sun *et al.*, 1980; Francisco *et al.*, 1986). A hydroquinone with a sesquiterpenoid side-chain (**265**) from *Dictyopteris undulata* shows antibiotic activity with fungi (Ochi *et al.*, 1979a). The side-chain is closed to a ring in dictyochromenol (**266**). It occurs in the same alga and has ichthyotoxic activity (Dave *et al.*, 1984). The cyclization of the side-chain bound with a phenol is often observed. In the case of zonarol (**275**) from *Dictyopteris undulata*, a decahydronaphthalene ring is formed (Fenical

Cytotoxic and other activities	Found in species
Inhibit SUED (4)	*Dictyota linearis* (4)
	Dictyota linearis (11)
	Dictyota linearis (11)

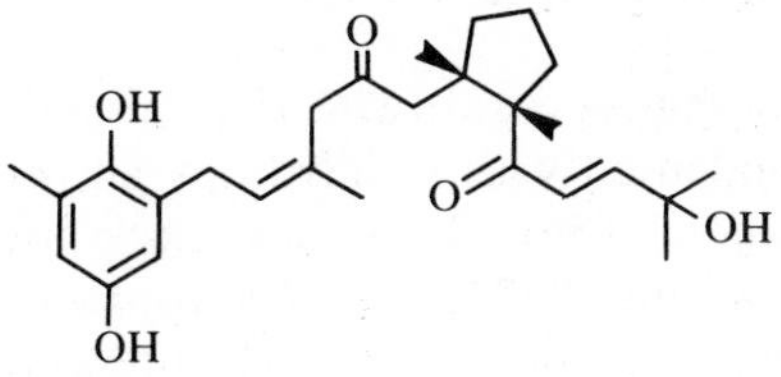

262 bifurcarenone

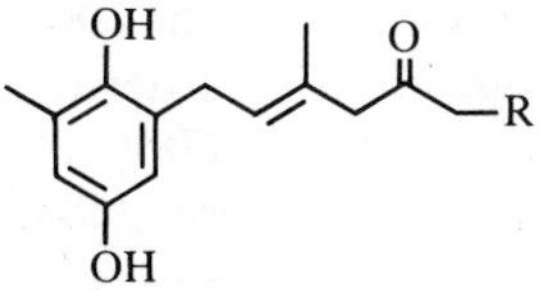

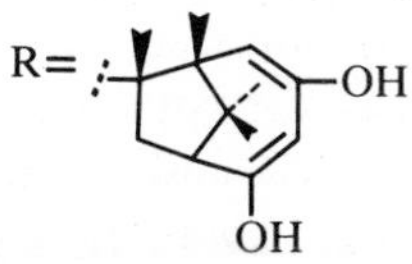

263 mediterraneol A 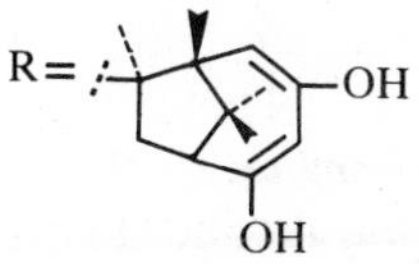   **264** mediterraneol B

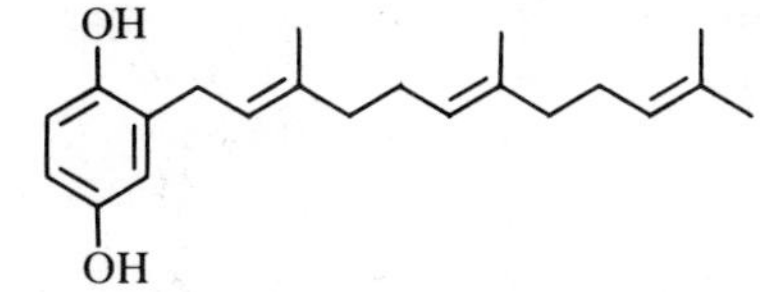

265 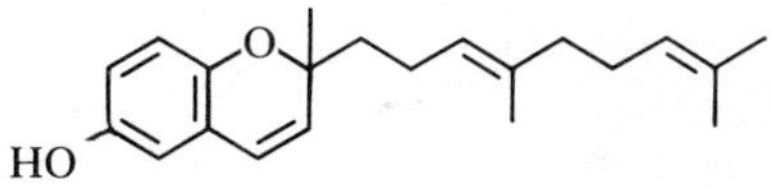   **266** dictyochromenol

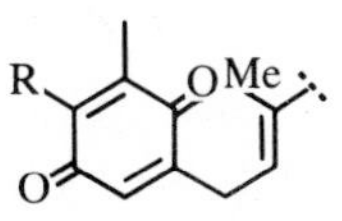

| **267** | R = H | **269** | R = H | **271** | R = H | **273** | R = H |
| **268** | R = Me | **270** | R = Me | **272** | R = Me | **274** | R = Me |

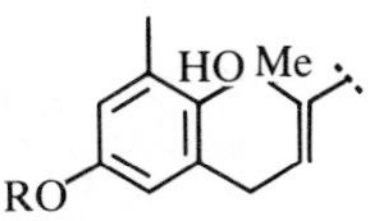

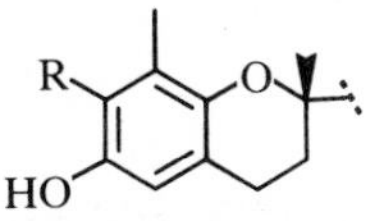

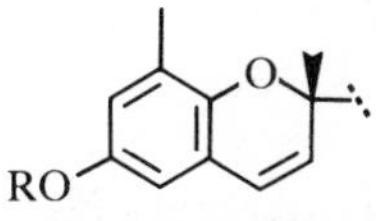

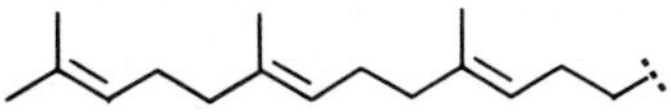

a 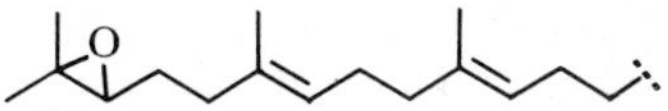   b

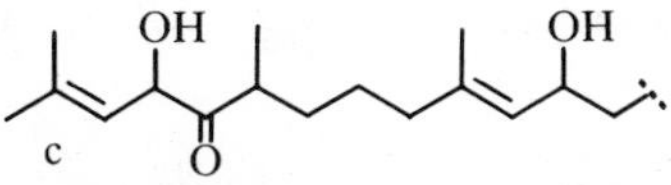

c

for further examples see Faulkner 1984

Fig. 7.33 Isoprenylated methylhydroquinones from brown algae.

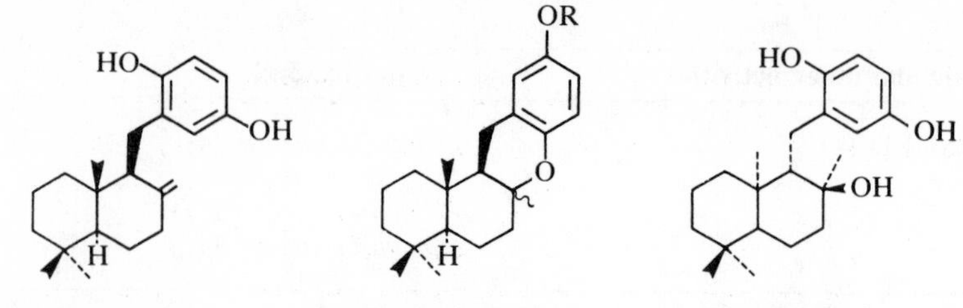

275 zonarol   **276** chromazonarol   **277** yahazunol

Fig. 7.34 Zonarol and analogues from brown algae.

et al., 1973), while in chromazonarol (**276**; Fenical and McConnell, 1975), the hydroquinone is condensed with a dihydropyran ring. Both substances are ichthyotoxic (Faulkner, 1986). Such substances also occur in sponges (*Disidea pallescens*). Yahazunol (**277**) from *Dictyopteris undulata* has strong antimicrobial activity against some yeasts and differs from zonarol stereochemically and by the addition of H_2O to the exocyclic methylene group (Ochi *et al.*, 1979b).

Acylderivatives of phloroglucinols (**278–283**) or resorcinols (**284–287**; Faulkner 1984) are often found alone or together with prenylated derivatives. They occur in the dictyotaceaens *Zonaria farlowii* (**278**), *Z. diesingiana* (**278, 279**; Gerwick and Fenical, 1982), *Z. tournefortii* (**280**; Tringali and Piattelli, 1982), *Lobophora papenfussii* (**283**; Gerwick and

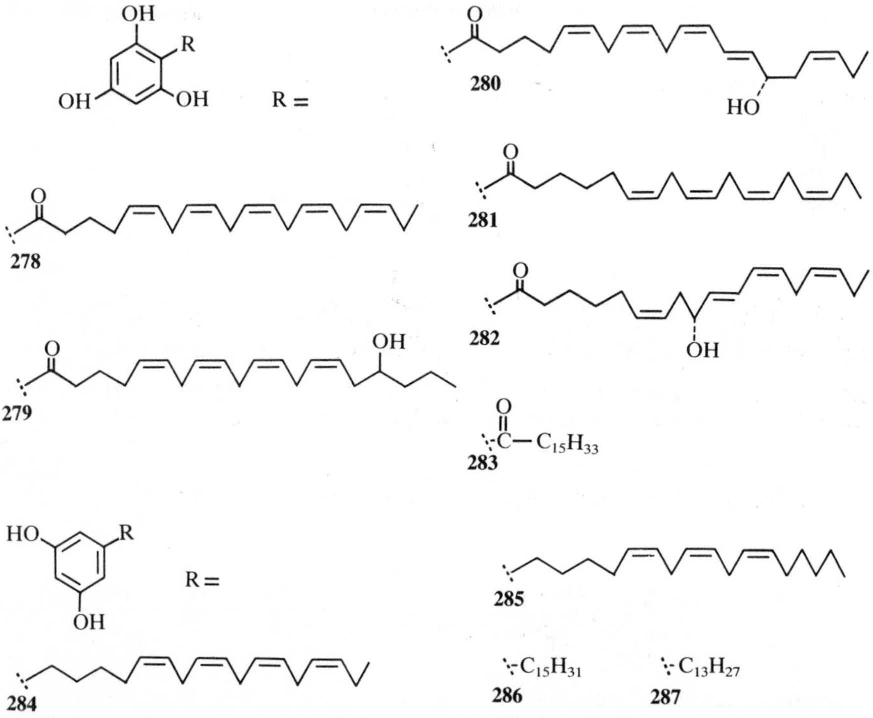

Fig. 7.35 Acylderivatives of phloroglucinols and resorcinols from brown algae.

Fenical, 1982) and also in *Cystophora torulosa* (methylated **281**, **285**; Gregson *et al.*, 1977).

The active anti-inflammatory principle of the sargassacean *Caulocystis cephalornithos* is a substance named 6-tridecylsalicylic acid (**288**) which is accompanied by the decarboxylation product 3-tridecylphenol (**289**) and the lacton (**290**) which can be formed by intramolecular addition to an unsaturated analogue of **288** (Kazlauskas *et al.*, 1980). Nothing is known about the activity of the last two.

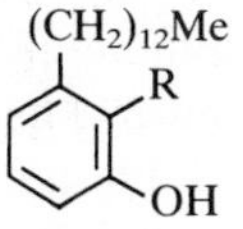

(CH₂)₁₂Me

288 R=CO₂H
289 R=H

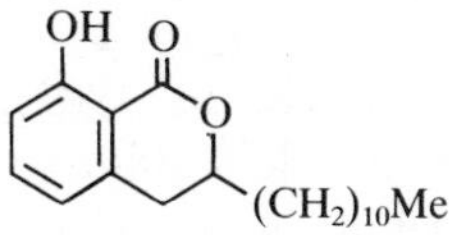

290

Fig. 7.36 Tridecylsalicylic acid and derivatives from brown algae.

Stypoldione

Freshly collected plants of *Stypopodium zonale* release compounds toxic to the Caribbean herbivorous fish *Eupomacentrus leucostictus*, the major component of which was the *o*-quinone, stypoldione (**291**; Gerwick *et al.*, 1979). It is formed from the pyrocatechol stypotriol (**292**) which is approximately five times more toxic but unstable. It is accompanied by similar substances such as stypodiol (**293**), which is nontoxic, and epistypodiol

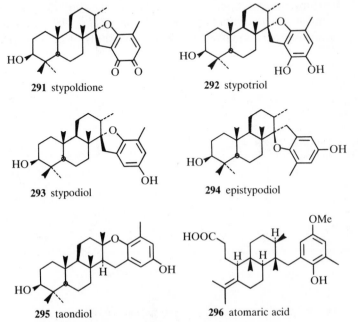

291 stypoldione   292 stypotriol

293 stypodiol   294 epistypodiol

295 taondiol   296 atomaric acid

Fig. 7.37 Prenylated hydroquinones from *Stypopodium zonale*.

(294), taondiol (295) and atomaric acid (296; Gerwick and Fenical, 1981), which produce distinct lethargic behaviour and narcosis. Both 295 and 296 were first isolated from *Taonia atomaria* (González *et al.*, 1971, 1974a). Substance 291 is a potent inhibitor of the first cleavage of fertilized sea urchin eggs (Jacobs *et al.*, 1981). It shows no evidence of acute toxicity in mice given doses up to 100 mg kg^{-1} but has interesting antitumour effects. It seems to inhibit microtubule polymerization *in vitro* by a novel mechanism (O'Brien *et al.*, 1984). In other studies it has been shown that 291 is inactive when added late in the cycle beyond metaphase (White and Jacobs, 1983) and has kinetics similar to mitotic spindle poisons, such as vinca alkaloids (White and Jacobs, 1983). Compound 291 has been shown to bind stoichiometrically to a low affinity site on the tubulin dimer and thus reduce the rate and extent of microtubule polymerization. This contrasts with known microtubule assembly inhibitors. Thus, this agent appears to have the potential of being a new and useful probe for elucidating steps in the cell cycle once its exact mechanism is known (Jacobs *et al.*, 1985).

Dinophyceae and Desmophyceae

Antibiotic effects from the single-celled Dinophyceae have occasionally been observed (Ragan, 1984). Cytotoxic, antimitotic, ichthyotoxic or haemolytic activities have been found during research on dinophyceaen toxins and in other research. Bass *et al.* (1983) isolated a haemolysin from *Gonyaulax monilata* which behaves like an unsaturated glycolipid. Yasumoto *et al.* (1987) succeeded in isolating two haemolytic galactoglycerolipids (297, 298) from *Amphidinium carteri*. A macrolide amphidinolide A (301) was found in an *Amphidinium* sp. (Kobayashi *et al.*, 1986) which showed an *in vitro* antitumour activity towards several kinds of cells. *Ptychodiscus*

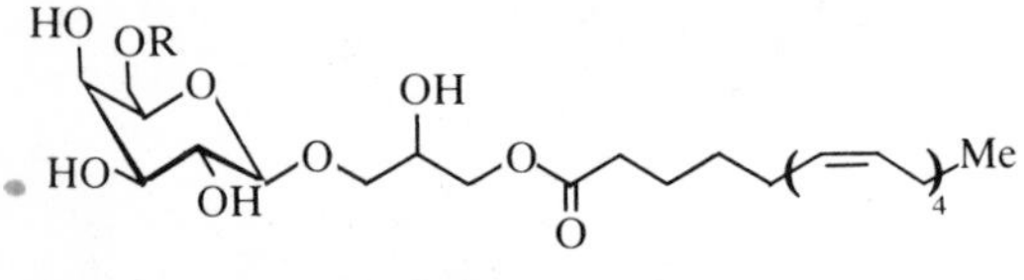

297   hemolysin I   R = H
298   hemolysin II   R = α-galactosyl

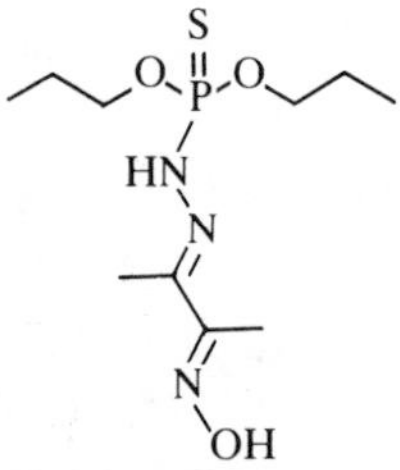

299 Gb-4-toxin

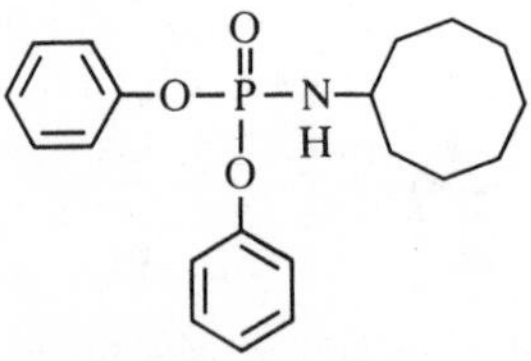

300 Pb-1-toxin

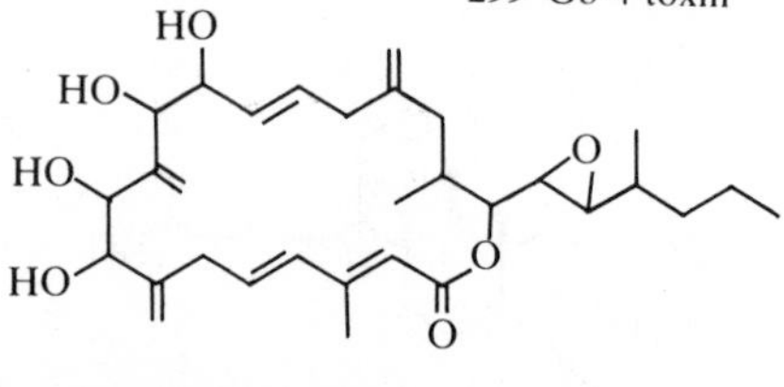

301 amphidinolide-A

Fig. 7.38 Bioactive substances of lesser importance from Dinophyceae.

brevis contains the ichthyotoxic phosphorus substances PB-1 (**300**; DiNovi *et al.*, 1983) and Gb-4 (**299**; Alam *et al.*, 1982).

The reports about dinoflagellate toxins, which can be dangerous for humans and animals, are of more significance and interest. Some of the Dinophyceae have a tendency to form 'blooms' or 'red tides'. The last term is preferred only by journalists and not by most experts in the field. Only some blooming organisms contain toxins and have effects on the ecosystem including large-scale fish mortality. The toxic dinoflagellates contain various toxins. According to the symptoms of poisoning, one distinguishes between paralytic shellfish poison (PSP), diarrhetic shellfish poison (DSP), neuro-toxins and others. A great number of excellent review articles about dinophyceaen toxins have been published: Shimizu (1978), Schantz (1981), Shimizu (1982, 1984), Steidinger and Baden (1984), Yasumoto (1985), Carmichael (1986), Metting and Pyne (1986).

Paralytic shellfish poison (PSP)

Toxicity of this type was first recognized in the form of intoxication caused by the ingestion of clams. The poisoning begins with a feeling of numbness on the lips and difficulties in articulation. The numbness progresses to the fingers and feet and finally the arms and legs in advancing paralysis and, in severe cases, death occurs 2–12 hours later due to respiratory arrest. Sommer *et al.* (1937) were the first to clarify the relationship between the toxicity in clams and the population of dinoflagellates on which they had fed. The PSPs are formed from various representatives of the genera *Gonyaulax, Protogonyaulax, Pyrodinium* and *Gymnodinium* (Table 7.6). The toxicity and toxin composition varies widely among *Protogonyaulax* isolates, even if they are related or conform to the same morphotype (Cembella *et al.*, 1987). Thirteen PSP toxins (**302–314**) have been isolated from clams and dinoflagellates. Compounds **302** and **303** comprise class I, class II is characterized by an α- or β-OSO_3^--group at C–11 (**304–307**), class III consists of sulphocarbamoyl derivatives (**308, 309**) and class IV has two sulphate groups (**310–313**). Class V has no carbamoyl-C (Steidinger and Baden, 1984). The highest mouse toxicity is produced by **302** and **306**; but **308–311** are only weakly effective, with toxicity amounting to 10% or less than that of **302** (Genenah and Shimizu 1981; Steidinger and Baden, 1984; Fig. 7.39) in the mouse bioassay. Compound **382** has been found as aphan-toxin in addition to **308** in *Aphanizomenon flos-aquae* (Alam *et al.*, 1978) and in the red alga *Jania* (Kotaki *et al.*, 1985). The less toxic sulphamates (**308–311**) are converted by bacteria (Kotaki *et al.*, 1985), weak acid hydrol-ysis (Nishio *et al.*, 1982) or enzymatically (Hall and Reichardt, 1984) into the more poisonous carbamates (**302** and **303**). The mechanisms of action of the PSPs have been intensively investigated. As a rule, **302**, whose effect is similar to tetrodotoxin in many respects, is used for this. The gonyau-toxins in principle have a similar mode of action. The toxins block the influx of sodium through the excitable nerve membranes, acting as a plug in the sodium channel (Hille, 1975) and thus preventing the formation of action

Table 7.6   Toxic dinoflagellates

Class: Dinophyceae

Gymnodiniales
   Ptychodiscus brevis   hemolysins, PbTx
   Gymnodinium catenatum   PSP
   Amphidinium carteri   hemolysins

Peridiniales
   Protogonyaulax catenella   PSP
   P. acatenella   PSP
   P. tamarensis   PSP
   Gonyaulax polyedra   PSP
   G. catenatum   PSP
   Pyrodinium bahamense var. *compressa*   PSP

Dinophysiales
   Dinophysis fortii   DSP
   D. acuminata   DSP
   Gambierdiscus toxicus   ciguatoxin

Class: Desmophyceae

   Prorocentrum lima   DSP

PSP, Paralytic shellfish poison; DSP, Diarrhetic shellfish poison.

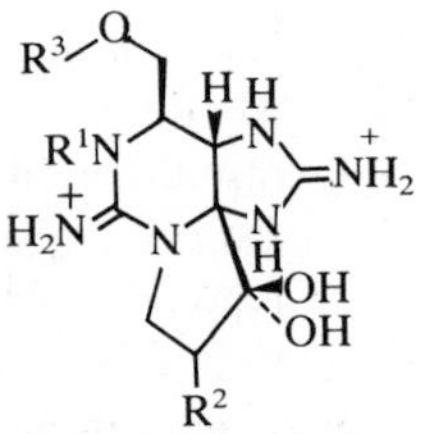

		R^1	R^2	R^3	related toxicity ref. (19)	ref. for first report
302	saxitoxin	$-H$	$-H$	$-CONH_2$	1	(15)
303	neosaxitoxin	$-OH$	$-H$	$-CONH_2$	0,50	(12)
304	gonyautoxin-I	$-OH$	$-\alpha OSO_3^-$	$-CONH_2$	0,80	(16)
305	gonyautoxin-II	$-H$	$-\alpha OSO_3^-$	$-CONH_2$	0,39	(16)
306	gonyautoxin-III	$-H$	$-\beta OSO_3^-$	$-CONH_2$	1,09	(16)
307	gonyautoxin-IV	$-OH$	$-\beta OSO_3^-$	$-CONH_2$	0,33	(12)
308	gonyautoxin-V (=B1)	$-H$	$-H$	$-CONHSO_3^-$	0,17	(12, 7)
309	gonyautoxin-VI (=B2)	$-OH$	$-H$	$-CONHSO_3^-$	0,09	(11, 9)
310	gonyautoxin-VIII (=C2)	$-H$	$-\beta OSO_3^-$	$-CONHSO_3^-$	0,14	(7, 8, 18)
311	epigonyautoxin-VII (=Cl)	$-H$	$-\alpha OSO_3^-$	$-CONHSO_3^-$	$8,3.10^{-3}$	(7, 8)
312	C3	$-OH$	$-\alpha OSO_3^-$	$-CONHSO_3^-$		(10)
313	C4	$-OH$	$-\beta OSO_3^-$	$-CONHSO_3^-$		(10)
314	decarbamoylsaxitoxin	$-H$	$-H$	$-H$		

Fig. 7.39   Structure, relative toxicity and occurrence of saxitoxins and gonyautoxins. Reference list for Fig. 7.39 see Fig. 7.41.

potentials (Narahashi, 1972). It is assumed that the formation of a hemiketal binding between the C-12-ketogroup and the receptor is necessary (Hille, 1975). The toxins have not found use in therapy but have particular importance as pharmacological tools because of their specific activity against sodium channels.

Bioactive polyether compounds

Diarrhetic shellfish poisoning (DSP)

Diarrhetic shellfish poisoning is distinctly different from PSP in both symptomatology and etiology. The predominant symptoms after ingestion of shellfish are gastrointestinal disturbances. It does not appear to be fatal but a high mobility rate and worldwide occurrence has been observed (Yasumoto et al., 1985). The dinoflagellate *Dinophysis fortii* was identified as the organism from which the toxin was produced and transmitted to shellfish (Yasumoto et al., 1980). From digestive glands of the scallop *Patinopecten yessoensis*, Yasumoto et al. (1985) isolated the dinophysistoxin 1 (DTX$_1$, **316**) and 3 (DTX$_3$, **317**) and five pectenotoxins (PTX$_1$–PTX$_5$) from which three (**318–320**) were elucidated (Yasumoto et al., 1985; Murata et al., 1986). Compound **316** is 35(S)-methylokadaic acid. Okadaic acid (**315**) is a potent cytotoxic component first isolated from the black sponge *Halichondria okadai* (Tachibana et al., 1981) and then from the dinoflagellate *Prorocentrum lima* (Murakami et al., 1982) which also contains a 5-methylene-6-hydroxy-2-hexenol ester of **316** (Yasumoto et al., 1987) and was also identified from a toxic mussel *Mytilus edulis* which had fed on *Dinophysis acuminata* (Yasumoto et al., 1985).

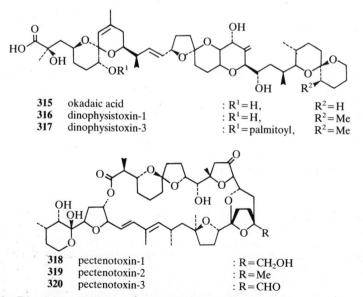

315	okadaic acid	: R^1=H,	R^2=H
316	dinophysistoxin-1	: R^1=H,	R^2=Me
317	dinophysistoxin-3	: R^1=palmitoyl,	R^2=Me

318	pectenotoxin-1	: R=CH$_2$OH
319	pectenotoxin-2	: R=Me
320	pectenotoxin-3	: R=CHO

Fig. 7.40 Diarrhetic shellfish poisons (DSP) from Dinophyceae.

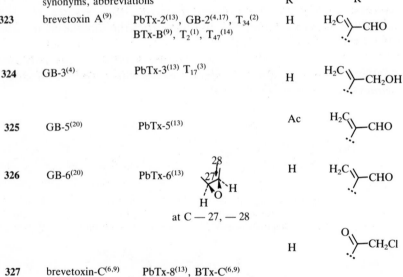

synonyms, abbreviations

321 brevetoxin A[9]   PbTx-1[13], T$_{46}$[14], Gb-1[20],   R = CHO
322 GB-7[20]   PbTx-7[13],   R = CH$_2$OH

	synonyms, abbreviations		R^1	R^2
323	brevetoxin A[9]	PbTx-2[13], GB-2[4,17], T$_{34}$[2] BTx-B[9], T$_2$[1], T$_{47}$[14]	H	H$_2$C⟩—CHO
324	GB-3[4]	PbTx-3[13] T$_{17}$[3]	H	H$_2$C⟩—CH$_2$OH
325	GB-5[20]	PbTx-5[13]	Ac	H$_2$C⟩—CHO
326	GB-6[20]	PbTx-6[13]	H	H$_2$C⟩—CHO
327	brevetoxin-C[6,9]	PbTx-8[13], BTx-C[6,9] BTx I[21]	H	⟩—CH$_2$Cl

at C — 27, — 28

Fig. 7.41 Structure and denomination of brevetoxins. Reference list for Figs 7.39 and 7.41: (1) Alam *et al.* (1975); (2) Baden *et al.* (1981); (3) Baden and Mende (1982); (4) Chou and Shimizu (1982); (5) Chou *et al.* (1985); (6) Golik *et al.* (1982); (7) Hall *et al.* (1980); (8) Kobayashi and Shimizu (1981); (9) Lin *et al.* (1981); (10) Noguchi *et al.* (1983); (11) Oshima *et al.* (1976); (12) Oshima *et al.* (1977); (13) Poli *et al.* (1986); (14) Risk *et al.* (1979); (15) Schantz *et al.* (1957); (16) Shimizu *et al.* (1975); (17) Shimizu (1978); (18) Shimizu (1980); (19) Shimizu (1984); (20) Shimizu *et al.* (1986); (21) Whitefleet-Smith *et al.* (1986).

Brevetoxins: Dinophyceaen neurotoxins

The blooms occurring on the coasts of Florida are often caused by *Ptychodiscus brevis* (formerly *Gymnodinium breve*) (Steidinger, 1983). It is responsible for massive fish kills, mollusc poisoning and human food poisoning. It also causes respiratory disorders when inhaled in an aerosol form (Pierce, 1986), perhaps due to bronchoconstriction (Baden *et al.*, 1982). This alga forms haemolytic agents and neurotoxins, which are cyclic polyethers with a long chained C-framework like the DSPs and are called brevetoxins (Steidinger and Baden, 1984). A survey of the compounds that have been isolated and the names used by the various research groups is given in Fig. 7.41.

In common with ouabain, compound **323** was found to have a strong inotropic effect on an isolated heart (Shimizu, 1982). The brevetoxins are potent neurotoxins, altering the membrane properties of excitable cell types in a way that activates sodium channels at fairly negative potentials and by inhibiting the fast sodium inactivation (Steidinger and Baden, 1984). Tetrodotoxin abolishes the effect caused by **329** (Wu *et al.*, 1985). Compound **329** causes a dose-dependent depolarization which is half-maximal at about 1 nM. Competition experiments using natural toxin probes specific for sites 1–4 of the voltage-dependent sodium channel have illustrated that **329** does not bind to any of the previously described sites. A fifth site is proposed (Poli *et al.*, 1986).

Ciguatera–scaritoxin–maitotoxin

Ciguatera is a disease of humans occurring in tropical and subtropical areas from eating certain eels and reef fish. Eating results in nausea, vomiting, metallic taste and many other symptoms. The toxins apparently originate from *Gambierdiscus toxicus* (Bagnis *et al.*, 1980), from which the poison, called maitotoxin from the surgeonfish *Ctaenochaetus striatus* (Tahitian name 'maito'), also originates (Yasumoto *et al.*, 1976). Ciguatera toxin and maitotoxin, however, have different pharmacological action (Ohizumi, 1987). None of the fish appear to be direct plankton feeders but they are predators of smaller fish that feed on plankton (Schantz, 1981). The ciguatoxin acts slowly and it is rarely fatal to man even though it has a high toxicity (LD_{50}, 0.45 μg kg^{-1}, i.p., mice). The consumers survive because of its extremely low concentration in fish flesh (Tachibana *et al.*, 1987).

In spite of intensive efforts it has not been possible to obtain a sufficient amount of material for structure analysis. Proton NMR studies with 1.3 mg chromatographically pure toxin from 75 kg of toxic eel viscera have shown that it is a polar and highly oxygenated molecule belonging to a class of polyethers like okadaic acid and brevetoxins (Tachibana *et al.*, 1987). There also seem to be similarities to the toxins from *Scarus gibbus* (e.g. scaritoxin) which were often investigated in vain (Tachibana *et al.*, 1987). Precise conclusions cannot be drawn from the various findings (Anderson and Lobel, 1987).

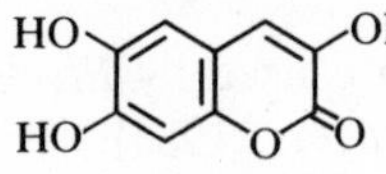

328 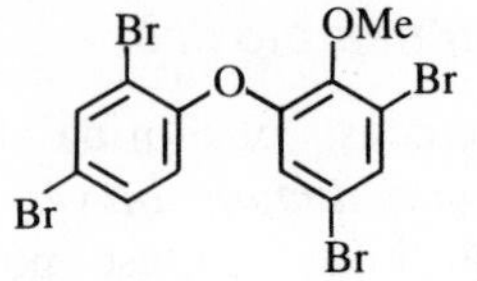   **329** 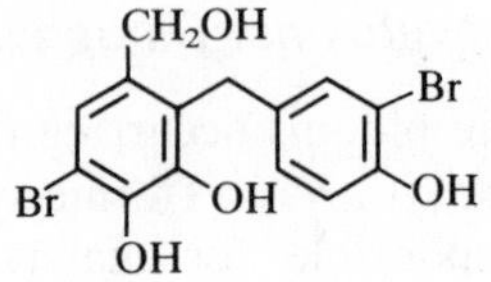   **330** avrainvilleol

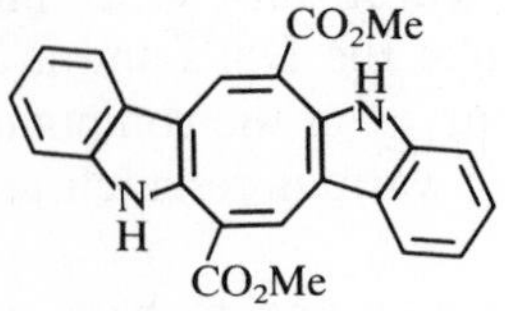

331 caulerpin 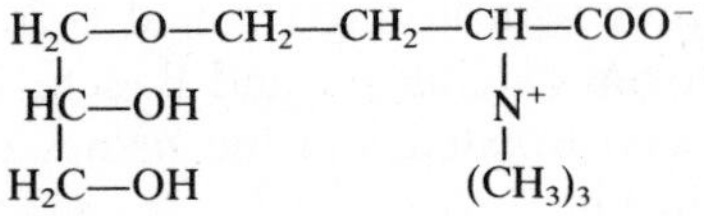   **332** ulvaline

Fig. 7.42 Various bioactive substances from green algae.

Chlorophyceae

Nitrogen-containing compounds

Several betaines have been detected in green algae (review: Blunden and Gordon, 1986). The reduction of the blood cholesterol level obtained with extracts of *Monostroma nitidum* could be caused by ulvaline (**332**; Abe and Kaneda, 1975) and β-alaninebetaine (**8**; Abe and Kaneda, 1973). For several other betaines found in green algae an activity was known. *Chlorella*, *Spirulina* and *Scenedesmus* contain a glycoprotein (mol. wt. = 121 000; sugar:protein = 1:1) which after treatment with α-amylase gives an anti-tumour product (Shinho, 1984). From the same algae a lipid mixture with activity against sarcoma-180 in mice was found (Watanabe and Fujita,1985). An L-asparaginase has been purified from a marine *Chlamydomonas* species, which possesses limited antitumour activity in an antilymphoma assay *in vivo* (Paul, 1982). The nitrogenous metabolite caulerpin (**331**; Maiti *et al.*, 1978) has been isolated from several species of *Caulerpa* and is considered to be responsible for the feeding deterrent activity of extracts of *C. prolifera*, but it causes little, if any, effect in comparison with the sesquiterpenoids (**333**, **335–338**, **343**, **346**; Paul *et al.*, 1987).

Carboxylic acids from green algae

The first studies aimed at identification of antimicrobial effects were carried out with culture solutions of various unicellular Chlorophyceae, especially *Chlorella*. It was often observed that the antimicrobial activities increased with the age of the cultures and were light dependent. An active fraction isolated from such cultures was called chlorellin (Pratt *et al.*, 1944). In spite of many efforts, no definite structure for the effective principles has been published. They are presumed to be a mixture of fatty acids, in which the

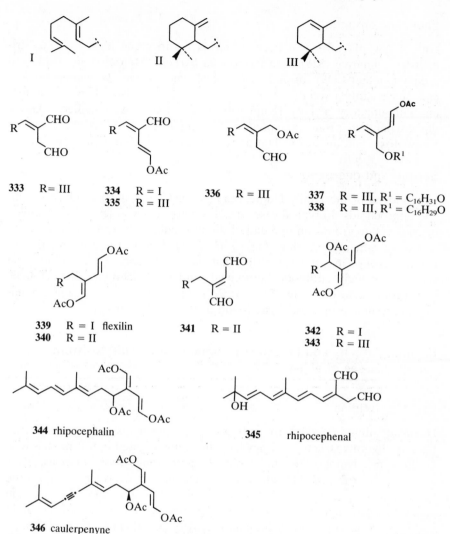

Fig. 7.43 Bioactive sequiterpenoids from green algae.

more unsaturated compounds are converted by photooxidation into anti-bacterially active derivatives (Spoehr *et al.*, 1949). From *Scenedesmus obliquus* a hexadecatetraenoic acid was isolated (Akhunov *et al.*, 1978), but it is known that the Chlorophyceae contain not only this but a number of highly unsaturated fatty acids (review: Pohl and Zurheide, 1979). The fatty acids from *Chaetomorpha minima* are even held responsible for the ichthyotoxic (*Oryzias latipes*, killifish) and haemolytic effect of this alga (Fusetani *et al.*, 1976). An antithrombotic activity was described for eicosapentaenoic acid-rich glycolipids from *Chlorella minutissima* (Seto *et al.*, 1984). Green algae sometimes contain dimethylpropiothetine (**184**) in substantial amounts (Glombitza, 1979; Iida *et al.*, 1985), from which the antibiotically effective acrylic acid (**185**) can be released after enzymatic cleavage.

Phenolic compounds

Only a few phenols have been found so far in green algae. A 3,6,7-tri-hydroxycoumarine (**328**) from *Cymopolia barbata* (Menzel *et al.*, 1983) and a diphenylether (**329**), brominated several times, from *Cladophora fascicularis* have antibacterial activity, and the latter also shows a potent anti-inflammatory activity (Kuniyoshi *et al.*, 1985). Avrainvilleol (**330**) a diphenylmethan from *Avrainvillea longicaulis* has an antibacterial, an ichthyotoxic and an antifeedant activity against fish (Sun *et al.*, 1983b).

Sesqui- and diterpenoids

Green algae of the family Udoteaceae (order Caulerpales) produce sesqui- and diterpenoids. In typical cases the substances carry at least two aldehyde or enolacetate groups on one end of the molecule. One end can be cyclized to a 5- or 6-membered ring (Figs. 7.43, 7.44). Due to the accumulation of reactive groups it is not surprising that these substances possess biological activities, namely antibiotic, antifungal, antilarval, ichthyotoxic, cytotoxic and feeding deterrent effects (Table 7.7). *In situ* the effects against predators predominate in these algae found in the Caribbean sea.

Haemagglutinins and bioactive polymeric carbohydrates from various classes of algae

Haemagglutinins

Haemagglutinins (lectins) are proteins or glycoproteins having specificities for carbohydrate structures. They have been investigated in test models with erythrocytes, lymphocytes and leukemia cells from various animals. They are widely distributed throughout higher plants and invertebrates, but little is known about their distribution in marine algae (reviews: Shimizu and Kamiya, 1983; Ingram, 1985). From 24 species of Puerto Rican algae, 11 had more or less specific agglutination activity (Boyd *et al.*, 1966). From an initial screening of 100 British algae, 19 were positive against human erythrocytes to various degrees; this was later studied in more detail and partially modified (Rogers *et al.*, 1980). Hori *et al.* (1986c) found haemagglutinating effects in the green algae *Ulva pertusa*, *Enteromorpha linza*, *Codium fragile*, *Boodlea coacta* and the red algae *Carpopeltis flabellata*, *Hypnea japonica* and *Gracilaria bursa-pastoris* and they investigated the carbohydrate and protein content of the haemagglutinins, their heat and pH stability, the influence of various cations and the specificity in the binding of various carbohydrates. There are also many haemagglutinating red algae on the Spanish coast (Fabregas *et al.*, 1985). Similar effects of several brown algae are possibly caused mainly by polyphenols (Rogers and Loveless, 1985) of the phloro-tannin type which would explain the cause of the low specificity of several Fucacean extracts (Wagner and Wagner, 1978). A follow-up examination of earlier positive findings often revealed that the activity found in the earlier collection was produced by polyphenols which may have been

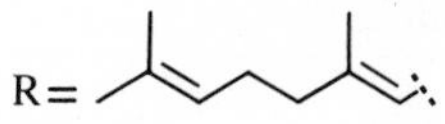

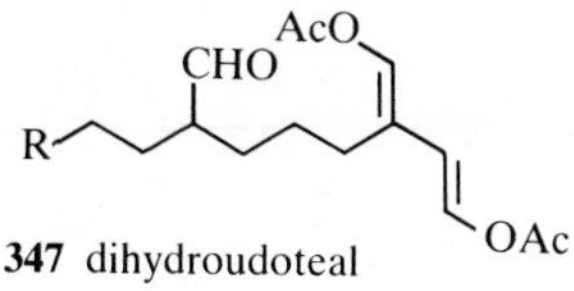

R=

347 dihydroudoteal

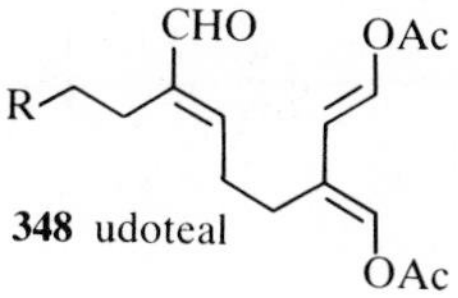

348 udoteal

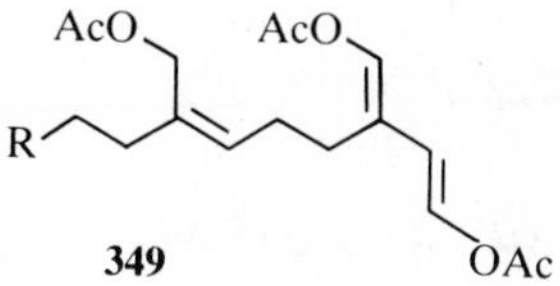

349

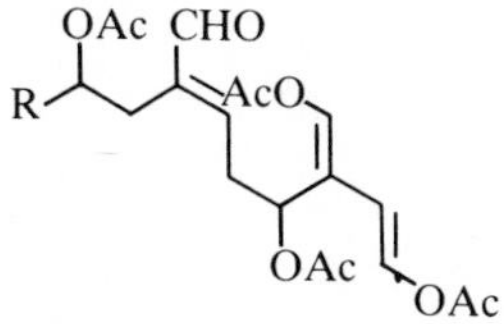

350

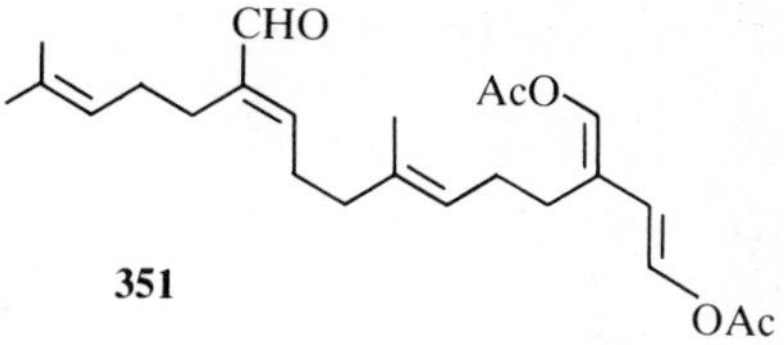

351

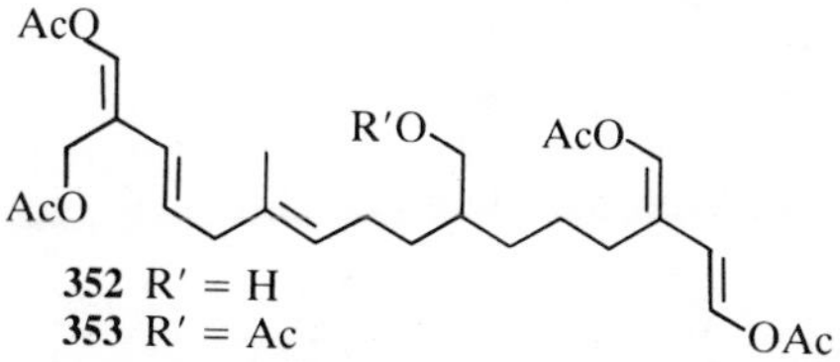

352 R′ = H
353 R′ = Ac

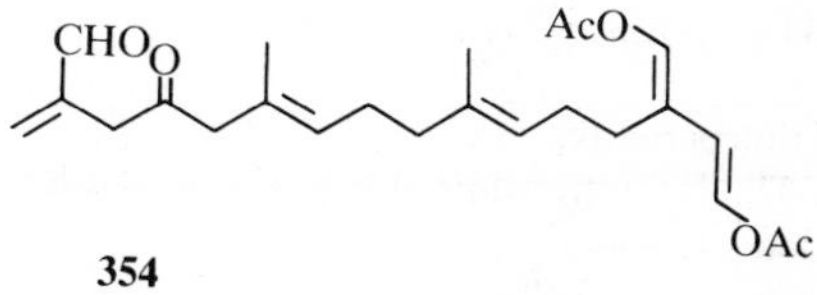

354

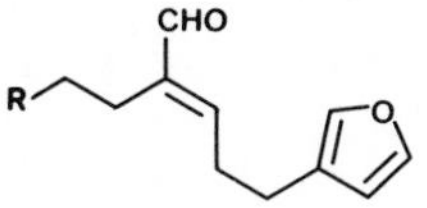

355 udoteafuran

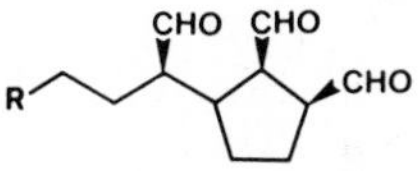

356 udoteatrial

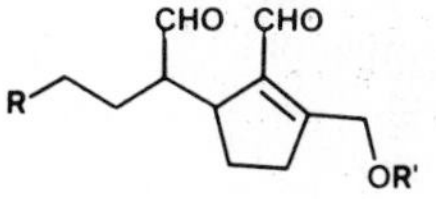

357 R′ = Ac
358 R′ = H

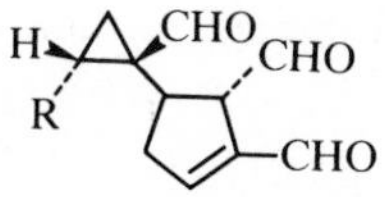

359 halimedatrial

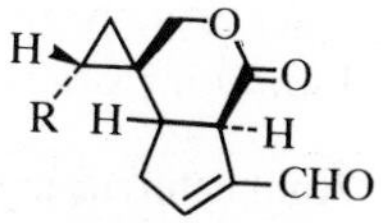

360 halimedalactone

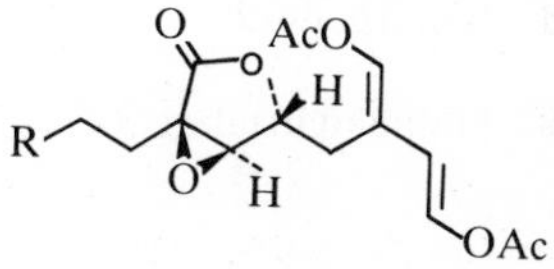

361

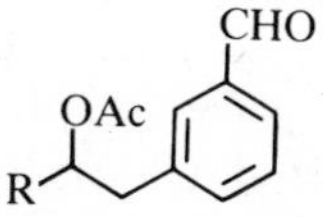

362

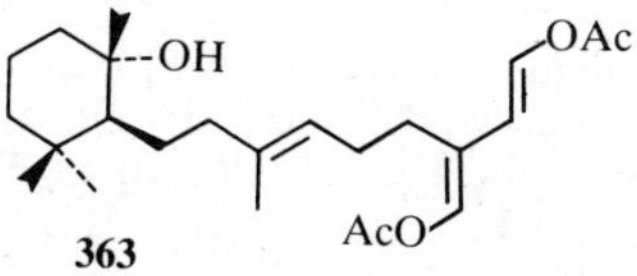

363

Fig. 7.44 Bioactive diterpenoids from green algae.

Table 7.7 Bioactivity of sesqui- and diterpenoids from green algae

Substance	Antimicrobial activity
Sesquiterpenoids	
334 'Aldehyde'	S.m., V.s., V.h., V.l., Le., Lu., Al. (2, 7)
339 Flexilin	S.a., D.h., Lu. (2, 7)
342 'Triacetate'	S.a., B.s., S.m., V.s., V.h., V.l. (2, 7)
344 Rhipocephalin	
345 Rhipocephenal	
341 'Dialdehyde'	
340 'Diacetate'	
343	SK (10)
335	L.c., V.l., V.p. SK (10)
333	V.l., V.p., SK (10)
336	V.l., V.p., SK (10)
346 Caulerpenyne	V.l. (10)
Diterpenoids	
347 Dihydroudoteal	S.a., V.s., V.h., V.l., D.h., Lu. (2, 7)
348 Udoteal	
349 'Diterpene diacetate'	S.a., B.c., S.m., V.s., V.h., V.l., C.a. (2, 7)
350 'Tetraacetate'	S.a., B.s., V.h., Lu. Al. (2, 8)
351 'Diterpenoid'	V.s. (2)
352 'Bisenolacetate'	S.a., B.s., V.h., V.l. (9)
353 'Pentaacetate'	S.a., B.s. (9)
354 'Aldehyde'	S.a., B.s. (9)
355 Udoteafuran	S.a. (2)
356 Udoteatrial	antimicrobial (3), S.a. (2)
357 'Diterpenoid'	S.m., V.s., V.h., V.l. (7)
358 'Diterpenoid'	S.m., V.s., V.h., V.l., D.h., Lu., Al. (2, 7)
359 Halimedatrial	S.m., V.h., V.l., V.s. (6), S.a., B.s,. Le., Lu., D.h., C.a. (2)
360 Halimedalactone	Antibacterial, antifungal (8)
361 'Epoxylactone'	S.a., B.s. (9)
362 'Bis-nor-diterpenoid'	Antibacterial, antifungal (8)
363 'Alcohol'	S.a., B.s., V.h., V.l., V.a., E.c., (9)

Reference list: (1) Blackman and Wells (1978); (2) Fenical and Paul (1984); (3) Nakatsu *et al.* (1981); (4) Paul and Fenical (1982); (5) Paul *et al.* (1982); (6) Paul and Fenical (1983); (7) Paul and Fenical (1984a); (8) Paul and Fenical (1984b); (9) Paul and Fenical (1985); (10) Paul *et al.* (1987); (11) Stallard and Faulkner (1974); (12) Sun and Fenical (1979).

Cytotoxic and other activities	Found in species
Inhibit SUED, 2 μg ml^{-1} (2, 6), ichthyotoxic (7)	Penicillus capitatus,
Inhibit SUED (7)	Udotea cyathiformis (7)
Inhibit SUED, 2 μg ml^{-1} (2, 7)	Caulerpa flexilis (1)
	Penicillus capitatus,
	Udotea cyathiformis (7)
Ichthyotoxic (P.c.), DL 2 μg ml^{-1} (7, 12), antifeedant (E.l.), 100–300 ppm (12), inhibit SUED 8 μg ml^{-1} (2)	Rhipocephalus phoenix (12)
Ichthyotoxic (P.c.), DL 10 μg ml^{-1}, antifeedant (E.l.) 100–300 ppm (12)	Rhipocephalus phoenix (12)
Ichthyotoxic (P.p.), 5 μg ml$^-$, inhibit SUED (L.p.), ED$_{50}$ 1 μg ml^{-1} (4)	Caulerpa bikinensis (4)
Ichthyotoxic (P.p.), 10 μg/ml^{-1} (4), inhibit SUED (L.p.), ED$_{50}$ 2 μ g/ml^{-1} (4)	Caulerpa bikinensis (4)
Ichthyotoxic (P.c.), 10 μg/ml^{-1} (10)	Caulerpa ashmeadii (10)
Ichthyotoxic (P.c.), 2.5 μg ml^{-1} (10)	Caulerpa ashmeadii (10)
Ichthyotoxic (P.c.), 2.5 μg ml^{-1} (10)	Caulerpa ashmeadii (10)
Ichthyotoxic (P.c.), 5.0 μg ml^{-1} (10)	Caulerpa ashmeadii (10)
Ichthyotoxic (P.c.), 20 μg ml^{-1} (10)	Caulerpa ashmeadii (10)
Inhibit SUED, 16 μg ml^{-1} (2, 7)	Penicillus dumetosus (7)
Inhibit SUED, 2 μg ml^{-1} (2), antifeedant (E.l.) 800 ppm (5), less ichthyotoxic	Udotea flabellum (5)
Inhibit SUED, 16 μg ml^{-1} (2, 7), ichthyotoxic (7)	Penicillus dumetosus (7)
Inhibit SUED, 16 μg ml^{-1} (8,11)	Halimeda (various sp.) (8)
Inactive against SUED (2)	Penicillus dumetosus (7)
Inhibit SUED, 8 μg ml^{-1} (9)	Tydemania expeditionis (9)
	Tydemania expeditionis (9)
Inhibit SUED, 8 μg ml^{-1} (9)	Chlorodesmis fastigiata (9)
	Udotea flabellum (3)
	Udotea flabellum (3)
Inhibit SUED, 1 μg ml^{-1} (2, 7), ichthyotoxic (7)	Udotea flabellum (7)
Inhibit SUED, 1 μg ml^{-1} (2, 7), ichthyotoxic (7)	Udotea flabellum (7)
Ichthyotoxic (E.p., D.a.), 5 μg ml^{-1} (6), antifeedant (E.p.) 2000 ppm (6), inhibit SUED, 1 μg ml^{-1} (2, 6)	Halimeda (various sp.) (6)
Inhibit SUED, 16 μg ml^{-1} (8)	Halimeda tuna, H. scabra (8)
Inhibit SUED, 8 μg ml^{-1} (9)	Udotea argentea (9)
Inhibit SUED (8)	Halimeda scabra and others (8)
	Caulerpa brownii (9)

present in larger amounts than in a sample prepared in a special manner later (Blunden *et al.*, 1986). The haemagglutinins were enriched from the total extracts but were more closely investigated in only a few cases (Table 7.8). Lectins have been shown to be extremely useful reagents for the study of erythrocyte and tumour cell surfaces. It would appear that marine algae

Table 7.8 Haemagglutinins from algae

Species	Molecular sizes, Da	Remarks	Reference
Rhodophyceae			
Agardhiella tenera	Fraction la 12 000–13 000	Amino acids and glucose analysed	(19)
Carpopeltis flabellata	25 000	'Carnin', glycoprotein, agglutinates rabbit, mouse, horse, weakly human, not sheep, chicken erythrocytes; amino acids analysed	(5)
Cystoclonium purpureum	CPA-1 12 500 subunits with 6000	Amino acids and sugars analysed	(8)
Palmaria palmata	43 000 2 subunits	Glycoprotein	(10)
Soliera chordalis	35 000 various oligomeric fractions	Anti-sialic acid agglutinin	(18)
Ptilota plumosa	PP-I 65 000 PP-II 170 000	Human B blood-group cells and D-galactose specific	(15) (14, 16)
Chlorophyceae			
Boodlea coacta	Boonin A 17 500 monomeric Boonin B 17 500 monomeric Boonin C 17 000 monomeric Boonin D 20 000 monomeric	Specific for rabbit erythrocytes, amino acids and sugars analysed	(4)
Codium fragile	3 different	Sulphated glycoprotein distinguished leukemia L5178Y and L1210 cells	(11)
Codium fragile ssp. *atlanticum* and ssp. *tomentosoides*	60 000 2 different, with 2-mercaptoethanol in subunits of 15 000 separable	Mainly A blood-group preference	(17) (12)

Species	Molecular sizes, Da	Remarks	Reference
Hypnea japonica	Hypnin A 4200 monomeric	Amino acids analysed (no sugars)	.
	Hypnin B, C dimeric or trimeric of A	Amino acids analysed (no sugars)	
	Hypnin D 12 000 monomeric	Amino acids analysed	(3)
Monospora sp.	At least two different agglutinins	Specific for sheep and human O erythrocytes	(9)
Phaeophyceae			
Fucus vesiculosus	2×10^6	Specific for sheep erythrocytes	(1)

For reference list see Table 7.9

are proving to be a potentially fruitful but unexploited source of these interesting and important molecules (Rogers and Blunden, 1980).

Bioactive polymeric carbohydrates

Numerous practical, industrial, and pharmaceutical applications have been found for algal carbohydrates (Hoppe, 1979), but they will not be listed in this review article. Polymeric carbohydrates are probably often the reason for antiviral properties of algal extracts (e.g. Ehresmann *et al.*, 1977; Richards *et al.*, 1978). Their properties as *in vivo* binders of metal pollutants have been repeatedly described (Hoppe, 1979; Fukuda, 1986). The enumeration presented here is deliberately restricted to antitumour activity, blood coagulating activity and some effects of the carrageenans.

Antitumour activity of algal carbohydrates

Jolles *et al.* (1963) were the first to report the influence of degraded sulphated laminaran on experimental tumour growth. During a 12-year screening of 150 Chlorophyceaen, Rhodophyceaen and Phaeophyceaen algae from the Sea of Japan and the Pacific Ocean (Nakazawa *et al.*, 1974) several algae were found with activity on Ehrlich ascites carcinoma. From the dialysed extract and from an ammonium sulphate precipitate of several *Sargassum* species as well as from *Codium pugniformis* (Nakazawa *et al.*, 1976b), they obtained a polysaccharide, a raw protein and a lower molecular weight fraction, each of which inhibited Ehrlich ascites carcinoma as well as sarcoma-180 cells. The findings were confirmed and led to a more exact investigation of several brown algae (see Table 7.9). The substances have

Table 7.9   Brown algal polysaccharides with effects on experimental tumour growth

Species	Proved with	Chemical analysis	Reference
Ecklonia cava	L-1210 leukaemia	Crude fucoidan of low activity	(22)
Eisenia bicyclis	L-1210 leukaemia	Active fucoidan, inactive laminaran	(22)
Laminaria angustata var. *longissima*	L-1210 leukaemia	Crude fucoidan	(22)
Laminaria japonica	L-1210 leukaemia	Crude fucoidan of low activity	(22)
Laminaria japonica var. *ochotensis*	L-1210 leukaemia	Crude fucoidan of low activity	(22)
Sargassum fulvellum	Sarcoma 180	Na-alginate	(2, 20)
Sargassum horneri	Ehrlich ascites	Polysaccharides in sugar protein fractions	(13)
Sargassum kjellmanianum	Sarcoma 180	Polymer carbohydrate, active alginate and sulphated polysaccharides, slightly active laminaran	(21)
Sargassum thunbergii	Ehrlich ascites	Not dialysable acidic polysaccharides	(6, 7)

Reference list for Tables 7.8 and 7.9: (1) Ferreiros and Criado (1983); (2) Fujihara *et al.* (1984); (3) Hori *et al.* (1986a): (4) Hori *et al.* (1986b); (5) Hori *et al.* (1987); (6) Iizima-Mizui *et al.* (1985); (7) Ito and Sugiura (1976); (8) Kamiya *et al.* (1980); (9) Kamiya (1982); (10) Kamiya *et al.* (1982); (11) Kobayashi and Shimizu (1976); (12) Loveless and Rogers (1985); (13) Nakazawa *et al.* (1976a); (14) Rogers *et al.* (1977); (15) Rogers and Blunden (1980); (16) Rogers *et al.* (1982); (17) Rogers *et al.* (1984); (18) Rogers and Topliss (1983); (19) Shiomi *et al.* (1979); (20) Yamamoto *et al.* (1977); (21) Yamamoto *et al.* (1981); (22) Yamamoto *et al.* (1984).

no direct cytotoxic activity, but they activate the immune system (Nakazawa *et al.*, 1976b).

Carbohydrates with blood-anticoagulating activity

The occurrence of substances in algae which inhibit blood clotting has been known for a long time (Elsner *et al.*, 1938) and has been intensively investigated recently. Abdel-Fattah *et al.* (1974) investigated a probably non-uniformly sulphatized polysaccharide (SPS) from *Sargassum linifolium*, sargassan, and identified high anticoagulant activity. Also fucose-containing SPS from various species of the genera *Ecklonia*, *Laminaria*, *Hizikia*, *Lessonia*, *Myagropsis* and other *Sargassum* species (Nishino and Nagumo,

1987) as well as *Fucus vesiculosus* influence various blood coagulating test models. This effect could also be proven for SPS from Chlorophyceae (e.g. *Ulva pertusa*; Fusetani and Hashimoto, 1975) and Rhodophyceae (Deacon-Smith and Rogers, 1982; Efimov *et al.*, 1983) (e.g. the carrageenans from *Chondrus crispus*; Hawkins and Leonard, 1963). The activity is linked to the sulphate ester group because inactive laminaran obtains anticoagulant activity through esterification with sulphuric acid (Hawkins and Leonard, 1958).

Bioactivity of carrageenans

The red alga *Chondrus crispus* is called Irish moss or Carrageen because it is found in great quantities in Carragheen (Waterford, Ireland). The mixture of polymeric SPS isolated from this alga is called carrageenan. Substances with a similar composition and activity have been found in many other red algae. They are used in many forms in industry to produce a gel (Klose and Glicksman, 1968) whose absorption and fate after an oral dose have been repeatedly investigated. Low molecular weight carrageenans are suspected of damaging the mucous membrane of the intestine producing ulcerative colitis and mucosal erosions; however, the high molecular weight species do not produce ulceration (review: Collins, 1978) and oral doses have proved harmless (Bekemeier and Giessler, 1987). Intraperitoneal doses are dangerous; they have local irritant activity and can be fatal for laboratory animals (DiRosa, 1972). The use of carrageenan as an irritant to induce oedema formation in the rat foot (Winter *et al.*, 1962) has become popular as a widely used test for anti-inflammatory activity. The results are only comparable with one another if it is indicated whether κ-carrageenan or the more effective λ-carrageenan from *Chondrus crispus* or ι-carrageenan from *Eucheuma* or other fractions were used (Thomson and Horne, 1976). Carrageenan is also able to induce inflammatory exudates in sites other than the foot; pleurisy (Ghiara *et al.*, 1986), in the brain (Gamache *et al.*, 1986), subcutaneous abscesses and development of a granulomatous tissue (DiRosa, 1972). Carrageenans are able to release kinins, inhibit models of acute non-immune inflammation, depress peptic activity and acidity of gastric juice, suppress ulcerogenesis in rats, inhibit the C1-component of complement system, inhibit delayed hypersensitivity when administered to tuberculous-sensitive guinea pigs, show immunogenic properties and have selective toxicity to macrophages (DiRosa, 1972). The anticoagulant activity is similar to that of other SPSs and heparin.

Cyanophyceae

The prokaryotic Cyanophyceae are found in freshwater or the sea. They often occur in toxic water blooms and the toxins belong to various classes of substances.

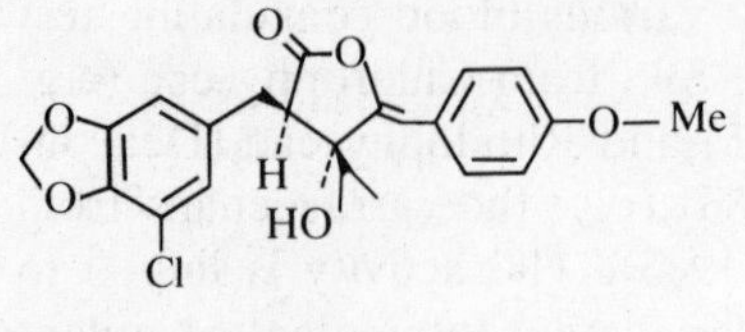

364 cyanobacterin

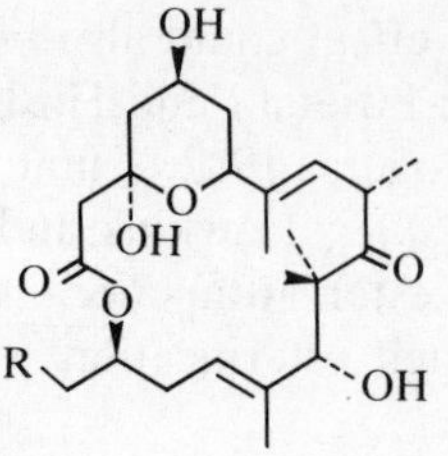

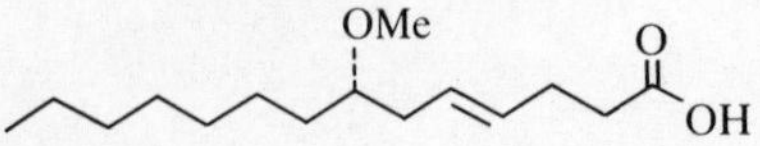

365

366   malyngolide

367 R = CH₃CH₂CH₂CH₂   acutiphycin

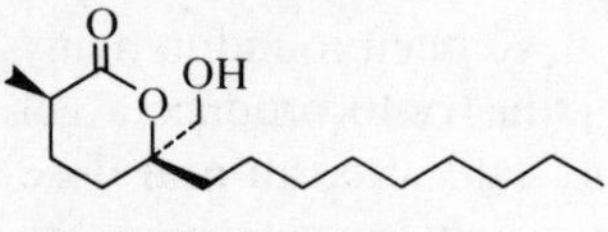

368 R = (H₃C, H)C=C(CH₂, H)   didehydroacutiphycin

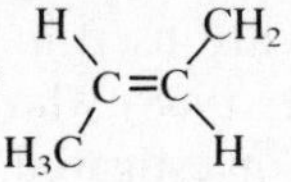

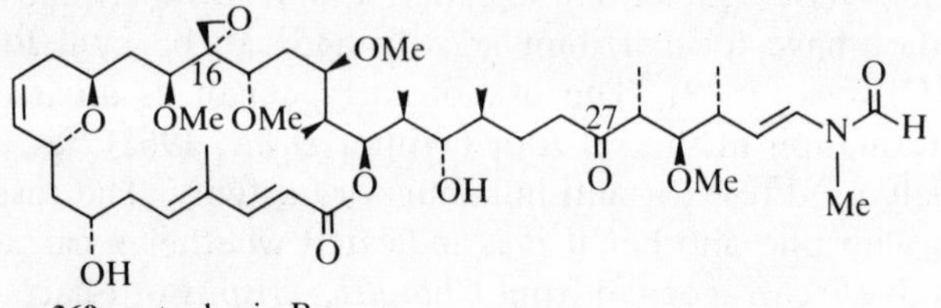

369 scytophycin B

370 A: same as B except at C27

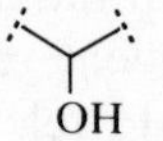

371 C: same as B except at C16

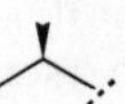

372 D: same as B except at C16

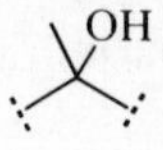

373 E: same as B except at C16

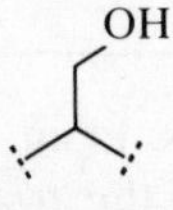

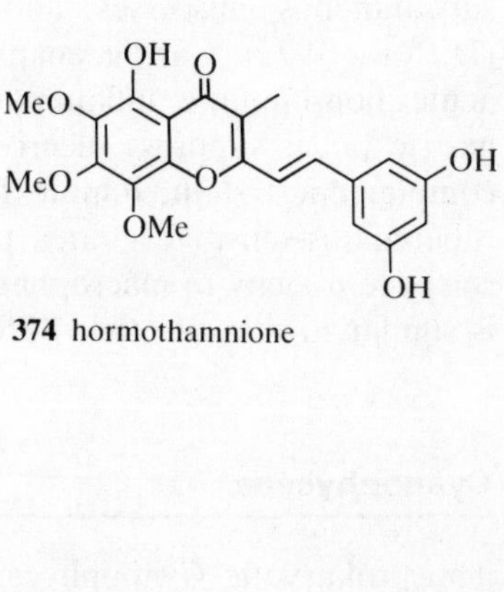

374 hormothamnione

Fig. 7.45 Lactones, acids and phenols from Cyanophyceae.

Bioactive lactons, phenols and acids from Cyanophyceae

The freshwater cyanobacterium *Scytonema hofmanni* produces a secondary metabolite cyanobacterin (Mason *et al.*, 1982) which is chlorinated and has a γ-ylidene-γ-butyrolactone with proven absolute configuration (**364**; Gleason *et al.*, 1986a). Cyanobacterin is a strong inhibitor of algae and higher plants but not of photosynthetic bacteria (Gleason and Paulson, 1984). It inhibits photosynthetic electron transport in the Hill reaction in photosystem II at a concentration (5 μM) generally lower than that of the known photosystem II inhibitor 3-(3,4-dichlorophenyl)-1,1-dimethylurea (DCMU; Gleason and Case, 1986), but at a unique site which is different from that of DCMU (Gleason *et al.*, 1986b). As **364** does not have any influence on bacteria, it is of no interest therapeutically but was patented as a herbicide (Gleason, 1986).

Barchi *et al.* (1984) isolated two new macrolides from the freshwater cyanobacterium *Oscillatoria acutissima*: acutiphycin (**367**) and 20,21-dide-hydroacutiphycin (**368**) with high cytotoxicity against KB and NIH/3T3 cells and significant antineoplastic activity *in vivo* against murine Lewis lung carcinoma. The highly cytotoxic and antimycotic scytophycines A, B, C, D, and E, are macrolides with a C_{21}-lacton ring (**369–373**). They were isolated from a terrestrial cyanobacterium *Scytonema pseudohofmanni* (Moore *et al.*, 1986b).

Lyngbya majuscula is a marine cyanobacterium with different constituents in shallow or deep water varieties which show interesting biological properties (Kashiwagi *et al.*, 1980). It is responsible for a severe erythematous, papulovesicular dermatitis, known as 'swimmers' itch' in Hawaiian waters (Serdula *et al.*, 1982). The most interesting inflammatory and vesicatory substances are the lyngbyatoxins, aplysiatoxins and oscillatoxins, which are described with other tumour promotors in the next section. Substances of less importance are the lipophilic malyngamide A (Cardellina *et al.*, 1979b), B and C (Cardellina *et al.*, 1978) which occur in shallow water varieties and malyngamides D and E which were found in deep water varieties (Mynderse and Moore, 1978a). The *O*-methyl acid (**365**) has antimicrobial activity to Gram-positive bacteria (Gerwick *et al.*, 1987) and the lacton (-)-malyngolide (**366**) is the major antibiotic in the lipid extract of a shallow water variety (Cardellina *et al.*, 1979c).

A γ-pyron hormothamnione (**374**) was isolated from *Hormothamnion enteromorphoides* and was found to be a potent cytotoxic agent which appears to inhibit RNA synthesis (Gerwick *et al.*, 1986). Four of the brominated phenols typical of red algae (e.g. **114, 117**) have also been found in *Calothrix brevissima* or in the culture filtrates, respectively. (Pedersén and DaSilva, 1973), but to date this work has not been verified by others.

Nitrogen-containing substances from Cyanophyceae

A malyngamide D (**375**) from *Lyngbya majuscula*, which is not identical to the malyngamide D described by Mynderse and Moore (1978a) is mildly cytotoxic to KB cells in tissue culture (Gerwick *et al.*, 1987). In a shallow water variety the non-toxic lipodipeptides majusculamides A and B were found but a deep water variety contains the depsipeptide majusculamide C (**377**) that controls the growth of a number of fungal plant pathogens (Moore and Mynderse, 1982; Carter *et al.*, 1984).

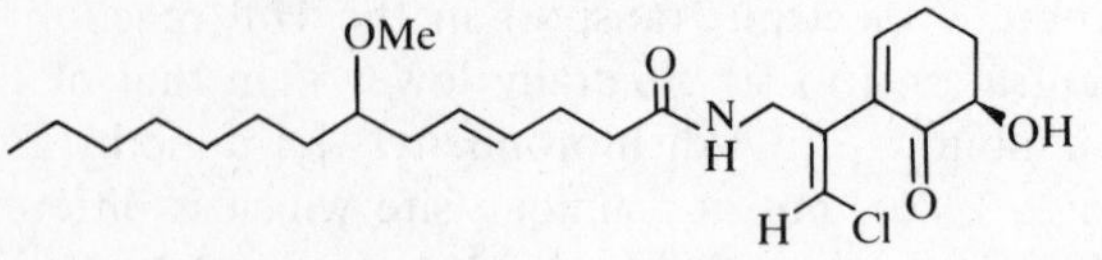

375 malyngamide D

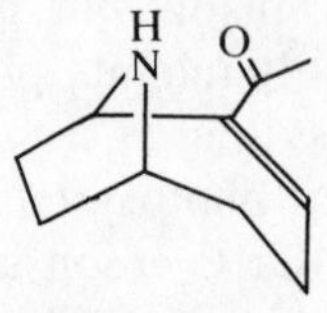

376 anatoxin-a

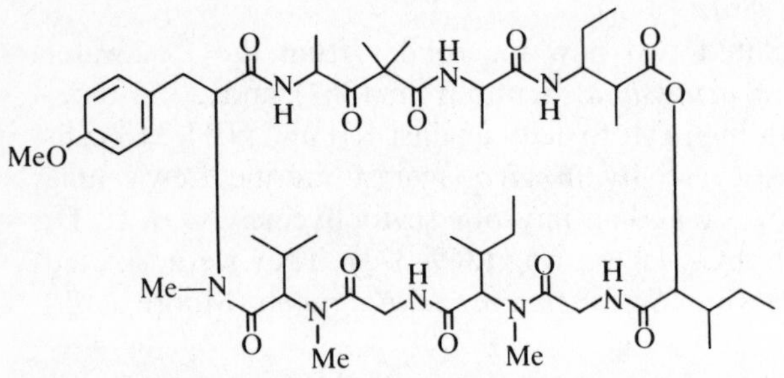

377 majusculamide C

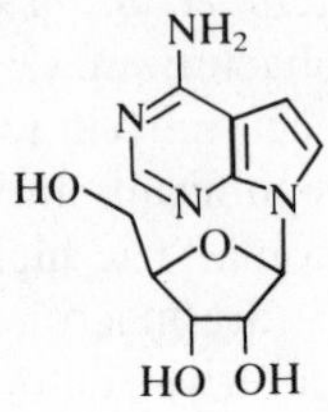

378 tubercidin

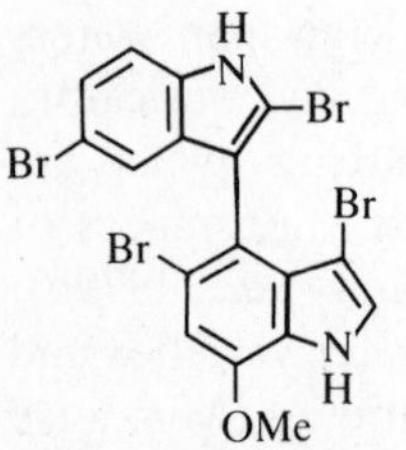

379 rivularin D₃   380   hapalindole A 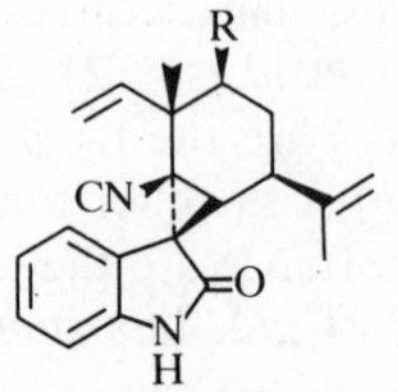   381   hapalindolenine A, R = Cl
                                                          382   hapalindolenine B, R = H

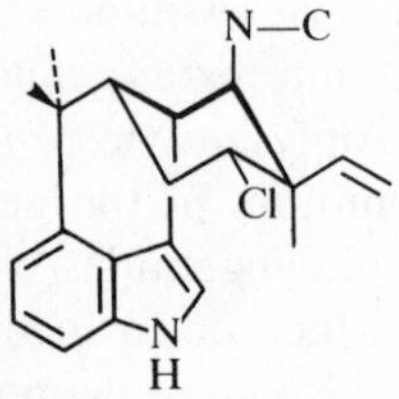

Fig. 7.46 Nitrogen-containing substances from Cyanophyceae.

Anabaena flos-aquae is responsible for numerous poisoning cases of livestock and other animals. This occurs frequently in the lakes of western Canada and western USA. A *very fast death factor* was proved, isolated and resolved from cultures of *A. flos-aquae* (Huber, 1972) and was identified as the alkaloid (+)-anatoxin-a (**376**). It is a potent and stereospecific neurotoxin, which depolarizes neuromuscular blocking and possesses both muscarinic and nicotinic activity (Carmichael *et al.*, 1979). The natural (+)-**376** detects two nicotine high affinity binding sites with a 40-fold difference in affinity constants, while the synthetic (-)-anatoxin-a binds only to one low affinity site (Zhang *et al.*, 1987). Compound **376** is a more potent agonist than acetylcholine or carbamylcholine because of a higher affinity for the acetylcholine receptor. It activates the appearance of channels with the same conductances as acetylcholine-induced channels but with a shorter channel lifetime (Swanson *et al.*, 1986). Different strains of *A. flos-aquae* have been shown to produce different signs of toxicity (Carmichael, 1982); **376** is found in clone number NRC 44–1 while NRC 525–17, for example, produces anatoxin-a(s), a toxin which causes salivation, lacrimation and may act as an anticholinesterase (Mahmood and Carmichael, 1986).

From the six biindoles isolated from the marine *Rivularia firma* (Norton and Wells, 1982), rivularin D_3 (**379**) had anti-inflammatory activity. Moore *et al.* (1984b) isolated an indolalkaloid hapalindole A (**380**) with a Cl and an isonitrile group from an extract of the edaphic cyanobacterium *Hapalosiphon fontinalis* (Stigonemataceae) with antialgal and antimycotic properties. In the meantime, 19 indoles occurring in minor amounts have been identified and the whole group, especially **380**, has been patented due to their antibacterial and antifungal activities (Moore and Patterson, 1986). The indolenines **381** and **382** from *Fischerella* (ATCC 53558) have similar structures but act as inhibitors of ^{3}H-arginine vasopressin binding to kidney tissue (Schwartz *et al.*, 1987).

The cytotoxic antibiotic tubercidin (**378**) already known from *Streptomyces tubercidicus*, was also found in *Tolypothrix byssoidea* (Scytonemataceae) and is responsibie for the cytostatic activity of the algal extract (Barchi *et al.*, 1983). Aphantoxin, which is a mixture containing saxitoxin (**302**), known from the marine dinoflagellate *Gonyaulax* (Ikawa *et al.*, 1982), was isolated from *Aphanizomenon flos-aquae*, which is the cause of many toxic water blooms (Sasner *et al.*, 1981). Several Cyanophyceae produce toxic proteins (Carmichael, 1982). The most common are the hepatotoxins of *Microcystis aeruginosa* but some of the strains, especially those from North America, are neurotoxic. A large number of blooms of *Microcystis* have been implicated in the loss of livestock and wild animals in several countries throughout the world (Carmichael *et al.*, 1985). Reports on *Microcystis* poisoning are often confusing. This may be attributed to possible differences in the effect of various strains, the use of extracts from mixed blooms or differences in investigation methods (review: Carmichael, 1986). All the studies of these toxins report that the toxins are small (about 500–2800 Da), usually cyclic, peptides. Histological examinations of affected livers reveal extensive centrilobular haemorrhagic necrosis. Damaged cells have extensive fragmentations and vesiculations of the membrane (Runnegar *et al.*, 1981). These toxins have been designated type C toxins (microcystis-type-C) or anatoxin from *A. flos-aquae* (Carmichael, 1982), or aerugosin (Gregson and Lohr, 1983), or only microcystis toxin. Other anatoxins than a (b, d) are distinguishable on the basis of pathogenic symptoms in animals (Carmichael and Gorham, 1978). Botes *et al.* (1984) recently suggested that the peptidic hepatotoxins should be called cyanoginosins with a two letter suffix to indicate the two variant amino acid residues in the cyclic heptapeptides. Cyanoginosins-XY are proposed to be cyclo-D-Ala-L-X-erythro-β-methyl-D-isoAsp-L-Y-Adda-D-isoGlu-*N*-methyldehydroAla where X and Y represent variable amino acids and Adda is 3-amino-9-methoxy-2,6,8-

trimethyl-10-phenyl-deca-4,6-dienoic acid. Cyanoginosin-LA was shown to be

D-Ala-L-Leu-erythro-β-Me-O-isoAsp-L-Ala-Adda-D-isoGlu-N-MedhAla

Further examples of this group have been elucidated (Botes *et al.*, 1985). Hepatotoxic effects have also been shown in many other cases but the toxins have not yet been identified (e.g. *Oscillatoria agardhii*; Berg and Søli, 1985).

Tumour promoters from Cyanophyceae

Tumour promoters are substances which are not carcinogenic themselves but which can set off tumour development either in the presence of subliminal doses of a carcinogen or after previous initiation (cocarcinogens). In a process which is well-known but not completely understood tumour formation takes place (Moore, 1982). The effect was first discovered with phorbol esters from *Croton tiglium* (Euphorbiaceae). The highly inflammatory 12-*O*-tetradecanoylphorbol-13-acetate (TPA) serves as a reference substance. Suitable test systems are: irritation of mouse ear, induction of ornithine decarboxylase in mouse skin, adhesion of human promyelocytic leukaemia (HL-60) cells, the two-step mouse skin carcinogenesis, inhibition of differentiation of Friend erythroleukaemia cells induced by dimethyl sulphoxide, induction of early antigen complex formation in Raji cells and aggregation of human blood lymphocytes (review: Moore, 1982; Fujiki *et al.*, 1984a).

During a systematic examination of numerous irritant substances, it turned out that teleocidin and its derivatives from *Streptomyces mediocidicus* (Takashima *et al.*, 1962), lyngbyatoxin A and its derivatives, aplysiatoxin and debromoaplysiatoxin from *Lyngbya majuscula* (Sugimura, 1982), oscillatoxin A and derivatives from *Schizothrix calcicola* and *Oscillatoria nigroviridis* (Mynderse and Moore 1978b) are potent tumour promoters. Lyngbyatoxin A (**383**) was isolated from a *L. majuscula* strain which had been collected leeward of Oahu (Hawaii) (Cardellina *et al.*, 1979a). Compound **383** is identical to teleocidin A and has very similar structural characteristics to teleocidin B (**385**). In **385** the monoterpenoid side-chain is connected to the aromatic ring of the indole under formation of a cyclohexene ring. Teleocidin B and its tetrahydroderivative (**384**) have an activity comparable to dihydroteleocidin B and TPA on the above mentioned test system (Fujiki *et al.*, 1981). Aplysiatoxin (**387**) is a substance with antineoplastic activity which was first isolated from *Stylocheilus longicauda* (Kato and Scheuer, 1974), which feeds preferentially on *L. majuscula*. Aplysiatoxin and debromoaplysiatoxin (**386**) have been found in a deep water variety of *L. majuscula* from Enewetak (Cardellina *et al.*, 1979a). They were identified together with 21-bromooscillatoxin A (**391**), 19,21-dibromooscillatoxin A (**392**), 19-bromoaplysiatoxin (**388**) and

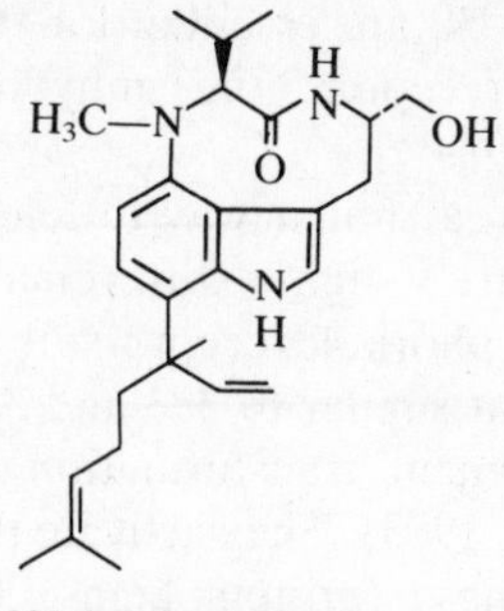

383 lyngbyatoxin A
teleocidin A

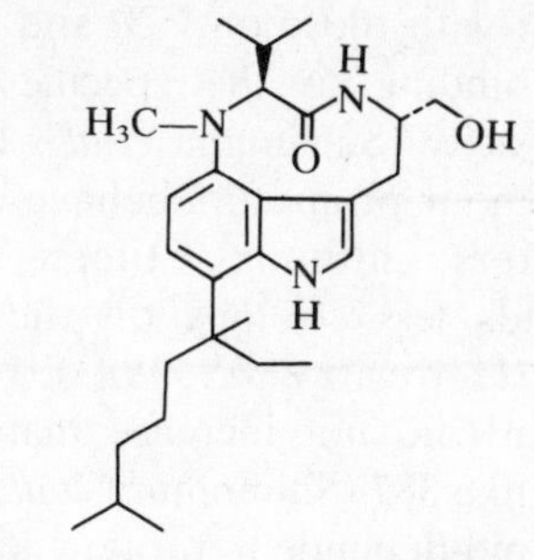

384 tetrahydrolyngbyatoxin A

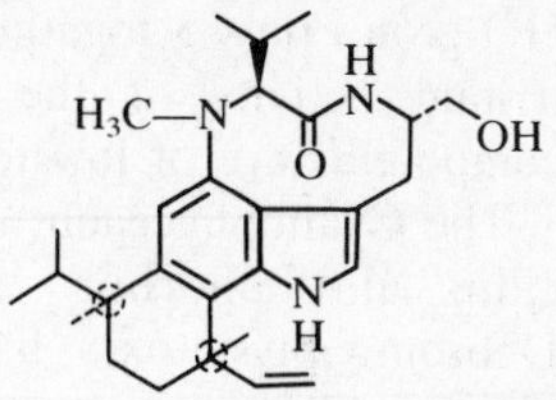

385 teleocidin B

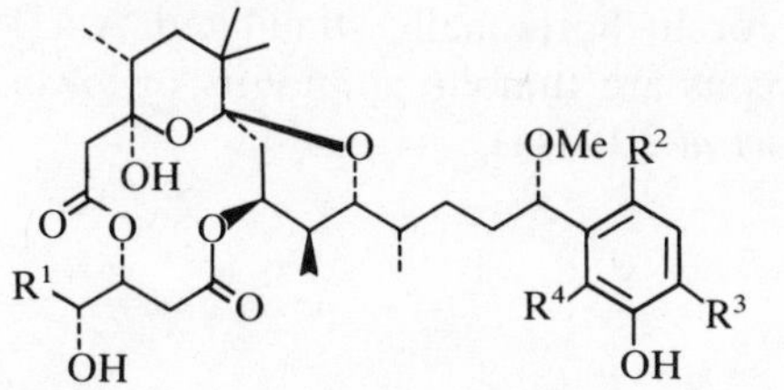

	R¹	R²	R³	R⁴
386 debromoaplysiatoxin	Me	H	H	H
387 aplysiatoxin	Me	Br	H	H
388 19-bromoaplysiatoxin	Me	Br	Br	H
389 19, 21-dibromoaplysiatoxin	Me	Br	Br	Br
390 oscillatoxin A	H	H	H	H
391 21-bromooscillatoxin A	H	Br	H	H
392 19, 21-dibromooscillatoxin A	H	Br	Br	H

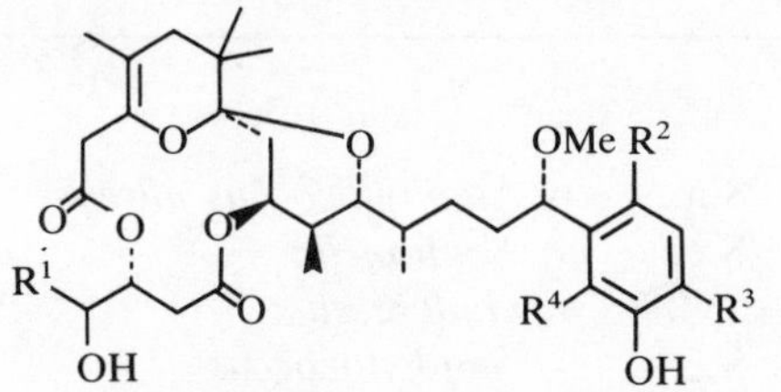

	R¹	R²	R³	R⁴
393 debromoanhydroaplysiatoxin	Me	H	H	H
394 anhydroaplysiatoxin	Me	Br	H	H
395 19-bromoanhydroaplysiatoxin	Me	Br	Br	H
396 anhydrooscillatoxin A	H	H	H	H

Fig. 7.47 Tumour promoters from Cyanophyceae.

other minor constituents of this type (Entzeroth *et al.*, 1985) in *Oscillatoria nigroviridis* and *Schizothrix calcicola* (Mynderse and Moore 1978b). The absolute configuration depicted in Fig. 7.47 was ascertained for these substances (Moore *et al.*, 1984a). Aplysiatoxins and oscillatoxins easily change into the corresponding anhydro-compounds (**393–396**). As the

HO-group on C3 together with those on C20 and C30 are essential for the irritant activity or the binding to the specific receptor, the anhydro-compounds are of low toxicity (Suganuma *et al.*, 1984).

The Cyanophyceaen tumour promotors behave in a similar way to teleo-cidin and phorbol esters, although there are slight differences. Debromoaplysiatoxin binds less strongly on the phorbol receptors than phorbol-12,13-dibutyrate (Horowitz *et al.*, 1983), but similar to **383** and **387** (Moore *et al.*, 1986a) and did not increase malignant transformation or stimulate DNA syntheses like **387** (Shimomura *et al.*, 1983). They activate the Ca^{2+}-activated phospholipid-dependent protein kinase (protein kinase C) (Fujiki *et al.*, 1984b) depending on the Ca^{2+}-concentration (Arcoleo and Weinstein, 1985). They also blocked the tyrosine-specific phosphorylation of the epidermal growth factor receptor in hormonally stimulated A-431 cells (Friedman *et al.*, 1984). Oscillatoxins are tumour promotors of lower potency than the aplysiatoxins (Fujiki *et al.*, 1984a).

Acknowledgements

We thank Mrs. Neu, Kleine and Gassen for their aid in literature supply, drawing the formulae and writing the manuscript and Mrs. de Voy for trans-lating the manuscript.

List of abbreviations used in tables

Bacteria

B.s.	= *Bacillus subtilis*	S.a.	= *Staphylococcus aureus*
B.sph.	= *Bacillus sphaericus*	S.c.	= *Salmonella*
C.f.	= *Citrobacter freundii*		*choleraesius*
D.h.	= *Dreschleria haloides*	S.e.	= *Staphylococcus*
E.c.	= *Escherichia coli*		*epidermidis*
En.c.	= *Enterobacter cloacae*	S.m.	= *Serratia marinorubra/*
M.s.	= *Mycobacterium*		*marcescens*
	smegmatis	SK	= SK–13, unidentified
K.p.	= *Klebsiella pneumoniae*		Gram-positive
L.c.	= *Lagenidium callinectis*	St.f.	= *Streptococcus faecalis*
Le.	= *Leptosphaeria* sp.	St.h.	= *Streptococcus*
Lu.	= *Lulworthia* sp.		*haemolyticus*
M.l.	= *Micrococcus luteus*	V.a.	= *Vibrio anguillarum*
P.a.	= *Pseudomonas*	V.h	= *Vibrio harveyi*
	aeruginosa	V.l.	= *Vibrio leiognathi*
P.s.	= *Proteus* sp.	V.p.	= *Vibrio phosphoreum*
P.v.	= *Proteus vulgaris*	V.s.	= *Vibrio splendida*

Fish

C.au.	= *Carassius auratus*
D.a.	= *Dascyllus aruanus*
E.l.	= *Eupomacentrus leucostictus* (reef-dwelling fish)
E.p.	= *Eupomacentrus planiformis* (damselfish)
G.p.	= *Gambusia patrelis*
P.c.	= *Pomacentrus coeruleus* (damselfish)
P.p.	= *Pomacentrus phillipinus*

Other animals

B.g.	= *Biomphalaria glabrata*
L.p.	= *Lytechinus pinctus* (sea urchin)
SUED	= Sea urchin egg division

Fungi

Al.	= *Alternaria* sp.
A.n.	= *Aspergillus niger*
C.c.	= *Cladosporium cucumerinum*
C.a.	= *Candida albicans*
M.r.	= *Mucor racemosus*

Plants

H.v.	= *Hordeum vulgare*

References

Abbott I A, Williamson E H 1974 *Limu: an Ethnobotanical Study of some Edible Hawaiian Seaweeds*, Pacific Tropical Botanical Garden, Hawaii

Abdel-Fattah A F, Hussein M M-D, Salem H M 1974 Studies of the purification and some properties of Sargassan, a sulphated heteropolysaccharide from *Sargassum linifolium*. *Carbohydr. Res.* **33**: 9–17

Abe S, Kaneda T 1973 Studies on the effect of marine products on cholesterol metabolism in rats–9. Effect of betaines on plasma and liver cholesterol levels. *Bull. Jpn. Soc. Sci. Fish.* **39**: 391–3

Abe S, Kaneda T 1975 Studies on the effect of marine products on cholesterol metabolism in rats–11. Isolation of a new betaine, ulvaline, from a green laver *Monostroma nitidum* and its depressing effect on plasma cholesterol levels. *Bull. Jpn. Soc. Sci. Fish.* **41**: 567–71

Akhunov A A, Gusakova S D, Taubaev T T, Umarov A U 1978 Isolation and antibiotic properties of cis-4,7,10,13-hexadecatetraenoic acid from the alga *Scenedesmus obliquus* UA-2-6. *Khim. Prir. Soedin. (Tashk)* **3**: 379–85

Alam M, Sanduja R, Hossain M B, Helm van der D 1982 *Gymnodinium breve* toxins. 1. Isolation and X-ray structure of O,O-dipropyl(E)-2-(1-methyl-2-oxopropylidene)phosphorohydrazidothioate(E)-oxime from the red tide dinoflagellate *Gymnodinium breve*. *J. Am. Chem. Soc.* **104**: 5232–4

Alam M, Shimizu Y, Ikawa M, Sasner J J 1978 Reinvestigation of the toxin from *Aphanizomenon flos-aquae* by high performance chromatographic methods. *J. Env. Sci. Health* Part A, **A13**: 493–9

Alam M, Trieff N M, Ray S M, Hudson J E 1975 Isolation and partial characterization of toxins from the dinoflagellate *Gymnodinium breve* Davis. *J. Pharm. Sci.* **64**: 865–7

Alvarado A B, Gerwick W H 1985 Dictyol H, a new tricyclic diterpenoid from the brown seaweed *Dictyota dentata*. *J. Nat. Prod.* **48**: 132–4

Amico V, Chillemi R, Oriente G, Piattelli M, Sciuto S, Tringali C 1982 Atti conv. Unita Oper. affer. sottoprog. Ris. Biol. Inquin. Mar. CNR Rome. p 267

Amico V, Oriente G, Piattelli M, Tringali C, Fattorusso E, Magno S, Mayol L 1980 Diterpenes based on the dolabellane skeleton from *Dictyota dichotoma*. *Tetrahedron* **36**: 1409–14

Anderson D M, Lobel P S 1987 The continuing enigma of ciguatera. *Biol. Bull.* **172**: 89–107

Andersson L, Bohlin L 1984 Studies of Swedish marine organisms 3. Procedure for the isolation of the bioactive principle, histamine, from the red alga *Furcellaria lumbricalis* (Huds.) Lamour. *Acta Pharm. Suec.* **21**: 373–6

Arcoleo J P, Weinstein I B 1985 Activation of protein kinase C by tumor promoting phorbol esters, teleocidin and aplysiatoxin in the absence of added calcium. *Carcinogenesis (Lond.)* **6**: 213–17

Aubert M, Aubert J, Gauthier M 1979 Antibiotic substances from marine flora. In Hoppe H A, Levring T, Tanaka Y (eds) *Marine Algae in Pharmaceutical Science*, Walter de Gruyter, Berlin, pp 267–91

Baden D G, Mende T J 1982 Toxicity of two toxins from the Florida red tide marine dinoflagellate, *Ptychodiscus brevis*. *Toxicon* **20**: 457–61

Baden D G, Mende T J, Bikhazi G, Leung I 1982 Bronchoconstriction caused by Florida red tide toxins. *Toxicon* **20**: 929–32

Baden D G, Mende T J, Lichter W, Wellham L L 1981 Crystallization and toxicology of T34: a major toxin from Florida's red tide organism *(Ptychodiscus brevis)*. *Toxicon* **19**: 455–62

Bagnis R, Chanteau S, Chungue E, Hurtel J M, Yasumoto T, Inoue A 1980 Origins of ciguatera fish poisoning: a new dinoflagellate, *Gambierdiscus toxicus* Adachi and Fukuyo, definitively involved as a causal agent. *Toxicon* **18**: 199–208

Baker J T 1984 Seaweeds in pharmaceutical studies and application. *Hydrobiologia* **116/117**: 29–40

Balansard G, Pellegrini M, Cavalli C, Timon-David P, Gasquet M 1983 Diagnosis and anthelminthic action of *Alsidium helminthocorton* (corsican moss), of *Jania rubens* Lamour. and of *Corallina officinalis* Kützg. *Ann. Pharm. Fr.* **41**: 77–86

Banaigs B, Francisco C, González E, Codomier L 1984 Lipids from the brown alga *Cystoseira barbata*. *Phytochemistry* **23**: 2951–2

Banaigs B, Francisco C, González E, Fenical W 1983 Diterpenoid metabolites from the marine alga *Cystoseira elegans*. *Tetrahedron* **39**: 629–38

Barchi J J Jr, Moore R E, Patterson G M L 1984 Acutiphycin and 20,21-didehydroacutiphycin, new antineoplastic agents from the cyanophyte *Oscillatoria acutissima*. *J. Am. Chem. Soc.* **106**: 8193–7

Barchi J J Jr, Norton T R, Furusawa E, Patterson G M L, Moore R E 1983 Identification of cytotoxin from *Tolypothrix byssoidea* as tubercidin. *Phytochemistry* **22**: 2851–2

Bass E L, Pinion J P, Sharif M E 1983 Characteristics of a hemolysin from *Gonyaulax monilata* Howell. *Aquat. Toxicol. (NY)* **3**: 15–22

Bates P, Blunt J W, Hartshorn M P, Jones A J, Munro M H G, Robinson W T, Selwyn C Y 1979 Halogenated metabolites of the red alga *Plocamium cruciferum*. *Aust. J. Chem.* **32**: 2545–54

Bekemeier H, Giessler A J 1987 Thrombosis induction by different carrageenans in rats and mice. *Naturwissenschaften* **74**: 345–6

Berdy J, Aszalos A, McNitt K L 1985 CRC Handbook of Antibiotic Compounds, Antibiotics from Higher Forms of Life, Vol 12 CRC Press, Florida, pp 539–89

Berg K, Søli N E 1985 Effects of *Oscillatoria agardhii*-toxins on blood pressure and isolated organ preparations. *Acta Vet. Scand.* **26**: 374–84

Berland B R, Bonin D J, Cornu A L , Maestrini S Y, Marino J P 1972 The antibacterial substances of the marine alga *Stichochrysis immobilis* (Chrysophyta). *J. Phycol.* **8**: 383–92

Bhakuni D S, Silva M 1974 Biodynamic substances from marine flora. *Botanica Mar.* **17**: 40–51

Biard J F, Verbist J F 1981 Agents antineoplasiques des algues marines: substances cytotoxiques de *Colpomenia peregrina* Sauvageau (Scytosiphonacees). *Plant. Med. Phytother.* **15**: 167–71

Biard J F, Verbist J F, Letourmeux Y, Floch R 1980 Cétols diterpeniques à activité antimicrobienne de *Bifurcaria bifurcata*. *Planta Med.* **40**: 288–94

Biscoe T J, Evans R H, Headley P M, Martin M, Watkins J C 1975 Domoic and quisqualic acids as potent amino acid excitants of frog and rat spinal neurones. *Nature (Lond.)* **255**: 166–7

Bittner M L, Silva M, Paul V J, Fenical W 1985 A rearranged chamigrene derivative and its potential biogenetic precursor from a new species of the marine red algal genus *Laurencia* (Rhodomelaceae). *Phytochemistry* **24**: 987–90

Biziere K, Slevin J T, Zaczek R, Collins J C, Coyle J T 1981 Kainic acid: neurotoxicity and receptor interactions. *Advances in Pharmacology, Proceedings of International Congress*, pp 271–6

Blackman A J, Wells R J 1978 Flexilin and trifarin, terpene 1,4-diacetoxybuta-1,3-dienes from two *Caulerpa* species (Chlorophyta). *Tetrahedron Lett.* **19**: 3063–4

Blunden G, Gordon S M 1986 Betaines and their sulphonio analogues in marine algae. In Round R E, Chapman D J (eds) *Progress in Phycological Research*, Vol 4, Biopress Ltd, pp 39–80

Blunden G, Rogers D J, Loveless R W, Patel A V 1986 Haemagglutinins in marine algae – lectins or phenols? In Bog-Hansen T C, Driessche van E (eds) *Lectins*, Vol 5, Walter de Gruyter, Berlin, pp 139–45

Blunt J W, Hartshorn M P, McLennan T J, Munro M H G, Robinson W T, Yorke S C 1978 Thyrsiferol: a squalene derived metabolite of *Laurencia thyrsifera*. *Tetrahedron Lett.* **19**: 69–72

Bonotto S 1979 List of multicellular algae of commercial use. In Hoppe H A, Levring T, Tanaka Y (eds) *Marine Algae in Pharmaceutical Science*, Walter de Gruyter, Berlin, pp 121–37

Botes D P, Tuinman A A, Wessels P L, Viljoen C C, Kruger H, Williams D H, Santikarn S, Smith R J, Hammond S J 1984 The structure of cyanoginosin-LA, a cyclic heptapeptide toxin from the cyanobacterium *Microcystis aeruginosa*. *Chem. Soc. Perkin. Trans. 1* pp 2311–8

Botes D P, Wessels P L, Kruger H, Runnegar M T C, Santicarn S, Smith R J, Barna J C J, Williams D H 1985 Structural studies on cyanoginosins-LR, -YR, -YA and -YM, peptide toxins from *Microcystis aeruginosa*. *J. Chem. Soc. Perkin. Trans. 1* pp 2747–8

Boyd W C, Almodovar L R, Boyd L G 1966 Agglutinins in marine algae for human erythrocytes. *Transfusion* **6**: 82–3

Brennan M R, Erickson K L 1978 Polyhalogenated indoles from the marine alga *Rhodophyllis membranacea* Harvey. *Tetrahedron Lett.* **19**: 1637–40

Brennan M R, Erickson K L, Minott D A, Pascoe K O 1987 Chamigrane metabolites from a Jamaican variety of *Laurencia obtusa*. *Phytochemistry* **26**: 1053–7

Burreson B J, Moore R E, Roller P 1976 Volatile halogen compounds in the alga *Asparagopsis taxiformis* (Rhodophyta). *J. Agric. Food Chem.* **24**: 856–61

Burreson B J, Woolard F X, Moore R E 1975 Evidence for the biogenesis of halogenated myrcenes from the red alga *Chondrococcus hornemanni*. *Chem. Lett.* **4**: 1111–14

Caccamese S, Azzolina R, Furnari G, Cormaci M, Grasso S 1980 Antimicrobial

and antiviral activities of extracts from mediterranean algae. *Botanica Mar.*
23: 285–8

Caccamese S, Toscano R M, Cerrini S, Gavuzzo E 1982 Laurencianol, a new halogenated diterpenoid from the marine alga *Laurencia obtusa*. *Tetrahedron Lett.*
23: 3415–8

Cardellina II J H, Dalietos D, Marner F J, Mynderse J S, Moore R E 1978 (-)-*Trans*-
7(*S*)-methoxytetradec-4-enoic acid and related amides from the marine cyanophyte
Lyngbya majuscula. *Phytochemistry* **17**: 2091–5

Cardellina II J H, Marner F J, Moore R E 1979a Seaweed dermatitis: structure of
lyngbyatoxin A. *Science* **204**: 193–5

Cardellina II J H, Marner F J, Moore R E 1979b Malyngamide A, a novel chlorinated metabolite of the marine cyanophyte *Lyngbya majuscula*. *J. Am. Chem. Soc.*
101: 240–2

Cardellina II J H, Moore R E, Arnold E V, Clardy J 1979c Structure and absolute
configuration of malyngolide, an antibiotic from the marine blue–green alga
Lyngbya majuscula Gomont. *J. Org. Chem.* **44**: 4039–42

Carmichael W W 1982 Chemical and toxicological studies of the toxic freshwater
cyanobacteria *Microcystis aeruginosa*, *Anabaena flos-aquae* and *Aphanizomenon
flos-aquae*. *S. Afr. J. Sci.* **78**: 367–72

Carmichael W W 1986 Algal toxins. *Adv. Bot. Res.* **12**: 47–101

Carmichael W W, Biggs D F, Peterson M A 1979 Pharmacology of anatoxin-a,
produced by the freshwater cyanophyte *Anabaena flos-aquae* NRC-44-1. *Toxicon*
17: 229–36

Carmichael W W, Gorham P R 1978 Anatoxins from clones of *Anabaena flos-aquae*
isolated from lakes of western Canada. *Mitt. Int. Verein. Theor. Angew. Limnol.*
21: 285–95

Carmichael W W, Jones C L A, Mahmood N A, Theiss W C 1985 Algal toxins and
water-based diseases. *CRC Crit. Rev. Env. Control.* **15**: 275–313

Carter D C, Moore R E, Mynderse J S, Niemczura W P, Todd J S 1984 Structure
of majusculamide C, a cyclic depsipeptide from *Lyngbya majuscula*. *J. Org.
Chem.* **49**: 236–41

Carter G T, Rinehart K L Jr, Li L H, Kuentzel S L, Connor J L 1978 Brominated
indoles from *Laurencia brongniartii*. *Tetrahedron Lett.* **19**: 4479–82

Cembella A D, Sullivan J J, Boyer G L, Taylor F J R, Anderson R J 1987 Variation
in paralytic shellfish toxin composition within the *Protogonyaulax tamarensis/catenella*
species complex; red tide dinoflagellates. *Biochem. Syst. Ecol.* **15**: 171–86

Chapman V J 1979 Seaweeds in pharmaceuticals and medicine: a review. In Hoppe
H A, Levring T, Tanaka Y (eds) *Marine Algae in Pharmaceutical Science*, Walter
de Gruyter, Berlin, pp 139–47

Chou H-N, Shimizu Y 1982 A new polyether toxin from *Gymnodinium breve* Davis.
Tetrahedron Lett. **23**: 5521–4

Chou H-N, Shimizu Y, Duyne G V, Clardy J 1985 Isolation and structures of two
new polycyclic ethers from *Gymnodinium breve* Davis (= *Ptychodiscus brevis*).
Tetrahedron Lett. **26**: 2865–8

Cimino G, Stefano de D, Minale L 1975 *ent*-Chromazonarol a chromansesquiter-
penoid from the sponge *Disidea pallescens*. *Experientia* **31**: 1117–18

Codomier L, Segot M, Combaut G 1981 Influence de composés organiques halogènes
sur la croissance d'*Asparagopsis armata* (Rhodophycée, Bonnemaisoniale).
Botanica Mar. **24**: 509–13

Collins M 1978 Algal toxins. *Microbiol. Rev.* **42**: 725–46

Crews P, Hanke F J, Naylor S, Hogue E R, Kho E, Braslau R 1984a Halogen regi-
ochemistry and substituent stereochemistry determination in marine monoterpenes
by 13C NMR. *J. Org. Chem.* **49**: 1371–7

Crews P, Kho E 1974 Cartilagineal. An unusual monoterpene aldehyde from a marine
alga. *J. Org. Chem.* **39**: 3303–4

Crews P, Kho E 1975 Plocamene B, a new cyclic monoterpene skeleton from a red
marine alga. *J. Org. Chem.* **40**: 2568–70

Crews P, Kho-Wiseman E, Montana P 1978 Halogenated alicyclic monoterpenes from the red alga, *Plocamium. J. Org. Chem.* **43**: 116–20

Crews P, Klein T E, Hogue E R, Myers B L 1982 Tricyclic diterpenes from the brown marine algae *Dictyota divaricata* and *Dictyota linearis. J. Org. Chem.* **47**: 811–5

Crews P, Myers B L, Naylor S, Clason E L, Jacobs R S, Staal G B 1984b Bio-active monoterpenes from red seaweeds. *Phytochemistry* **23**: 1449–51

Crews P, Ng P, Kho-Wiseman E, Pace C 1976 Halogenated monoterpenes of the red alga *Microcladia. Phytochemistry* **15**: 1707–11

Dave M N, Kusumi T, Ishitsuka M, Iwashita T, Kakisawa H 1984 A piscicidal chromanol and a chromenol from the brown alga *Dictyopteris undulata. Heterocycles* **22**: 2301–7

Davies L P, Jamieson D D, Baird-Lambert J A, Kazlauskas R 1984 Halogenated pyrrolopyrimidine analogues of adenosine from marine organisms: pharmacological activities and potent inhibition of adenosine kinase. *Biochem. Pharmacol.* **33**: 347–56

Deacon-Smith R, Rogers D J 1982 Anti-haemostatic activities of British marine algae. Winter Meeting of the British Phycological Society 1982. *Br. Phycol. J.* **17**: 231

Diaz-Pifferer M 1979 Contributions and potentialities of Caribbean marine algae in pharmacology. In Hoppe H A, Levring T, Tanaka Y (eds) *Marine Algae in Pharmaceutical Science*, Walter de Gruyter, Berlin, pp 149–64

DiNovi M, Trainor D A, Nakanishi K, Sanduja R, Alam M 1983 The structure of PB-1, an unusual toxin isolated from the red tide dinoflagellate *Ptychodiscus brevis. Tetrahedron Lett.* **24**: 855–8

DiRosa M 1972 Biological properties of carrageenan. *J. Pharm. Pharmacol.* **24**: 89–102

Efimov V S, Usov A I, Olskaya T S, Balyunis A I, Rozkin M Y 1983 Comparative study of anticoagulant activity of sulfate polysaccharides obtained from Red Sea algae. *Farmakol. Toksikol.* **46**: 61–7

Ehresmann D W, Deig E F, Hatch M T, DiSalvo L H, Vedros N A 1977 Antiviral substances from California marine algae. *J. Phycol.* **13**: 37–40

Elsner M P, Liedmann A, Oppers K 1938 Tierversuche mit einem gerinnungshemmenden Algenstoff aus *Delesseria sanguinea* Lam. *Naunyn-Schmiedeberg's Arch. Exp. Pathol. Pharmacol.* **190**: 510–15

Enoki N, Tsuzuki K, Omura S, Ishida R, Matsumoto T 1983 New antimicrobial diterpenes, dictyol F and epidictyol F from the brown alga *Dictyota dichotoma. Chem. Lett.* **12**: 1627–30

Entzeroth M, Blackman A J, Mynderse J S, Moore R E 1985 Structures and stereochemistries of oscillatoxin B, 31-noroscillatoxin B, oscillatoxin D, and 30-methyloscillatoxin D. *J. Org. Chem.* **50**: 1255–9

Erickson K L 1983 Constituents of *Laurencia*. In Scheuer P J (ed) *Marine Natural Products, Chemical and Biological Perspectives*, Vol 5, Academic Press, New York, pp 132–257

Esping U 1957a A factor inhibiting fertilization of sea urchin eggs from extracts of the alga *Fucus vesiculosus*. 1. The preparation of the factor inhibiting fertilization. *Ark. Kemi* **11**: 107–15

Esping U 1957b A factor inhibiting fertilization of sea urchin eggs from extracts of the alga *Fucus vesiculosus*. 2. The effect of the factor inhibiting fertilization on some enzymes. *Ark. Kemi* **11**: 117–27

Fabregas J, Llovo J, Muñoz A 1985 Hemagglutinins in red seaweeds. *Botanica Mar.* **28**: 517–20

Farber J L 1981 The role of calcium in cell death. *Life Sci.* **29**: 1289–95

Fattorusso E, Magno S, Mayol L, Santacroce C, Sica D 1976 Oxocrinol and crinitol, novel linear terpenoids from the brown alga *Cystoseira crinita*. *Tetrahedron Lett.* **17**: 937–40

Faulkner D J 1976 3β-Bromo-8-epicaparrapi oxide, the major metabolite of *Laurencia obtusa*. *Phytochemistry* **15**: 1993–4

Faulkner D J 1977 Interesting aspects of marine natural products chemistry. *Tetrahedron* **33**: 1421–43

Faulkner D J 1978 Antibiotics from marine organisms. *Top. Antibiot. Chem.* **2**: 9–58

Faulkner D J 1984 Marine natural products: Metabolites of marine algae and herbivorous marine molluscs. *Nat. Prod. Rep.* **1**: 251–80

Faulkner D J 1986 Marine natural products. *Nat. Prod. Rep.* **3**: 1–31

Fenical W 1974 Polyhaloketones from the red seaweed *Asparagopsis taxiformis*. *Tetrahedron Lett.* **15**: 4463–6

Fenical W 1975 Halogenation in the Rhodophyta, a review. *J. Phycol.* **11**: 245–59

Fenical W 1978 Diterpenoids. In Scheuer P J (ed) *Marine Natural Products. Chemical and Biological Perspectives*. Vol 2, Academic Press, New York, pp 174–245

Fenical W, McConnell O J 1975 Chromazonarol and isochromazonarol, new chromanols from the brown seaweed *Dictyopteris undulata (zonarioides)*. *Experientia* **31**: 1004–5

Fenical W, McConnell O J 1976 Simple antibiotics from the red seaweed *Dasya pedicellata* var. *stanfordiana*. *Phytochemistry* **15**: 435–6

Fenical W, McConnell O J, Stone A 1979 Antibiotics and antiseptic compounds from the family Bonnemaisoniaceae (Florideophyceae). *International Seaweed Symposium Proceedings* **9**: 387–400

Fenical W, Norris J N 1975 Chemotaxonomy in marine algae: chemical separation of some *Laurencia* species (Rhodophyta) from the gulf of California. *J. Phycol.* **11**: 104–8

Fenical W, Paul V J 1984 Antimicrobial and cytotoxic terpenoids from tropical green algae of the family Udoteaceae. *Hydrobiologia* **116/117**: 135–40

Fenical W, Sims J J 1974 Cycloeudesmol, an antibiotic cyclopropane containing sesquiterpene from the marine alga *Chondria oppositiclada* Dawson. *Tetrahedron Lett.* **15**: 1137–40

Fenical W, Sims J J, Squatrito D, Wing R M, Radlick P 1973 Zonarol and isozonarol, fungitoxic hydroquinones from the brown seaweed *Dictyopteris zonarioides*. *J. Org. Chem.* **38**: 2383–6

Ferreiros C M, Criado M T 1983 Purification and partial characterization of a *Fucus vesiculosus* agglutinin. *Rev. Esp. Fisiol.* **39**: 51–60

Findlay J A, Patil A D 1984 Antibacterial constituents of the diatom *Navicula delognei*. *J. Nat. Prod.* **47**: 815–18

Findlay J A, Patil A D 1986 Antibacterial constituents of the red alga *Cystoclonium purpureum*. *Phytochemistry* **25**: 548–50

Finer J, Clardy J, Fenical W, Minale L, Riccio R, Battaile J, Kirkup M, Moore R E 1979 Structures of dictyodial and dictyolactone, unusual marine diterpenoids. *J. Org. Chem.* **44**: 2044–7

Francisco C, Banaigs B, Teste B, Cave A 1986 Mediterraneols: A novel biologically active class of rearranged diterpenoid metabolites from *Cystoseira mediterranea* (Phaeophyta). *J. Org. Chem.* **51**: 1115–20

Francisco C, Combaut G, Teste J, Prost M 1978 Eléganolone, nouveau cétol diterpenique linéaire de la phéophycée *Cystoseira elegans*, *Phytochemistry* **17**: 1003–5

Friedman B A, Frackelton A R Jr, Ross A H, Connors J M, Fujiki H, Sugimura T, Rosner M R 1984 Tumor promoters block tyrosine-specific phosphorylation of the epidermal growth factor receptor. *Proc. Nat. Acad. Sci. USA* **81**: 3034–8

Fujihara M, Iizima N, Yamamoto I, Nagumo T 1984 Purification and chemical and physical characterization of an antitumor polysaccharide from the brown seaweed *Sargassum fulvellum*. *Carbohydr. Res.* **125**: 97–106

Fujiki H, Mori M, Nakayasu M, Terada M, Sugimura T, Moore R E 1981 Indole alkaloids: dihydroteleocidin B, teleocidin, and lyngbyatoxin A as members of a new class of tumor promoters. *Proc. Nat. Acad. Sci. USA* **78**: 3872–6

Fujiki H, Suganuma M, Tahira T, Yoshioka A, Nakayasu M, Endo Y, Shudo K, Takayama S, Moore R E, Sugimura T 1984a New classes of tumor promoters: teleocidin, aplysiatoxin, and palytoxin. *Proc. Int. Symp. Princess Takamatsu Cancer Res. Fund* 1983 pp 37–45

Fujiki H, Tanaka Y, Miyake R, Kikkawa U, Nishizuka Y, Sugimura T 1984b Activation of the calcium-activated, phospholipid-dependent protein kinase (protein kinase C) by new classes of tumor promoters: teleocidin and debromoaplysiatoxin. *Biochem. Biophys. Res. Commun.* **120**: 339–43

Fukuda K (Shimizu Chemical K K) 1986 Food from seaweed having ion-exchange capability. *Eur. Pat. Appl.* EP 205.174 JP Appl. 85/129.028 1985. Ref.: C.A. 1987, **106**: 155085n

Fukuyama Y, Miura I, Kinzyo Z, Mori H, Kido M, Nakayama Y, Takahashi M, Ochi M 1985 Eckols, novel phlorotannins with a dibenzo-*p*-dioxin skeleton possessing inhibitory effects on a α2-macroglobulin from the brown alga *Ecklonia kurome* Okamura. *Chem. Lett.* **14**: 739–42

Fukuzawa A, Miyamoto M, Kumagai Y, Masamune T 1986 Ecdysone-like metabolites, 14α-hydroxypinnasterols from the red alga *Laurencia pinnata*. *Phytochemistry* **25**: 1305–7

Fusetani N, Hashimoto Y 1975 Structures of two water soluble hemolysins isolated from the green alga *Ulva pertusa*. *Agric. Biol. Chem.* **39**: 2021–5

Fusetani N, Ozawa C, Hashimoto Y 1976 Studies on marine toxins. 53. Fatty acids as ichthyotoxic constituents of a green alga *Chaetomorpha minima*. *Bull. Jpn. Soc. Sci. Fish.* **42**: 941

Gamache D A, Povlishock J T, Ellis E F 1986 Carrageenan-induced brain inflammation. Characterization of the model. *J. Neurosurg.* **65**: 679–85

Garthwaite J, Garthwaite G, Hajos F 1986 Amino acid neurotoxicity relationship to neuronal depolarization in rat cerebellar slices. *Neuroscience* **18**: 449–60

Gauthier M J, Bernard P, Aubert M 1978a Modification de la fonction antibiotique de deux diatomées marines, *Asterionella japonica* (Cleve) et *Chaetoceros lauderi* (Ralfs), par le dinoflagelle *Prorocentrum micans* (Ehrenberg). *J. Exp. Mar. Biol. Ecol.* **33**: 37–50

Gauthier M J, Bernard P, Aubert M 1978b Production d'un antibiotique lipidique photo-sensible par la diatomée marine *Chaetoceros lauderi* (Ralfs). *Ann. Microbiol.* **129B**: 63–70

Genenah A A, Shimizu Y 1981 Specific toxicity of paralytic shellfish poisons. *J. Agric. Food Chem.* **29**: 1289–91

Gerwick W H, Fenical W 1981 Ichthyotoxic and cytotoxic metabolites of the tropical brown alga *Stypopodium zonale* (Lamouroux) Papenfuss. *J. Org. Chem.* **46**: 22–7

Gerwick W H, Fenical W 1982 Phenolic lipids from related marine algae of the order Dictyotales. *Phytochemistry* **21**: 633–7

Gerwick W H, Fenical W 1983 Spatane diterpenoids from the tropical marine algae *Spatoglossum schmittii* and *Spatoglossum howleii* (Dictyotaceae). *J. Org. Chem.* **48**: 3325–9

Gerwick W H, Fenical W, Engen van D, Clardy J 1980 Isolation and structure of spatol, a potent inhibitor of cell replication from the brown seaweed *Spatoglossum schmittii*. *J. Am. Chem. Soc.* **102**: 7991–3

Gerwick W H, Fenical W, Fritsch N, Clardy J 1979 Stypotriol and stypoldione; ichthyotoxins of mixed biogenesis from the marine alga *Stypopodium zonale*. *Tetrahedron Lett.* **20**: 145–8

Gerwick W H, Fenical W, Sultanbawa M U S 1981 Spatane diterpenoids from the tropical marine alga *Stoechospermum marginatum* (Dictyotaceae). *J. Org. Chem.* **46**: 2233–41

Gerwick W H, Lopez A, Duyne van G D, Clardy J, Ortiz W, Baez A 1986 Hormothamnione, a novel cytotoxic styrylchromone from the marine cyanophyte *Hormothamnion enteromorphoides* Grunow. *Tetrahedron Lett.* **27**: 1979–82

Gerwick W H, Reyes S, Alvarado B 1987 Two malyngamides from the Caribbean cyanobacterium *Lyngbya majuscula*. *Phytochemistry* **26**: 1701–4

Ghiara P, Bartalini M, Tagliabue A, Boraschi D 1986 Antiinflammatory activity of IFN-β in carrageenan-induced pleurisy in the mouse. *Clin. Exp. Immunol.* **66**: 606–14

Gleason F K 1986 Cyanobacterin herbicide. *U.S.* US 4.626.271 1986, Appl. 776.842 1985. Ref.: C.A. 1987, **106**: 45736t

Gleason F K, Case D E 1986 Activity of the natural algicide, cyanobacterin, on angiosperms. *Pl. Physiol.* **80**: 834–7

Gleason F K, Case D E , Sipprell K D, Magnuson T S 1986b Effect of the natural algicide, cyanobacterin, on a herbicide-restistant mutant of *Anacystis nidulans* R2. *Pl. Sci.* **46**: 5–10

Gleason F K, Paulson J L 1984 Site of action of the natural algicide, cyanobacterin, in the blue–green alga, *Synechoccus sp. Arch. Microbiol.* **128**: 273–7

Gleason F K, Porwoll J, Flippen-Anderson J L, George C 1986a X-ray structure determination of the naturally occurring isomer of cyanobacterin. *J. Org. Chem.* **51**: 1615–16

Glombitza K W 1979 Antibiotics from algae. In Hoppe H A, Levring T, Tanaka Y (eds) *Marine Algae in Pharmaceutical Science*, Walter de Gruyter, Berlin, pp 303–42

Glombitza K W, Gerstberger G 1985 Phlorotannins with dibenzodioxin structural elements from the brown alga *Eisenia arborea*. *Phytochemistry* **24**: 543–51

Glombitza K W, Stoffelen H 1972 2,3-Dibrom-5-hydroxybenzyl-l′,4-disulfat (Dikaliumsalz) aus Rhodomelaceen. *Planta Med.* **22**: 391–5

Glombitza K W, Vogels H P 1985 Antibiotics from algae. 35. Phlorotannins from *Ecklonia maxima*. *Planta Med.* **51**: 308–12

Golik J, James J C, Nakanishi K, Lin Y-Y 1982 The structure of brevetoxin C. *Tetrahedron Lett.* **23**: 2535–8

González A G, Arteaga J M, Martín J D, Rodriguez M L , Fayos J, Martines-Ripolls 1978 Two new polyhalogenated monoterpenes from the red alga *Plocamium cartilagineum*. *Phytochemistry* **17**: 947–8

González A G, Darias J, Diáz A, Fourneron J D, Martín J D, Peréz C 1976 Evidence for biogenesis of halogenated chamigrenes from the red alga *Laurencia obtusa*. *Tetrahedron Lett.* **17**: 3051–4

González A G, Darias J, Martín J D 1971 Taondiol, a new component from *Taonia atomaria*. *Tetrahedron Lett.* **12**: 2729–32

González A G, Darias J, Martín J D 1973 Caespitol, a new halogenated sesquiterpene from *Laurencia caespitosa*. *Tetrahedron Lett.* **14**: 2381–4

González A G, Darias J, Martín J D, Norte M 1974a Atomaric acid, a new component from *Taonia atomaria*. *Tetrahedron Lett.* **15**: 3951–4

González A G, Darias J, Martín J D, Peréz C 1974b Revised structure of caespitol and its correlation with isocaespitol. *Tetrahedron Lett.* **15**: 1249–50

González A G, Darias V, Estévez E 1982 Chemotherapeutic activity of polyhalogenated terpenes from Spanish algae. *Planta Med.* **44**: 44–6

González A G, Martín J D, Martin V S, Norte M 1979 Carbon-13 NMR application to *Laurencia* polyhalogenated sesquiterpenes. *Tetrahedron Lett.* **20**: 2719–22

Gregson R P, Kazlauskas R, Murphy P T, Wells R J 1977 New metabolites from the brown alga *Cystophora torulosa*. *Aust. J. Chem.* **30**: 2527–32

Gregson R P, Lohr R R 1983 Isolation of peptide hepatotoxins from the blue–green alga *Microcystis aeruginosa*. *Comp. Biochem. Physiol.* **74C**: 413–17

Gregson R P, Marwood J F, Quinn R J 1979 The occurrence of prostaglandins PGE$_2$ and PGF$_{2\alpha}$ in a plant – the red alga *Gracilaria lichenoides*. *Tetrahedron Lett.* **20**: 4505–6

Güven K C, Bora A, Sunam G 1970 Hordenine from the alga *Phyllophora nervosa*. *Phytochemistry* **9**: 1893

Hall S, Reichardt P B 1984 Cryptic paralytic shellfish toxins. In Ragelis E D (ed) *Seafood Toxins*. Am. Chem. Soc. Symposium Series no 262, Washington DC, pp 113–23

Hall S, Reichardt P B, Neve R A 1980 Toxins extracted from an Alaskan isolate of *Protogonyaulax* sp. *Biochem. Biophys. Res. Commun.* **97**: 649–53

Hansen J A 1973 Antibiotic activity of the Chrysophyte *Ochromonas malhamensis*. *Physiol. Plant.* **29**: 234–8

Hashimoto Y 1979 Marine toxins and other bioactive metabolites. *Jpn. Sci. Soc. Press, Tokyo*, p 269

Hawkins W W, Leonard V G 1958 The physiological activity of laminarine sulphate. *Can. J. Biochem. Physiol.* **36**: 161–70

Hawkins W W, Leonard V G 1963 The antithrombic activity of carrageenan in human blood. *Can. J. Biochem. Physiol.* **41**: 1325–7

Higa T 1981 Phenolic substances. In Scheuer P J (ed) *Marine Natural Products, Chemical and Biological Perspectives*, Vol 4, Academic Press, New York, pp 93–145

Higgs M D 1981 Antimicrobial components of the red alga *Laurencia hybrida* (Rhodophyta, Rhodomelaceae). *Tetrahedron* **37**: 4255–8

Higgs M D, Vanderah D J, Faulkner D J 1977 Polyhalogenated monoterpenes from *Plocamium cartilagineum* from the British coast. *Tetrahedron* **33**: 2775–80

Hille B 1975 The receptor for tetrodotoxin and saxitoxin: a structural hypothesis. *Biophys. J.* **15**: 615–9

Hirschfeld D R, Fenical W, Lin G H Y, Wing R M, Radlick P, Sims J J 1973 Marine natural products. 8. Pachydictyol A, an exceptional diterpene alcohol from the brown alga *Pachydictyon coriaceum*. *J. Am. Chem. Soc.* **95**: 4049–50

Hodgkin J H, Craigie J S, McInnes A G 1966 The occurrence of 2,3-dibromobenzyl alcohol-4,5-disulfate, dipotassium salt, in *Polysiphonia lanosa*. *Can. J. Chem.* **44**: 74–8

Hoppe H A 1979 Marine algae and their products and constituents in pharmacy. In Hoppe H A, Levring T, Tanaka Y (eds) *Marine Algae in Pharmaceutical Science*, Walter de Gruyter, Berlin, Vol 1, 25–119

Hoppe H A 1982 Marine algae: their products and constituents. In Hoppe H A, Levring T (eds) *Marine Algae in Pharmaceutical Science*, Vol 2, Walter de Gruyter, Berlin, pp 1–48

Hori K, Matsuda H, Miyazawa K, Ito K 1987 A mitogenic agglutinin from the red alga *Carpopeltis flabellata*. *Phytochemistry* **26**: 1335–8

Hori K, Miyazawa K, Fusetani N, Hashimoto K, Ito K 1986a Hypnins, low molecular weight peptidic agglutinins isolated from a marine red alga, *Hypnea japonica*. *Biochim. Biophys. Acta* **873**: 228–36

Hori K, Miyazawa K, Ito K 1986b Isolation and characterization of glucoconjugate-specific isoagglutinins from a marine green alga *Boodlea coacta* (Dickie) Murray et De Toni. *Botanica Mar.* **29**: 323–8

Hori K, Miyazawa K, Ito K 1986c Preliminary characterization of agglutinins from seven marine algal species. *Bull. Jpn. Soc. Sci. Fish.* **52**: 323–31

Horowitz A D, Fujiki H, Weinstein I B, Jeffery A, Okin E, Moore R E, Sugimura T 1983 Comparative effects of aplysiatoxin, debromoaplysiatoxin, and teleocidin on receptor binding and phospholipid metabolism. *Cancer Res.* **43**: 1529–35

Huber C S 1972 The crystal structure and absolute configuration of 2,9-diacetyl-9-azabicyclo(4,2,1)non-2,3-ene. *Acta Crystallogr.* **B28**: 2577

Iida H, Nakamura K, Tokunaga T 1985 Dimethyl sulfide and dimethyl-β-propiothetine in sea algae. *Bull. Jpn. Soc. Sci. Fish.* **51**: 1145–50

Iizima-Mizui N, Fujihara M, Himeno J, Komiyama K, Umezawa I, Nagumo T 1985 Antitumor activity of polysaccharide fractions from the brown seaweed *Sargassum kjellmanianum*. *Kitasato Arch. Exp. Med.* **58**: 59–71

Ikawa M, Thomas V M Jr, Buckley L J, Uebel J J 1973 Sulfur and the toxicity of the red alga *Ceramium rubrum* to *Bacillus subtilis*. *J. Phycol.* **9**: 302–4

Ikawa M, Wegener K, Foxall T L, Sasner J J Jr 1982 Comparison of the toxins of the blue–green alga *Aphanizomenon flos-aquae* with the *Gonyaulax* toxins. *Toxicon* **20**: 747–52

Imanaka Y, Kato K (Teijin Ltd) 1986 Novel triterpenes and antitumor agents. *Jpn. Kokai Tokyo Koho* JP 61.152.687 (86.152.687) (Cl C07D493/04) 1986. Ref.: C.A. 1987, **106**: 18883 g

Ingram G A 1985 Lectins and lectin-like molecules in lower plants – 1. Marine algae. *Dev. Comp. Immunol.* **9**: 1–10

Irie T, Suzuki M, Hayakawa Y 1969 Constituents of marine plants. 12. Isolation of aplysin, debromoaplysin and aplysinol from *Laurencia okamurai*. *Bull. Chem. Soc. Jpn.* **42**: 843–4

Irie T, Suzuki M, Kurosawa E, Masamune T 1970 Laurinterol, debromolaurinterol and isolaurinterol, constituents of *Laurencia intermedia* Yamada. *Tetrahedron* **26**: 3271–7

Irie T, Suzuki M, Masamune T 1968 Laurencin, a constituent of *Laurencia glandulifera* Kützing. *Tetrahedron* **24**: 4193–205

Ito H, Sugiura M 1976 Antitumor polysaccharide fraction from *Sargassum thunbergii*. *Chem. Pharm. Bull.* **24**: 1114–5

Ito K, Ishikawa K, Tsuchiya Y 1976 Studies on the depressive factors in *Heterochordaria abietina* affecting the blood cholesterol level in rats. 3. Quantitative determination of iodoamino acids in marine algae and their effects of depressing the blood cholesterol level in rats. *Tohoku J. Agric. Res.* **27**: 53–61

Jacobs R S, Culver P, Langdon R, O'Brien T, White S 1985 Some pharmacological observations on marine natural products. *Tetrahedron* **41**: 981–4

Jacobs R S, White S, Wilson L 1981 Selective compounds derived from marine organisms: effects on cell division in fertilized sea urchin eggs. *Fed. Proc.* **40**: 26–9

Jamieson D D, Taylor K M 1979 Sedative and smooth muscle depressant effects of marine natural products. *Proc. Aust. Physiol. Pharmacol. Soc.* **10**: 275P

Jolles B, Remington M, Andrews P S 1963 Effects of sulphated degraded laminarin on experimental tumor growth. *Brit. J. Cancer*. **17**: 109–15

Kamiya H 1982 Properties of the hemagglutinins in the red alga *Monospora* sp. *Bull. Jpn. Soc. Sci. Fish.* **48**: 1365

Kamiya H, Ogata K, Hori K 1982 Isolation and characterization of a new agglutinin in the red alga *Palmaria palmata* (L.) O. Kuntze. *Botanica Mar.* **25**: 537–40

Kamiya H, Shiomi K, Shimizu Y 1980 Marine biopolymers with cell specificity. 3. Agglutinins in the red alga *Cystoclonium purpureum*: Isolation and characterization. *J. Nat. Prod.* **43**: 136–9

Kashiwagi M, Mynderse J S, Moore R E, Norton T R 1980 Antineoplastic evaluation of Pacific basin marine algae. *J. Pharm. Sci.* **69**: 735–8

Kato T, Kumanireng A S, Ichinose I, Kitahara Y, Kakinuma Y, Nishihira M, Kato M 1975 Active components of *Sargassum tortile* effecting the settlement of swimming larvae of *Coryne uchidai*. *Experientia* **31**: 433–4

Kato Y, Scheuer P J 1974 Aplysiatoxin and debromoaplysiatoxin, constituents of the marine mollusc *Stylocheilus longicauda*. *J. Am. Chem. Soc.* **96**: 2245–6

Kaul P N, Kulkarni S K, Kurosawa E 1978 Novel substances of marine origin as drug metabolism inhibitors. *J. Pharm. Pharmacol.* **30**: 589–90

Kawauchi H, Sasaki T 1978 Isolation and identification of hordenine, *p*-(2-dimethylamino)ethylphenol from *Ahnfeltia paradoxa*. *Bull. Jpn. Soc. Sci. Fish.* **44**: 135–7

Kazlauskas R, Mulder J, Murphy P T, Wells R J 1980 New metabolites from the brown alga *Caulocystis cephalornithos*. *Aust. J. Chem* **33**: 2097–101

Kazlauskas R, Murphy P T, Quinn R J, Wells R J 1976 New laurene derivatives from *Laurencia filiformis*. *Aust. J. Chem.* **29**: 2533–9

Kazlauskas R, Murphy P T, Quinn R J, Wells R J 1977 A new class of halogenated lactones from the red alga *Delisea fimbriata* (Bonnemaisoniaceae). *Tetrahedron Lett.* **18**: 37–40

Kazlauskas R, Murphy P T, Wells R J 1978 Two derivatives of farnesylacetone from brown alga *Cystophora moniliformis*. *Experientia* **34**: 156–7

Kazlauskas R, Murphy P T, Wells R J, Baird-Lambert J A, Jamieson D D 1983 Halogenated pyrrolo[2,3-d]pyrimidine nucleosides from marine organisms. *Aust. J. Chem.* **36**: 165–70

Kazlauskas R, Murphy P T, Wells R J, Blackman A J 1982 Macrocyclic enolethers containing an acetylenic group from the red alga *Phacelocarpus labillardieri*. *Aust. J. Chem.* **35**: 113–20

Kirkup M P, Moore R E 1983 Indole alkaloids from the marine red alga *Martensia fragilis*. *Tetrahedron Lett.* **24**: 2087–90

Klose R, Glicksman M 1968. In Furia F E (ed) *Handbook of Food Additives*. Chemical Rubber Co., Cleveland, Ohio, p 313

Kobayashi J, Ishibashi M, Nakamura H, Ohizumi Y, Yamasu T, Sasaki T, Hirata Y 1986 Amphidinolide-A, a novel antineoplastic macrolide from the marine dino-flagellate *Amphidinium* sp. *Tetrahedron Lett.* **27**: 5755–8

Kobayashi A, Shimizu Y 1976 Marine biopolymers with cell specificity. 1. Purification of agglutinins from *Codium fragile*. *Lloydia* **39**: 474

Kobayashi A, Shimizu Y 1981 Gonyautoxin 8, a cryptic precursor of paralytic shellfish poisons. *Chem. Commun.* pp 827–8

Komura T, Nagayama H 1983 Effect of unsaponifiable lipid components from the red alga *Porphyra tenera* on pancreatic carboxylesterase activity toward triacetin *in vitro*. *Agric. Biol. Chem.* **47**: 383–7

Komura T, Nagayama H, Wada S 1974 Isolation of phytol from the *Hizikia* unsaponifiable matter and its activating effect on pancreatic lipase activity. (Studies on the lipase activator in marine algae part 4). *Nippon Nôgeikagaku Kaishi* **48**: 459–66

Kotaki Y, Oshima Y, Yasumoto T 1985 Bacterial transformation of paralytic shellfish toxins. In Anderson D M, White A W, Baden D G (eds) *Toxic Dinoflagellates, Proc. Int. Conf.*, Elsevier Science Publishing. New York, pp 287–92

Kozakai H, Oshima Y, Yasumoto T 1982 Isolation and structural elucidation of hemolysin from the phytoflagellate *Prymnesium parvum*. *Agric. Biol. Chem.* **46**: 233–6

Kubo I, Matsumoto T, Ichikawa N 1985 Absolute configuration of crinitol, an acyclic diterpene insect growth inhibitor from the brown alga *Sargassum tortile*. *Chem. Lett.* **14**: 249–52

Kuniyoshi M, Yamada K, Higa T 1985 A biologically active diphenyl ether from the green alga *Cladophora fascicularis*. *Experientia* **41**: 523–4

Kurata K, Amiya T 1980 Bis(2,3,6-tribromo-4,5-dihydroxybenzyl)ether from the red alga, *Symphyocladia latiuscula*. *Phytochemistry* **19**: 141–2

Kurosawa E, Fukuzawa A, Irie T 1972 *Trans-* and *cis*-laurediol unsaturated glycols from *Laurencia nipponica* Yamada. *Tetrahedron Lett.* **13**: 2121–4

Kurosawa E, Fukuzawa A, Irie T 1973 Isoprelaurefucin, new bromo compound from *Laurencia nipponica* Yamada. *Tetrahedron Lett.* **14**: 4135–9

Kurosawa E, Suzuki M 1983 Metabolic products of red seaweed *Laurencia*. *Kagaku To* Seibutsu **21**: 23–32

Kutt E C, Martin D F 1975 Report on a biochemical red tide repressive agent. *Env. Lett.* **9**: 195–208

Leary J V, Kfir R, Sims J J, Fulbright D W 1979 The mutagenicity of natural products from marine algae. *Mutat. Res.* **68**: 301–5

Lin Y Y, Risk M, Ray S M, Engen van D, Clardy J, Gorlick J, James J C, Nakanishi K 1981 Isolation and structure of brevetoxin B from the red tide dinoflagellate *Gymnodinium breve*. *J. Am. Chem. Soc.* **103**: 6773–4

Lindequist U, Teuscher E 1987 Marine Wirkstoffe. *Pharmazie* **42**: 1–10

Loveless R W, Rogers D J 1985 Biochemical studies on the lectins from subspecies of *Codium fragile*. Winter Meeting of the British Phycological Society 1985. *Br. Phycol. J.* **20**: 188

Maeda M, Kodama T, Tanaka T, Ohfune Y, Nomoto K, Nishimura K, Fujita T 1984 Insecticidal and neuromuscular activities of domoic acid and its related compounds. *J. Pesticide Sci.* **9**: 27–32

Maeda M, Kodama T, Tanaka T, Yoshizumi H, Nomoto K, Takemoto T, Fujita T 1985 Structures of insecticidal substances isolated from a red alga *Chondria armata*. *Tennen Yuki Kagobutsu Toronkai Koen Yoshishu* **27**: 616–23

Maeda M, Kodama T, Tanaka T, Yoshizumi H, Takemoto T, Nomoto K, Fujita T 1986 Structures of isodomoic acids A, B and C, novel insecticidal amino acids from the red alga *Chondria armata*. *Chem. Pharm. Bull.* **34**: 4892–5

Mahmood N A, Carmichael W W 1986 The pharmacology of anatoxin-a(s), a neurotoxin produced by the freshwater cyanobacterium *Anabaena flos-aquae* NRC 525–17. *Toxicon* **24**: 425–34

Maiti B C, Thomson R H, Mahendran M 1978 The structure of caulerpin, a pigment from *Caulerpa* algae. *J. Chem. Res.* (S): 126–7

Manley S L, Chapman D J 1980 Metabolism of 4-hydroxybenzaldehyde, 3-bromo-4-hydroxybenzaldehyde and bromide by cell-free fractions of the marine red alga *Odonthalia floccosa*. *Phytochemistry* **19**:1453–7

Marderosian der A 1979 Marine pharmacology – focus on algae and microorganisms. In Hoppe H A, Levring T, Tanaka Y (eds) *Marine Algae in Pharmaceutical Science*, Walter de Gruyter, Berlin, pp 165–202

Martín J D, Darias J 1978 Algal sesquiterpenoids. In Scheuer P J (ed) *Marine Natural Products. Chemical and Biological Perspectives*, Vol 1, Academic Press, New York, pp 125–73

Mason C P, Edwards K R, Carlson R E, Pignatello J, Gleason F K, Wood J M 1982 Isolation of chlorine-containing antibiotic from the freshwater cyanobacterium *Scytonema hofmanni*. *Science* **215**: 400–2

McConnell O J, Fenical W 1977a Halogen chemistry of the red alga *Asparagopsis*. *Phytochemistry* **16**: 367–74

McConnell O J, Fenical W 1977b Polyhalogenated 1-octene-3-ones, antibacterial metabolites from the red seaweed *Bonnemaisonia asparagoides*. *Tetrahedron Lett.* **18**: 1851–4

McConnell O J, Fenical W 1978 Ochtodene and ochtodiol: Novel polyhalogenated cyclic monoterpenes from the red seaweed *Ochtodes secundiramea*. *J. Org. Chem.* **43**: 4238–41

McConnell O J, Fenical W 1980 Halogen chemistry of the red alga *Bonnemaisonia*. *Phytochemistry* **19**: 233–47

McGeer E G, Olney J W, McGeer P L (eds) 1978 *Kainic Acid as a Tool in Neurobiology*, Raven Press, New York

McLachlan J 1985 Macroalgae (seaweeds): industrial resources and their utilization. *Pl. Soil* **89**: 137–57

McLachlan J, Craigie J S 1966 Antialgal activity of some simple phenols. *J. Phycol.* **2**: 133–5

Menzel D, Kazlauskas R, Reichelt J 1983 Coumarins in the siphonalean green algal family Dasycladaceae Kützing (Chlorophyceae). *Botanica Mar.* **26**: 23–9

Metting B, Pyne J W 1986 Biologically active compounds from microalgae. *Enzyme Microbial Technol.* **8**: 386–94

Moore R E 1978 Algal nonisoprenoids. In Scheuer P J (ed) *Marine Natural Products. Chemical and Biological Perspectives*, Vol 1, Academic Press, New York, pp 44–124

Moore R E 1981 Constituents of blue–green algae. In Scheuer P J (ed) *Marine Natural Products. Chemical and Biological Perspectives*, Vol 4, Academic Press, New York, pp 1–52

Moore R E 1982 Toxins, anticancer agents, and tumor promoters from marine prokaryotes. *Pure Appl. Chem.* **54**: 1919–34

Moore R E, Blackman A J, Cheuk C E, Mynderse J S, Matsumoto G K, Clardy J, Woodard R W, Craig J C 1984a Absolute stereochemistries of the aplysiatoxins and oscillatoxin A. *J. Org. Chem.* **49**: 2484–9

Moore R E, Cheuk C, Patterson G M L 1984b Hapalindoles: new alkaloids from the blue–green alga *Hapalosiphon fontinalis. J. Am. Chem. Soc.* **106**: 6456–7

Moore R E, Mynderse J S 1982 Majusculamide C. *U.S.* US 4.342.751 Appl.241.812 1981. Ref.: C.A. 1982, **97**: 214751j

Moore R E, Patterson G M L 1986 Hapalindoles. *Eur. Pat. Appl.* EP 171.283, US Appl. 638.847, 1984. Ref.: C.A. 1986, **104**: 205530k

Moore R E, Patterson G M L, Entzeroth M, Morimoto H, Suganuma M, Hakii H, Fujiki H, Sugimura T 1986a Binding studies of ^{3}H lyngbyatoxin A and ^{3}H debromoaplysiatoxin to the phorbol ester receptor in a mouse epidermal particulate fraction. *Carcinogenesis (Lond)* **7**: 641–4

Moore R E, Patterson G M L, Mynderse J S, Barchi J J Jr, Norton T R, Furusawa E, Furusawa S 1986b Toxins from cyanophytes belonging to the Scytonemataceae. *Pure Appl. Chem.* **58**: 263–71

Murakami Y, Oshima Y, Yasumoto T 1982 Identification of okadaic acid as a toxic component of a marine dinoflagellate *Prorocentrum lima. Bull. Jpn. Soc. Sci. Fish.* **48**: 69–72

Murata M, Sano M, Iwashita T, Naoki H, Yasumoto T 1986 The structure of pectenotoxin-3, a new constituent of diarrhetic shellfish toxins. *Agric. Biol. Chem.* **50**: 2693–5

Mynderse J S 1975 Halogenated Monoterpenes from *Plocamium cartilagineum* Dixon and *Plocamium violaceum* Farlow. PhD Thesis, San Diego, California

Mynderse J S, Faulkner D J, Finer J, Clardy J 1975 (1R,2S,4S,5R)-1-Bromo-*trans*-2-chlorovinyl-4,5-dichloro-1,5-dimethyl cyclohexane, a new monoterpene skeletal type from the red alga *Plocamium violaceum. Tetrahedron Lett.* **16**: 2175–8

Mynderse J S, Moore R E 1978a Malyngamides D and E, two *trans*-7-methoxy-9-methylhexadec-4-enamides from a deep water variety of the marine cyanophyte *Lyngbya majuscula. J. Org. Chem.* **43**: 4359–63

Mynderse J S, Moore R E 1978b Toxins from blue–green algae, structures of oscillatoxin A and three related bromine-containing toxins. *J. Org. Chem.* **43**: 2301–3

Nadal N G M, Rodríguez L U, Dolagaray J I 1966 Low toxic effect of antimicrobial substances from marine algae. *Botanica Mar.* **9**: 62–3

Nadler J V, Perry B W, Cotman C W 1978 Intraventricular kainic acid preferentially destroys hippocampal pyramidal cells. *Nature (Lond.)* **271**: 676–7

Nakatsu T, Ravi B N, Faulkner D J 1981 Antimicrobial constituents of *Udotea flabellum. J. Org. Chem.* **46**: 2435–8

Nakayama M, Fukuoka Y, Nozaki H, Matsuo A, Hayashi S 1980 Structure of (+)-kjellmanianone, a highly oxygenated cyclopentenone from the marine alga *Sargassum kjellmanianum. Chem. Lett.* **9**: 1243–6

Nakazawa S, Abe F, Kuroda H, Kohno K, Higashi T, Umezaki I 1976a Antitumor effect of water-extracts from marine algae (3) *Codium pugniformis* Okamura. *Chemotherapy* **24**: 448–50

Nakazawa S, Abe F, Kuroda H, Kohno K, Higashi T, Umezaki I 1976b Antitumor effect of water-extracts from marine algae (2) *Sargassum horneri* (Turner) C. Agardh. *Chemotherapy* **24**: 443–7

Nakazawa S, Kuroda H, Abe F, Nishino T, Ohtsuki M, Umezaki I 1974 Antitumor effect of water-extracts from marine algae (1). *Chemotherapy* **22**: 1435–42

Narahashi T 1972 Mechanism of action of tetrodotoxin and saxitoxin on excitable membranes. *Fed. Proc.* **31**: 1124–31

Natsuno T, Mikami M, Saito K 1986 A potent inhibitor of bacterial growth from the seaweed, *Chondrus crispus. Shigaku* **74**: 412–21

Naylor S, Hanke F J, Manes L V, Crews P 1983 Chemical and biological aspects of marine monoterpenes. *Fortschr. Chem. Org. Naturst.* **44**: 189–241

Nishino T, Nagumo T 1987 Sugar constituents and blood-anticoagulant activities of fucose-containing sulfated polysaccharides in nine brown seaweed species. *Nippon Nôgeikagaku Kaishi* **61**: 362–3

Nishio S, Noguchi T, Onoue Y, Maruyama J, Hashimoto K, Seto H 1982. Isolation and properties of gonyautoxin-5, an extremely low-toxic component of paralytic shellfish poison. *Bull. Jpn. Soc. Sci. Fish.* **48**: 959–65

Noguchi T, Onoue Y, Maruyama J, Hashimoto K, Nishio S, Ikeda T 1983 The new paralytic shellfish poisons from *Protogonyaulax catenella. Bull. Jpn. Soc. Sci. Fish.* **49**: 1931

Norton R S, Wells R J 1982 A series of chiral polybrominated biindoles from the marine blue–green alga *Rivularia firma.* Application of ^{13}C NMR spin-lattice relaxation data and 13– ^{1}H coupling constants to structure elucidation. *J. Am. Chem. Soc.* **104**: 3628–35

Nozaki H, Fukuoka Y, Matsuo A, Soga O, Nakayama M 1980 Structure of sargassumlactam, a new β,γ-unsaturated-γ-lactam, from the marine alga *Sargassum kjellmanianum. Chem. Lett.* **9**: 1453–4

O'Brien E T, White S, Jacobs R S, Boder G B, Wilson L 1984 Pharmacological properties of a marine natural product, stypoldione, obtained from the brown alga *Stypopodium zonale. Hydrobiologia* **116/117**: 141–5

Ochi M, Asao K, Kotsuki H, Miura I, Shibata K 1986 Amijitrienol and 14-deoxy-isoamijiol, two diterpenoids from the brown seaweed *Dictyota linearis. Bull. Chem. Soc. Jpn.* **59**: 661–2

Ochi M, Kotsuki H, Inoue S, Taniguchi M, Tokoroyama T 1979a Isolation of 2-(3,7,11-trimethyl-2,6,10-dodecatrienyl)hydroquinone from the brown seaweed *Dictyopteris undulata. Chem. Lett.* **8**: 831–2

Ochi M, Kotsuki H, Muraoka K, Tokoroyama T 1979b The structure of yahazunol, a new sesquiterpene-substituted hydroquinone from the brown seaweed *Dictyopteris undulata* Okamura. *Bull. Chem. Soc. Jpn.* **52**: 629–30

Ohizumi Y 1987 Pharmacological actions of the marine toxins ciguatoxin and maitotoxin isolated from poisonous fish. *Biol. Bull.* **172**: 132–6

Ohta K 1977 Antimicrobial compounds in the marine red alga *Beckerella subcostatum. Agric. Biol. Chem.* **41**: 2105–6

Ohta K 1979 Chemical studies on biologically active substances in seaweeds. In Jensen A, Stein J R (eds) *Proc. Int. Seaweed Symp. 9.* Science Press. Princeton, pp 401–11

Ohta K, Takagi M 1977 Antimicrobial compounds of the marine red alga *Marginisporum aberrans. Phytochemistry* **16**: 1085–6

Okuda R K, Klein D, Kinnel R B, Li M, Scheuer P J 1982 Marine natural products: the past twenty years and beyond. *Pure Appl. Chem.* **54**: 1907–14

Oshima Y, Buckley L J, Alam M, Shimizu Y 1977 Heterogeneity of paralytic shellfish poisons. Three new toxins from cultured *Gonyaulax tamarensis* cells, *Mya arenaria*, and *Saxidomus giganteus. Comp. Biochem. Physiol.* **57c**: 31–4

Oshima Y, Fallon W E, Shimizu Y, Noguchi T, Hashimoto Y 1976 Toxins of the *Gonyaulax* sp. and infested bivalves in Owase bay. *Bull. Jpn. Soc. Sci. Fish.* **42**: 851–6

Otsuka Pharmaceutical Co. Ltd 1983a *Jpn Kokai Tokkyo Koho* JP 58.118.580 (83.118.580) (Cl. C07D319/24). Ref.: C.A. 1984, **100**: 12635g

Otsuka Pharmaceutical Co. Ltd 1983c *Jpn Kokai Tokkyo Koho* JP 58.118.591 (83.118.591) (Cl. C07D493/04). Ref.: C.A. 1984, **100**: 12637j

Otsuka Pharmaceutical Co. Ltd 1983b *Jpn Kokai Tokkyo Koho* JP 58.118.581 (83.118.581) (Cl. C07D319/24). Ref.: C.A. 1984. **100**: 12636h

Ozawa H, Gomi Y, Otsuki I 1967 Pharmacological studies on laminine monocitrate. *Yakugaku Zasshi* **87**: 935–9

Paul J H 1982 Isolation and characterization of a *Chlamydomonas* L-asparaginase. *Biochem. J.* **203**: 109–15

Paul V J, Fenical W 1982 Toxic feeding deterrents from the tropical marine alga *Caulerpa bikinensis* (Chlorophyta). *Tetrahedron Lett.* **23**: 5017–20

Paul V J, Fenical W 1983 Isolation of halimedatrial: Chemical defense adaptation in the calcareous reef-building alga *Halimeda. Science* **221**: 747–9

Paul V J, Fenical W 1984a Bioactive terpenoids from Caribbean marine algae of the genera *Penicillus* and *Udotea* (Chlorophyta). *Tetrahedron* **40**: 2913–8

Paul V J, Fenical W 1984b Novel bioactive diterpenoid metabolites from tropical marine algae of the genus *Halimeda* (Chlorophyta). *Tetrahedron* **40**: 3053–62

Paul V J, Fenical W 1985 Diterpenoid metabolites from Pacific marine algae of the order Caulerpales (Chlorophyta). *Phytochemistry* **24**: 2239–43

Paul V J, Littler M M, Littler D S, Fenical W 1987 Evidence for chemical defense in tropical green alga *Caulerpa ashmeadii* (Caulerpaceae: Chlorophyta): Isolation of new bioactive sesquiterpenoids. *J. Chem. Ecol.* **13**: 1171–85

Paul V J, McConnell O J, Fenical W 1980 Cyclic monoterpenoid feeding deterrents from the red marine alga *Ochtodes crockeri. J. Org. Chem.* **45**: 3401–7

Paul V J, Sun H H, Fenical W 1982 Udoteal, a linear diterpenoid feeding deterrent from the tropical green alga *Udotea flabellum. Phytochemistry* **21**: 468–9

Pedersén M, DaSilva E J 1973 Simple brominated phenols in the blue–green alga *Calothrix brevissima* West. *Planta* **115**: 83–6

Pedersén M, Fries E 1975 Bromophenols in *Fucus vesiculosus. Z. Pflanzenphysiol.* **74**: 272–4

Pesando D 1972 Etude chimique et structurale d'une substance lipidique antibiotique produite par une diatomée marine: *Asterionella japonica. Revue Intern. Océanogr. Méd.* **25**: 49–69

Pierce R H 1986 Red tide (*Ptychodiscus brevis*) toxin aerosols: a review. *Toxicon* **24**: 955–65

Pohl P, Zurheide F 1979 Fatty acids and lipids of marine algae and the control of their biosynthesis by environmental factors. In Hoppe H A, Levring T, Tanaka Y (eds) *Marine Algae in Pharmaceutical Science*, Walter de Gruyter, Berlin, pp 437–523

Poli M A, Mende T J, Baden D G 1986 Brevetoxins, unique activators of voltage-sensitive sodium channels, bind to specific sites in rat brain synaptosomes. *Mol. Pharmacol.* **30**: 129–35

Pratt R, Daniels T C, Eiler J B, Gunnison J B, Kumler W D, Oneto J F, Strait L A, Spoehr H A, Hardin G J, Milner H W, Smith J H C, Strain H H 1944 Chlorellin, an antibacterial substance from *Chlorella. Science* **99**: 351–2

Ragan M A 1984 Bioactivities in marine genera of atlantic Canada: The unexplored potential. *Proc. NS Inst. Sci.* **34**: 83–132

Ragan M A 1985 The high-molecular-weight polyphloroglucinols of the marine brown alga *Fucus vesiculosus* L.: degradative analysis. *Can. J. Chem.* **63**: 294–303

Ragan M A, Glombitza K W 1986 Phlorotannins, brown algal polyphenols. In Round R E, Chapman D J (eds) *Progress in Phycological Research*, Vol 4, Biopress Ltd, pp 129–41

Reiner E, Topliff J, Wood J D 1967 Hypocholesterolemic agents derived from sterols of marine algae. *Can. J. Biochem. Physiol.* **40**: 1401–6

Richards J T, Kern E R, Glasgow L A, Overall J C Jr, Deign E F, Hatch M T 1978 Antiviral acitivity of extracts from marine algae. *Antimicrob. Agents Chemother.* **14**: 24–30

Risk M, Lin Y Y, MacFarlane R D, Sadagopa-Ramanujam V M, Smith L L, Trieff N M 1979 Purification and chemical studies on a major toxin from *Gymnodinium breve.* In Taylor D L, Seliger H H (eds) *Toxic Dinoflagellate Blooms*, Elsevier Science Publishing, New York, pp 335–44

Rivera P, Astudillo L, Rovirosa J, San-Martín A 1987 Halogenated monoterpenes of the red alga *Shottera nicaensis. Biochem. Syst. Ecol.* **15**: 3–4

Roche J, Yagi Y, Michel R, Lissitstky S, Eysseric-Lafon M 1951 Sur la characterisation de la monobromotyrosine et de la thyroxine dans les Gorgonines. *Bull. Soc. Chem. Biol.* **33**: 526–31

Rogers D J, Blunden G 1980 Structural properties of the anti-B lectin from the red alga *Ptilota plumosa* (Huds.) C. Ag. *Botanica Mar.* **23**: 459–62

Rogers D J, Blunden G, Evans P R 1977 *Ptilota plumosa*, a new source of a bloodgroup B specific lectin. *Med. Lab. Sci.* **34**: 193–200

Rogers D J, Blunden G, Guiry M D, Northcott M J 1982 Evaluation of *Ptilota plumosa* from Ireland as a source of haemagglutinin. *Botanica Mar.* **25**: 399–400

Rogers D J, Blunden G, Topliss J A, Guiry M D 1980 A survey of some marine organisms for haemagglutinins. *Botanica Mar.* **23**: 569–77

Rogers D J, Loveless R W 1985 'Haemagglutinins' of the Phaeophyceae and nonspecific aggregation phenomena by polyphenols. *Botanica Mar.* **28**: 133–7

Rogers D J, Loveless R W, Northcott M J 1984 Specificity studies on lectin-type haemagglutinins from *Codium fragile*. *J. Pharm. Pharmacol.* **36**: 71P

Rogers D J, Topliss J A 1983 Purification and characterization of an anti-sialic acid agglutinin from the red alga *Solieria chordalis* (C. Ag.) J. Ag. *Bot. Mar.* **26**: 301–5

Roller P, Au K, Moore R E 1971 Isolation of *S*-(3-oxoundecyl) thioacetate, bis-(3-oxoundecyl) disulphide, (-)-3-hexyl-4,5-dithiacycloheptanone and *S*-(*trans*-3-oxoundec-4-enyl) thioacetate from *Dictyopteris*. *Chem. Commun.* pp 503–4

Rosa de S, Stefano de S, Macura S, Trivellone E, Zavodnik N 1984 Chemical studies of north Adriatic seaweeds – 1. New dolabellane diterpenes from the brown alga *Dilophus fasciola*. *Tetrahedron* **40**: 4991–5

Rose A F, Pettus J A Jr, Sims J J 1977 Marine natural products–13. Isolation and synthesis of some halogenated ketones from the red seaweed *Delisea fimbriata*. *Tetrahedron Lett.* **18**: 1847–50

Runnegar M T, Falconer I R, Silver J 1981 Deformation of isolated rat hepatocytes by a peptide from the hepatotoxin from the blue–green alga *Microcystis aeruginosa*. *Naunyn-Schmiedebergs Arch. Pharmacol.* **317**: 268–72

Sakemi S, Higa T, Jefford C W, Bernardinelli G 1986 Venustatriol. A new anti-viral triterpene tetracyclic ether from *Laurencia venusta*. *Tetrahedron Lett.* **27**: 4287–90

San-Martin A, Rovirosa J 1986 Variations in the halogenated monoterpene metabolites of *Plocamium cartilagineum* of the Chilean coast. *Biochem. Syst. Ecol.* **14**: 459–61

Sasner J J Jr, Ikawa M, Foxall T L, Watson W H 1981 Studies on aphantoxin from *Aphanizomenon flos-aquae* in New Hampshire. In Carmichael W W (ed) *The Water Environment – Algal Toxins and Health*, Plenum Press, New York, pp 389–404

Schantz E J 1981 Poisons produced by dinoflagellates – a review. In Carmichael W W (ed) *The Water Environment – Algal Toxins and Health*, Plenum Press, New York, pp 25–36

Schantz E J, Ghazarossian V E, Shnoes H K, Strong F M, Springer J P, Pezzanite J O, Clardy J 1975 The structure of saxitoxin. *J. Am. Chem. Soc.* **97**: 1238–9

Schantz E J, Mold J D, Stanger D W, Shavel J, Riel F J, Bowden P J, Lynch J M, Wyler R S, Riegel B R, Sommer H 1957 Paralytic shellfish poison. 6. A procedure for the isolation and purification of the poison from the toxic clams and mussel tissues. *J. Am. Chem. Soc.* **79**: 5230–5

Schlenk D, Gerwick W H 1987 Dilophic acid, a diterpenoid from the tropical brown seaweed *Dilophus guineensis*. *Phytochemistry* **26**: 1081–4

Schwartz R E, Hirsch C F, Springer J P, Pettibone D J, Zink D L 1987 Unusual cyclopropane-containing hapalindolinones from a cultured cyanobacterium. *J. Org. Chem.* **52**: 3704–6

Serdula M, Bartolini G, Moore R E, Gooch J, Wiebenga N 1982 Seaweed itch on windward Oahu. *Hawaii Med. J.* **41**: 200–1

Seshadri R, Sieburth J M 1971 Cultural estimation of yeasts on seaweeds. *Appl. Microbiol.* **22**: 507–12

Seto A, Kitagawa K, Yamashita M, Fujita T (Nisshin Oil Mills Ltd) 1984 Preparation of glycolipids rich in eicosapentaenoic acid from *Chlorella minutissima. Jpn. Kokai Tokkyo Koho* JP 61 63.624 (86 63.624). Ref.: C.A. 1986, **105**: 30041t

Shiba T, Wakamiya T (Ajinomoto Co. Inc) 1985 1-Amino-2-(guanidinomethyl) cyclopropane-1-carboxylic acid. *Jpn. Kokai Tokkyo Koho* JP 60.246.360 (85.246.360) Appl,. 84/102.281 1984. Ref.: C.A. 1986, **104**: 204131u

Shilo M 1981 The toxic principles of *Prymnesium parvum.* In Carmichael W W (ed) *The Water Environment – Algal Toxins and Health,* Plenum Press, New York, pp 37–47

Shimizu Y 1978 Dinoflagellate toxins. In Scheuer P J (ed) *Marine Natural Products, Chemical and Biological Perspectives,* Vol 1, Academic Press, New York, pp 1–44

Shimizu Y 1980 Red tide toxins – paralytic shellfish poisons produced by *Gonyaulax* organisms. *Kagaku To Seibutsu* **18**: 792–9

Shimizu Y 1982 Recent progress in marine toxin research. *Pure Appl. Chem.* **54**: 1973–80

Shimizu Y 1984 Paralytic shellfish poisons. In Zechmeister L, Herz W, Griesebach H, Kirby G W (eds) *Progress in the Chemistry of Organic Natural Products,* Vol 45, Springer, Wien, pp 237–64

Shimizu Y, Alam M, Oshima Y, Fallon W E 1975 Presence of four toxins in red tide infested clams and cultured *Gonyaulax tamarensis* cells. *Biochem. Biophys. Res. Commun.* **66**: 731–7

Shimizu Y, Chou H-N, Bando H, Duyne van G, Clardy J C 1986 Structure of brevetoxin A (GB-1 toxin), the most potent toxin in the Florida red tide organism *Gymnodinium breve (Ptychodiscus brevis). J. Am. Chem. Soc.* **108**: 514–15

Shimizu Y, Kamiya H 1983 Bioactive marine biopolymers. In Scheuer P J (ed) *Marine Natural Products. Chemical and Biological Perspectives,* Vol 5, Academic Press, New York, pp 391–427

Shimomura K, Mullinix M G, Kakunaga T, Fujiki H, Sugimura T 1983 Bromine residue at a hydrophilic region influences the biological activity of aplysiatoxin, a tumor promoter. *Science* **222**: 1242–4

Shinho K (Chlorella Ind Co Ltd) 1984 Antitumor glycoproteins from *Chlorella* and other species. *Jpn. Kokai Tokkyo Koho* JP 61 69.728 (86 69.728). Ref.: C.A. 1986, **105**: 30034t

Shinozaki H 1978 Discovery of novel actions of kainic acid and related compounds. In McGeer E G, Olney J W, McGeer P L (eds) *Kainic Acid as a Tool in Neurobiology,* Raven Press, New York

Shiomi K, Kamiya H, Shimizu Y 1979 Purification and characterization of an agglutinin in the red alga *Agardhiella tenera. Biochim. Biophys. Acta* **576**: 118–27

Sieburth J M 1961 Antibiotic properties of acrylic acid, a factor in the gastrointestinal antibiosis of polar marine animals. *J. Bact.* **82**: 72–9

Sieburth J M, Conover J 1965 *Sargassum* tannin, an antibiotic which retards fouling. *Nature (Lond.)* **208**: 52–3

Silva de S S M, Gamage S K T, Kumar N S, Balasubramaniam S 1982 Antibacterial activity of extracts from the brown seaweed *Stoechospermum marginatum. Phytochemistry* **21**: 944–5

Sims J J, Donnell M S, Leary J V, Lacy G H 1975 Antimicrobial agents from marine algae. *Antimicrob. Agents Chemother.* **7**: 320–1

Sims J J, Lin G H Y, Wing R M 1974 Marine natural products 11. Elatol, a halogenated sesquiterpene alcohol from the red alga *Laurencia elata. Tetrahedron Lett.* **15**: 3487–90

Snader K M, Higa T 1986 Antiviral chamigrene derivative. *PTC Int. Appl.* WO 8603,739, US Appl. 682,896 1984. Ref.: C.A. 1987, **106**: 12959q

Sommer H, Whedon W F, Kofoid C A, Stohler R 1937 Relation of paralytic shellfish poison to certain plankton organisms of the genus *Gonyaulax*. *A. M. A. Arch. Pathol.* **24**: 537–59

Spence I, Jamieson D D, Taylor K M 1979 Anticonvulsant activity of farnesylacetone epoxide – a novel marine natural product. *Experientia* **35**: 238–9

Sperk G, Lassmann H, Baran H, Seitelberger F, Hornykiewicz O 1985 Kainic acid-induced seizures: dose-relationship of behavioral, neurochemical and histopathological changes. *Brain Res.* **338**: 289–95

Spoehr H A, Smith J H C, Strain H H, Milner H W, Hardin G J 1949 Fatty acid antibacterials from plants. *Carnegie Inst. Washington Pub.* **586**: 1–67

Stallard M O, Faulkner D J 1974 Chemical constituents of the digestive gland of the sea hare *Aplysia californica* – 1. Importance of diet. *Comp. Biochem. Physiol.* **49B**: 25–35

Steidinger K A 1983 A re-evaluation of toxic dinoflagellate biology and ecology. In Round R E, Chapman D J (eds) *Progress in Phycological Research*, Vol 2, Elsevier Science Publishers, New York, pp 147–88

Steidinger K A, Baden D G 1984 Toxic marine dinoflagellates. In Spector D L (ed) *Dinoflagellates*, Academic Press, New York, pp 201–61

Stein J R, Borden C A 1984 Causative and beneficial algae in human disease conditions: a review. *Phycologia* **23**: 485–501

Stierle D 1977 *Halogenated Monoterpenes from the Red Alga* Plocamium. PhD Thesis, Univ. of Calif. Riverside

Stierle D B, Sims J J 1984 Plocamenone, a unique halogenated monoterpene from the red alga *Plocamium. Tetrahedron Lett.* **25**: 153–6

Stierle D B, Wing R M, Sims J J 1979 Marine natural products – 16. Polyhalogenated acyclic monoterpenes from the red alga *Plocamium* of Antarctica. *Tetrahedron* **35**: 2855–9

Suganuma M, Fujiki H, Tahira T, Cheuk C, Moore R E, Sugimura T 1984 Estimation of tumor promoting activity and structure-function relationship of aplysiatoxins. *Carcinogenesis (Lond.)* **5**: 315–18

Sugimura T 1982 Potent tumor promoters other than phorbol ester and their significance. *Gann* **73**: 499–507

Sun H H, Fenical W 1979 Rhipocephalin and rhipocephanal; toxic feeding deterrents from the tropical marine alga *Rhipocephalus phoenix. Tetrahedron Lett.* **20**: 685–8

Sun H H, Ferrara N M, McConnell O J, Fenical W 1980 Bifurcarenone, an inhibitor of mitotic cell division from the brown alga *Bifurcaria galapagensis. Tetrahedron Lett.* **21**: 3123–6

Sun H H, McEnroe F J, Fenical W 1983a Acetoxycrenulide, a new bicyclic cyclopropane-containing diterpenoid from the brown seaweed *Dictyota crenulata. J. Org. Chem.* **48**: 1903–6

Sun H H, Paul V J, Fenical W 1983b Avrainvilleol, a brominated diphenylmethane derivative with feeding deterrent properties from the tropical green alga *Avrainvillea longicaulis. Phytochemistry* **22**: 743–5

Suntory Ltd 1982 Domoic acid as an insecticide. *Jpn. Kokai Tokkyo Koho* JP 58.222.004 (83 222.004). Ref.: C.A. 1984, **100**: 116498f

Suzuki T, Kurosawa E 1978 New aromatic sesquiterpenoids from the red alga *Laurencia okamurai* Yamada (1). *Tetrahedron Lett.* **19**: 2503–6

Suzuki T, Takeda S, Suzuki M, Kurosawa E, Kato A, Imanaka Y 1987 Cytotoxic squalene-derived polyethers from the marine red alga *Laurencia obtusa* (Hudson) Lamouroux. *Chem. Lett.* **16**: 361–4

Swanson K L, Allen C N, Aronstam R S, Rapoport H, Albuquerque E X 1986 Molecular mechanisms of the potent and stereospecific nicotinic receptor agonist (+)-anatoxin-a. *Mol. Pharmacol.* **29**: 250–7

Tachibana K, Nukina M, Joh Y-G, Scheuer P J 1987 Recent developments in the molecular structure of ciguatoxin. *Biol. Bull.* **172**: 122–7

Tachibana K, Scheuer P J, Tsukitani Y, Kikuchi H, Engen van D, Clardy J, Gopic-

hand Y, Schmitz F J 1981 Okadaic acid, a cytotoxic polyether from two marine sponges of the genus *Halichondria. J. Am. Chem. Soc.* **103**: 2469–71

Takagi T, Asahi M, Itabashi Y 1985 Fatty acid composition of twelve algae from Japanese waters. *Yukagaku* **34**: 1008–12

Takashima M, Sakai H, Arima K 1962 A new toxic substance, teleocidin, produced by *Streptomyces*, part 3, Production, isolation and chemical characterization of teleocidin B. *Agric. Biol. Chem.* **26**: 660–8

Takemoto T, Daigo K, Takagi N 1965 Studies on the hypotensive constituents of marine algae. 3. Determination of laminine in Laminariaceae. *Yakugaku Zasshi* **85**: 37–40

Takemoto T, Daigo K 1960 Constituents of *Chondria armata* and their pharmacological effects. *Arch. Pharmacol.* **293**: 627–33

Tanaka J, Higa T 1984 Hydroxydictyodial, a new antifeedant diterpene from the brown alga *Dictyota spinulosa. Chem. Lett.* **13**: 231–2

Thomson A W, Horne C H W 1976 Toxicity of various carrageenans in the mouse. *Br. J. Exp. Path.* **57**: 455–9

Tringali C, Nicolosi G, Piattelli M, Rocco C 1984a Three further dolabellane diterpenoids from *Dictyota* sp. *Phytochemistry* **23**: 1681–4

Tringali C, Oriente G, Piattelli M, Nicolosi G 1984b Structure and conformation of two new dolabellane-based diterpenes from *Dictyota* sp. *J. Nat. Prod.* **47**: 615–9

Tringali C, Piattelli M 1982 Further metabolites from the brown alga *Zonaria tournefortii. Gazz. Chim. Ital.* **112**: 465–8

Tringali C, Piattelli M, Nicolosi G 1984c Structure and conformation of new diterpenes based on the dolabellane skeleton from a *Dictyota* species. *Tetrahedron* **40**: 799–803

Tringali C, Piattelli M, Nicolosi G, Hostettmann K 1986 Molluscicidal and antifungal activity of diterpenoids from brown algae of the family Dictyotaceae. *Planta Med.* **53**: 404–6

Ueno Y, Nawa H, Ueganagi J, Morimoto H, Nakamori R, Matsuoka T 1955 Studies on the active components of *Digenea simplex* Ag. and related compounds. *J. Pharmacol. Soc. Jpn.* **75**: 807–44

Valdebenito H, Bittner M, Sammes P G, Silva M, Watson W H 1982 A compound with antimicrobial activity isolated from the red seaweed *Laurencia chilensis. Phytochemistry* **21**: 1456–7

Wagner M, Wagner B 1978 Agglutinine in marinen Braunalgen. *Z. Allg. Mikrobiol.* **18**: 355–60

Wakamiya T, Nakamoto H, Shiba T 1984 Structural determination of carnosadine, a new cyclopropyl amino acid, from red alga *Grateloupia carnosa. Tetrahedron Lett.* **25**: 4411–12

Walker P D, McAllister II J P, 1986 Anterograde transport of horseradish peroxidase in the nigrostriatal pathway after neostriatal kainic acid lesions. *Exp. Neurol.* **93**: 334–47

Watanabe S, Fujita T (Nisshin Oil Mills Ltd) 1985 Immune adjuvants as antitumor agents from marine algae. *Jpn. Kokai Tokkyo Koho* JP 61 197.525 (86.197.526). Ref.: C.A. 1987, **106**: 38468d

White S J, Jacobs R S 1979 Inhibition of microtubular polymerization by the halogenated sesquiterpene elatone (elatol ketone). *Pharmacologist* **21**: 170

White S J, Jacobs R S 1981 Inhibition of cell division and of microtubule assembly by elatone, a halogenated sesquiterpene. *Mol. Pharmacol.* **20**: 614–20

White S J, Jacobs R S 1983 Effect of stypoldione on cell cycle progression, DNA and protein synthesis, and cell division in cultured sea urchin embryos. *Mol. Pharmacol.* **24**: 500–8

White S J, Tanalski R, Fenical W, Jacobs R S 1978 Inhibition of cell cleavage by a halogenated sesquiterpene (elatol) and the 9-ketone synthetic derivative (elatol ketone). *Pharmacologist* **20**: 210

Whitefleet-Smith J, Boyer G L, Schnoes H K 1986 Isolation and spectral characteristics of four toxins from the dinoflagellate *Ptychodiscus brevis*. *Toxicon* **24**: 1075–90

Winter C A, Risley E A, Nuss G W 1962 Carrageenan-induced in hind paw edema of rat as an assay for antiinflammatory drugs *Proc. Soc. Exp. Biol. Med.* **11**: 544–7

Wratten S J, Faulkner D J 1976 Cyclic polysulfides from the red alga *Chondria californica*. *J. Org. Chem.* **41**: 2465–7

Wratten S J, Faulkner D J 1977 Metabolites of the red alga *Laurencia subopposita*. *J. Org. Chem.* **42**: 3343–9

Wright J L C 1984 Biologically active marine metabolites: some recent examples. *Proc. NS Inst. Sci.* **34**: 133–61

Wu C H, Huang J M C, Vogel S M, Luke V S, Atchison W D, Narahashi T 1985 Actions of *Ptychodiscus brevis* toxins on nerve and muscle membranes. *Toxicon* **23**: 481–7

Yamamoto I, Nagumo T, Fujihara M, Takahashi M, Ando Y, Okada M, Kawai K 1977 Antitumor effect of seaweeds. 2. Fractionation and partial characterization of the polysaccharide with antitumor activity from *Sargassum fulvellum*. *Jpn. J. Exp. Med.* **47**: 133–40

Yamamoto I, Nagumo T, Takahashi M, Fujihara M, Suzuki Y, Iizima N 1981 Antitumor effect of seaweeds. 3. Antitumor effect of an extract from *Sargassum kjellmanianum*. *Jpn. J. Exp. Med.* **51**: 187–9

Yamamoto I, Takahashi M, Tamura E, Maruyama H, Mori H 1984 Antitumor activity of edible marine algae: Effect of crude fucoidan fractions prepared from edible brown seaweeds against L-1210 leukemia. *Hydrobiologia* **116/117**: 145–8

Yasumoto T 1985 Recent progress in the chemistry of dinoflagellate toxins. In Anderson D M, White A W, Baden D G (eds) *Toxic Dinoflagellates*, Elsevier Science Publishing. New York, pp 259–79

Yasumoto T, Bagnis R, Vernoux J P 1976 Toxicity of the surgeonfishes. 2. Properties of the principal water-soluble toxin. *Bull. Jpn. Soc. Sci. Fish.* **42**: 359–65

Yasumoto T, Murata M, Oshima Y, Sano M, Matsumoto G K, Clardy J 1985 Diarrhetic shellfish toxins. *Tetrahedron* **41**: 1019–25

Yasumoto T, Oshima Y, Sugawara W, Fukuyo Y, Oguri H, Igarashi T, Fujita N 1980 Identification of *Dinophysis fortii* as the causative organism of diarrhetic shellfish poisoning. *Bull. Jpn. Soc. Sci. Fish.* **46**: 1405–11

Yasumoto T, Seino N, Murakami Y, Murata M 1987 Toxins produced by benthic dinoflagellates. *Biol. Bull.* **172**: 128–31

Zhang X, Stjernlöf P, Adem A, Nordberg A 1987 Anatoxin-a, a potent ligand for nicotinic cholinergic receptors in rat brain. *Eur. J. Pharmacol.* **135**: 457–8

8

The genetic manipulation of cyanobacteria and its potential uses

O. Ciferri, O. Tiboni and A. M. Sanangelantoni

Introduction

Cyanobacteria are organisms of great importance, being important primary producers, and also prokaryotes endowed with unique, or almost so, characteristics such as chromatic adaptation and aerobic nitrogen fixation. However, little is known of their general biology. For instance, no formal genetics has been performed on these organisms and even their taxonomy is very confused.

In this chapter, we shall attempt to outline briefly the present status of our knowledge on the recombination mechanisms that exist in cyanobacteria. Furthermore, we shall describe the restriction enzymes that have been characterized as well as the plasmids that have been isolated and their possible utilization for genetic manipulation. Finally, a list of the genes which have been cloned and characterized will be presented, together with an appraisal of the perspectives offered by the genetic manipulation of cyanobacteria.

Given the uncertain state of cyanobacterial taxonomy, we shall employ only the generic name even for those isolates for which the species name is commonly utilized (e.g. *Spirulina* in place of *Spirulina platensis*). In general, the generic name most recently proposed has been utilized (e.g. *Synechococcus* for *Anacystis nidulans*) except in Table 8.2 where the specific name that gave rise to the acronym designating the restriction enzyme has been retained (e.g. *Agmenellum quadruplicatum* instead of *Synechococcus* in the case of endonuclease AquI). Finally, when available, the number of the Pasteur Institute Collection (PCC) has also been given (Rippka *et al.*, 1979).

'Classic' genetic recombination

A fundamental step in evolution is the ability of organisms to adapt genetically to environmental changes and to transmit these and other characteristics by genetic recombination.

There is no doubt that identification and characterization of the 'classic' mechanisms for genetic recombination in cyanobacteria has lagged tremendously in comparison to the other prokaryotes. This is in part due to the little information available on the general physiology of these organisms, the difficulties in culturing many isolates, the impossibility of growing single clones, the slow growth rate and the absence of well-defined, stable mutants such as those auxotrophic for amino acids that have been so important in the development of bacterial genetics. Indeed, for a long time cyanobacteria have been thought to be devoid of any gene transfer system, until Kumar (1962) reported the isolation of double-resistant colonies by mixing cultures of *Synechococcus* mutants resistant to either streptomycin or penicillin. The validity of these data was later challenged, when it was reported that gene transfer had probably not occurred. The apparent acquisition of the resistance to penicillin was due to the inactivation of the antibiotic in the selection medium or, more likely, to the presence in the inoculum of non-metabolizing (and therefore insensitive to penicillin) cells of the streptomycin-resistant strain (Pikalek, 1967). Similarly, the reports of the occurrence of genetic exchange in *Cylindrospermum* (Singh and Sinha, 1965) and *Anabaena* (Singh, 1967) have also been questioned.

The first, well-documented demonstration of the occurrence of genetic transfer was provided by isolating double-resistant mutants after co-culture of clonally-raised mutants of *Synechococcus* resistant to streptomycin or polymixin B (Bazin, 1968). However, in these experiments it was impossible to establish whether conjugation or transformation had occurred. Later reports demonstrated the existence in the unicellular forms of the latter mechanism only.

Following these reports, a number of papers have appeared claiming transformation or, in some cases, conjugation between different mutants of the filamentous cyanobacterium *Nostoc* for a variety of characters including nitrogen fixation and heterocyst formation, resistance to antibiotics, antimetabolites, herbicides and pesticides (Stewart and Singh, 1975; Trehan and Sinha, 1981; Prasad and Vaishampayan, 1984; Vaishampayan and Prasad, 1984). Similar results have not been reported for other filamentous cyanobacteria and subsequent investigations demonstrated that only transformation must have occurred since stable recombinant clones were also produced when purified DNA was utilized (Trehan and Sinha, 1982). Indeed, it was recently found that herbicide resistance in the unicellular diazotrophic cyanobacterium *Synechocystis* could be transferred to *Nostoc* in an efficient and stable manner, either in mixed cultures or by DNA-mediated transformation (Singh *et al.*, 1987).

The first clear evidence in favour of the occurrence of a DNA-mediated transformation in unicellular cyanobacteria was reported for *Synechococcus* 7943 (Shestakov and Khyen, 1970). The frequency of recombinant clones, although relatively low when compared to that of the bacterial systems, was 10^4-fold higher than the frequency of spontaneous mutation, and transformation was completely inhibited by DNase treatment.

The possibility of obtaining stable auxotrophic mutants of *Synechococcus*

6301 (Herdman and Carr, 1972) allowed the establishment with certainty of the occurrence of transformation in these cyanobacteria (Orkwiszewski and Kaney, 1974).

Another type of transformation, mediated by extracellular nucleic acids secreted into the growth medium, was also reported for *Synechococcus* 6301 (Herdman and Carr, 1971; Herdman, 1973a,b). In these experiments, transformation was found to be inactivated by pretreatment with either DNase or RNase, thus indicating the possible existence in this organism of two transformation systems, one mediated by purified DNA and sensitive to DNase only, and a second one mediated by an extracellularly-secreted DNA/RNA complex and sensitive to both DNase and RNase. Similar results were obtained in the case of *Synechocystis* 6308 (Devilly and Houghton, 1977). In these investigations, mutational events were frequently observed in association with recombination, suggesting that the process of recombination itself could be mutagenic. It is worth noticing that mutation during recombination has been reported in bacteria (e.g. Yoshikawa, 1966), but never at a frequency as high as that observed in *Synechococcus*. However, so far these data have not been extended to other cyanobacteria.

Until 1977, most data on genetic transformation had been reported for only two strains of the genus *Synechococcus*, except for one paper describing the phenomenon in *Synechocystis* 6714 (Astier and Espardellier, 1976). All attempts to transfer antibiotic resistance from a streptomycin-resistant strain of *Synechocystis* 6308 to the streptomycin-sensitive wild-type were constantly unsuccessful although, quite unexpectedly, the same DNA was able to transform to streptomycin resistance the wild-type *Synechococcus* (Devilly and Houghton, 1977). Pretreatment of *Synechocystis* recipient cells with calcium chloride allowed transformation even in this organism. This treatment was also efficient in the transfer of resistance to rifampicin from *Synechococcus* to *Synechocystis*. This seems to be the only indication of the usefulness of treatment with calcium chloride to achieve transformation in cyanobacteria. Indeed, in a detailed investigation on the conditions for an efficient transformation in *Synechococcus* 7002, treatment with calcium chloride was found to be ineffective (Stevens and Porter, 1980).

Transformation within the genus *Synechocystis* and *Synechococcus* was also utilized to ascertain the relatedness of different isolates. It was found, for instance, that DNA from certain strains (e.g. *Synechocystis* 6803, 6702 and 6714) could transform these strains, but not others (such as 6701 or *Synechococcus* 7942). This difference seemed to be correlated with a difference in the gross DNA composition as indicated by the mol% G+C content (Grigorieva and Shestakov, 1982), the compatible strains having values around 47.4 mol% G+C and the incompatible ones 35.7 and 55.6 mol% G+C. More recently, however, intergeneric transformation with chromosomal DNA within the *Synechococcus/Synechocystis* group did not appear to be correlated with the gross G+C content of the strains (Stevens and Porter, 1986).

In conclusion, the results so far reported confirm that transformation occurs in cyanobacteria, but that the phenomenon is not widespread.

Indirectly, the results re-emphasize the great confusion which exists in the taxonomy of cyanobacteria, and indicate that any conclusion on the presence and distribution of the mechanisms for genetic recombination in this group of prokaryotes, must await the establishment of a reliable, and realistic, classification system.

In the 1980s, with the advent of the techniques for the isolation and manipulation of plasmid DNA, efforts were made to construct suitable vectors and work out experimental conditions for efficient and reproducible transformation of the most commonly used strains of the *Synechococcus* and *Synechocystis* genera.

In order to establish, for example, the relations that may exist between photosynthesis and transformation, Golden and Sherman (1984) studied gene transfer in *Synechococcus* utilizing chiefly plasmids pCH1 (van den Hondel *et al.*, 1980) or pSG111 (Golden and Sherman, 1983). An enhancement of transformation was found when photosynthesis was inhibited either by incubation of cells in the dark during contact with donor DNA, or by treatment with chemicals, such as DCMU (3-(3,4-dichlorophenyl)-1,1-dimethylurea) and CCCP (carbonyl cyanide-*m*-chlorophenyl hydrazone). Similarly, under iron deficiency, which induces a reduction in the amount of photosynthetic membranes, phycobilisomes and carboxysomes, the efficiency of transformation appeared to be increased.

Recently, a system for transforming *Synechocystis* 6803 with plasmid DNA from heterologous sources, which does not require the construction of biphasic vectors and which does not rely on homologous recombination (to achieve transformation), has been developed (Dzelzkalns and Bogorad, 1986). Different plasmids, carrying multiple resistance markers for a number of antibiotics, were utilized and the system involved UV irradiation of cells prior to transformation. The effect of UV pretreatment appeared to be the induction of an array of biochemical processes similar to those of the UV-induced SOS response of *E. coli*, including a temporary alleviation of host-controlled restriction. In these experiments it was observed that only a segment of the introduced plasmids was recovered in the DNA of the transformed cells and that, in different transformants, the plasmid DNA was inserted at different sites in the chromosome. Furthermore, hybridization data suggested the occurrence of rearrangements or duplications of DNA sequences around the insertion site.

Successful transformation by unmodified *E. coli* plasmid pBR322 has been reported in *Synechococcus* 6301 (Daniell *et al.*, 1986, 1987) using either intact cells or highly permeable cells (permeaplasts) that have a good efficiency of cell wall regeneration and subsequent division. The resistance marker (ampicillin) was expressed in such cells after at least 18 h of contact with the donor DNA and the plasmid was recovered intact and not integrated into the chromosome. However, in other studies, it was found that even if the recovered plasmid seemed to have the same size of that used for transformation, it contained segments of the host chromosome (Stassi *et al.*, 1981).

In conclusion, at least so far, the only mechanism for genetic recombination that has been convincingly demonstrated in cyanobacteria is transformation. This mechanism has been shown to occur mostly in the unicellular strains classified as *Synechococcus* and *Synechocystis* that appear to possess a physiological mechanism for the uptake of DNA. Reports of transformation of the filamentous forms, including *Nostoc*, await confirmation.

Plasmids

The occurrence of extrachromosomal DNA in cyanobacteria was first observed by Asato and Ginoza in 1973 in the most commonly utilized unicellular strain, *Synechococcus* 6301. Such plasmids were subsequently characterized (Simon, 1978; Lau and Doolittle, 1979; van den Hondel *et al.*, 1979). Since then, as summarized in Table 8.1, plasmids have been found in, and characterized from, a large number of both unicellular or filamentous forms. The majority of cyanobacteria so far examined appear to contain several different plasmids with a size ranging from 1.9 to 990 Md. Obviously, the physical and genetic characterization of these elements is essential, not only for genetic analysis and manipulation of the organisms, but also for understanding the ecological and physiological significance of the plasmids. At the time of their discovery, it was reasonable to assume that these plasmids had functions comparable to those of the analogous elements found in bacteria, such as resistance to heavy metals and antibiotics, sexuality, degradation of aromatic compounds or toxin production. However, up to now, no proof for plasmid-encoded functions has been found (e.g. Lau and Doolittle, 1979; Lau *et al.*, 1980; Castets *et al.*, 1986). Their involvement in photosynthesis has often been postulated (Asato and Ginoza, 1973; Roberts and Koths, 1976), but organisms apparently lacking plasmids, such as *Synechococcus* 6715 (van den Hondel *et al.*, 1979), *Spirulina* (Tiboni unpublished) and others (Simon, 1978; Herdman, 1982) appear to possess a normal photosynthesis.

All plasmids so far characterized appear to be genetically cryptic, and their role in the physiology, ecology and evolution of cyanobacteria is still unknown. Nevertheless, there is good evidence, supported by restriction analysis and hybridization experiments, for the occurrence of similar if not identical plasmids in strains isolated independently and from different geographical locations, thus suggesting their possible transferable nature. Indeed, plasmids with identical molecular weights and patterns of digestion by restriction enzymes were found in the closely related *Synechococcus* 6301, 6908 (Lau and Doolittle, 1979; van den Hondel *et al.*, 1979) and 6311 (van den Hondel and van Arkel, 1980) but also in *Synechococcus* 6707, a strain which differs from the others in its chromosomal G+C content (67.7 mol% versus 55 mol%, Herdman *et al.*, 1979) and geographical origin (6301 from Texas versus 6707 from California, Rippka *et al.*, 1979). Furthermore, different plasmids in the same strain and different plasmids within different strains contain regions of sequence homology (Lau *et al.*, 1980; Potts, 1984)

Table 8.1 Occurrence of plasmids in cyanobacteria

Genus and strain (PCC)		G+C* Mol%	Number	Size in Md
UNICELLULAR				
Synechococcus	6301	55.1	2	4.7, 37
			2	5.1, 30.5
			2	5.3, 32.7
			2	5.3, 32.1
	6311	54.8	1	5.0
			2	5.3, 33
	6908	56.0	2	5.3, 32.4
			2	5.3, 30.5
	7942	55.6	2	5.3, 33
			2	5.2, 32
			2	5.3, 32.1
			3	5.2, 33.5, 660–990
	7943	–	2	5.3, 33
	6707	68.4	2	5.3, 33
	7002	49.1	6	3.0, 6.5, 10.3, 20.1, 24.7, 75
			5–7	3.5, 6.2, 10.6, 26, ≥50
			5	3.0, 10.3, 20.1, 24.7, 75
	73109	49.0	4–5	5.9, 10.6, ≥50
	7003	49.3	5	1.85, 2.9, 3.5, 33, ≥50
	7425	48.6	2	5.4, 24
	6312	50.2	1	15.9
	7335	47.4	2	21.1, 23.1
	7336	–	3	3.5, 21.1, 46.2
Synechocystis	6714	47.9	1	30
	7005	45.4	1	2.0
	6701	35.8	8	2–40
			8	3.3, 5.0, 8.9, 9.9 + 4 bands
			>10	2.7, 4.2, 5.5, 9.8, 11.9, 15, + others 23–39.6
	6803	47.5	4	1.5(×3), 3.4
			4	1.6, 3.4, 33, 66
	6808	36.0	5	2.5, 14.5, 17.1, 33, 39.6
Pseudoanabaena	6802	45.9	4	1.6, 24.4, 26.4, 66
FILAMENTOUS				
Pseudoanabaena	7409	–	6	1.8, 2.8, 4.3, 6.6, 85.8, 118.8
Calothrix	7601	–	4	24, 32, 40, 48
			>10	6.0, 9.8, 12.6, 59.4 + others ≤132[†]
			4	7.8, 9.9, 11, 13[†]
Nostoc	7524	39.0	3	4.0, 8.0, 28
	8009	–	3	5.5, 19.8, 39.6
			6	5.2, 19.8, 25.1, 26.4, 148.5, approx. 264
	6705	49.4	3	2.6, 14, 40
Anabaena	7938	–	5	14, 24, 40, 53, 74
	7936	–	3	5.7, 26, 36
	7937	–	2	25, 32
	7120	42.5	4	3.3, 33, 42, 74
	7118	41–43	2	3.6, 14

Reference

Simon (1978)
Lau and Doolittle (1979)
van den Hondel *et al.* (1979)
Engwall and Gendel (1985)
Friedberg and Seijffers (1979)
van den Hondel and van Arkel (1980)
van den Hondel *et al.* (1979)
Lau and Doolittle (1979)
van den Hondel *et al.* (1980)
Laudenbach *et al.* (1983)
Engwall and Gendel (1985)
Rebière *et al.* (1986)
van den Hondel and van Arkel (1980)
van den Hondel *et al.* (1979)
Roberts and Koths (1976)
Lau *et al.* (1980)
Rebière *et al.* (1986)
Lau *et al.* (1980)
Lau *et al.* (1980)
van den Hondel *et al.* (1979)
van den Hondel *et al.* (1979)
Rebière *et al.* (1986)
Rebière *et al.* (1986)
Lau *et al.* (1980)
van den Hondel *et al.* (1979)
Lau *et al.* (1980)
Anderson and Eiserling (1985)
Rebière *et al.* (1986)

Chauvat *et al.* (1986)
Rebière *et al.* (1986)
Rebière *et al.* (1986)
Rebière *et al.* (1986)

Rebière *et al.* (1986)
Simon (1978)
Rebière *et al.* (1986)

Rebière *et al.* (1986)
Reaston *et al.* (1980)
Lambert and Carr (1982)
Lambert *et al.* (1984)
Rebière *et al.* (1986)
Lambert and Carr (1982)
Lambert *et al.* (1984)
Simon (1978)
Simon (1978)
Simon (1978)
Simon (1978)
Lambert and Carr (1982)

Table 8.1 Cont.

Genus and strain (PCC)	G+C* Mol%	Number	Size in Md
Oscillatoria limnetica	–	1	5.1
LPP group 7606	46.0	2	2.1, 9.0
		3	0.9, 10, approx. 12
6306	48.8	1	9.4
		3	0.85, 2.6, 12
		3	0.9, 10, approx. 12
73110	49.9	3	0.9, 10, approx. 12
6402	47.4	3	3.0, 10, 30

* G+C Mol% of the genomic DNA reported by Herdman *et al.* (1979)
† Plasmids isolated from different subcultures of the same strain.

suggesting that these regions may be analogous to the transferable genetic elements carried by many bacterial plasmids (Lau *et al.*, 1980).

In conclusion, these data, together with the evidence for interspecific transfer, indicate that cyanobacterial plasmids, although genetically cryptic, may play a role in the ecology and evolution of this group of prokaryotes.

Restriction enzymes

Although there are more than 500 restriction enzymes exhibiting at least 100 different site specificities (Roberts, 1985), enzymes with new specificities are continuously sought to increase the possibilities of DNA manipulation. Many species of cyanobacteria, both filamentous and unicellular, appear to be a promising source of such enzymes.

Cyanobacteria contain different endonucleases in different combinations, some of which are isoschizomers even in unrelated strains. For example, the isoschizomers AquI and AvaI, that recognize and cleave the sequence CPyCGPuG, are produced by a unicellular strain (*Synechococcus*) and a filamentous one (*Anabaena*). Similarly, besides other restriction endonucleases, *Fremyella* and *Nostoc* contain enzymes with the same specificity as AvaII (Whitehead and Brown, 1985a) which was first identified in *Anabaena* (Hughes and Murray, 1980). This is not unusual as the same restriction enzyme has been found in different bacterial genera (Roberts, 1985). Such findings may indicate that some of the activities are plasmid- or phage-borne, although, at the moment, there is no evidence correlating restriction activity and presence of plasmids or phages. A list of the sequence-specific endonucleases so far found in cyanobacteria is given in Table 8.2.

Little is known about the *in vivo* function(s) of sequence-specific endonucleases. It is likely that one of the factors responsible for the difficulty in the development of a reliable gene transfer system in cyanobacteria is indeed the presence of restriction endonucleases. In fact, conjugation can

Reference

Friedberg and Seijffers (1979)
Simon (1978)
Potts (1984)
Simon (1978)
Lambert and Carr (1982)
Potts (1984)
Potts (1984)
Potts (1984)

occur in *Anabaena* only when biphasic plasmid DNA is modified either by eliminating (Wolk *et al.*, 1984) or by methylating specific sequences (Elhai, personal communication).

Genetic manipulation

The possibility of genetic manipulation of cyanobacteria offers tremendous opportunities for both basic and applied research.

As already reported, many species were found to contain one or more plasmids, but, so far, all of the plasmids isolated turned out to be devoid of any known, selectable genetic marker. Thus, for the development of a cloning system, it has been necessary to create appropriate vectors by introducing genetic markers into the native plasmids. The first report of the expression of a foreign marker in a cyanobacterium was provided by van den Hondel *et al.* (1980) who described the transformation of *Synechococcus* 7942 with a bacterial plasmid containing the transposon Tn901 conferring resistance to ampicillin. The authors demonstrated that the transposon was able to transpose at different sites onto the native cyanobacterial plasmid, pUH24, generating hybrid plasmids containing the ampicillin-resistance gene as well as a single restriction site suitable for cloning. Subsequently, because the translocation property of Tn901 could complicate the use of such plasmids as vectors, a smaller deletion plasmid, pUC1, with the immobilized transposon and an additional single restriction site was constructed. Unfortunately, these plasmids were unable to replicate in *E. coli* and great efforts were made to construct biphasic *E. coli*/cyanobacterial vectors capable of replicating in the bacterium and in the transformable strain *Synechococcus* 7942. This was accomplished by ligating to pUC1 the bacterial plasmid pACYC184 (Kuhlemeier *et al.*, 1981). The resulting recombinant plasmids, pUC104 and pUC105, carrying ampicillin and chloramphenicol resistance, differed in respect to the site of insertion and orientation of pACYC184, but were able to transform both *E. coli* and the cyanobacterium.

Table 8.2   Restriction enzymes identified in cyanobacteria

Organism*	Enzyme	Isoschizomer of
UNICELLULAR		
Agmenellum quadruplicatum	AquI	AvaI
Anacystis nidulans	AniI	
Aphanocapsa sp.	SecI	
	SecII	MspI
	SecIII	MstII
Aphanothece halophytica	AhaI	CauII
	AhaII	AcyI
	AhaIII	
Eucapsis sp.	EspI	
FILAMENTOUS		
Anabaena catanula	AcaI	
Anabaena cylindrica	AcyI	
Anabaena flos-aquae	AflI	AvaII
	AflII	
	AflIII	
Anabaena oscillarioides	AosI	MstI
	AosII	AcyI
Anabaena sp.	AocI	SauI
	AocII	
Anabaena sp., str. Waterbury	AstWI	AcyI
Anabaena subcylindrica	AsuI	
	AsuII	
	AsuIII	AcyI
Anabaena variabilis	AvaI	
	AvaII	
	AvaIII	
Anabaena variabilis UW	AvrI	AvaI
	AvrII	
Calothrix scopulorum	CscI	SacII
Fischerella sp.	FspI	MstI
	FspII	AsuII
Fremyella diplosiphon	FdiI	AvaII
	FdiII	MstI
Mastigocladus laminosus	MlaI	AsuII
Microcoleus sp.	MstI	
	MstII	SauI
Nostoc sp. 6705	NspBI	AsuII
	NspBII	
Nostoc sp. 7524	Nsp(7524)I	
	Nsp(7524)II	SduI
	Nsp(7524)III	AvaI
	Nsp(7524)IV	AsuI
	Nsp(7524)V	AsuII
Nostoc sp. 7413	NspHI	Nsp(7524)I
	NspHII	AvaII
Nostoc sp. 8009	NspMACI	BglII

Recognition sequence[†]	Reference
CPyCGPuG	Lau and Doolittle (1980)
	Karreman *et al.* (1986)
	Gallagher and Burke (1985)
CCNNGG	Calléja *et al.* (1985)
CCGG	
CCTNAGG	
CC(C/G)GG	Whitehead and Brown (1982)
GPu ↓ CGPyC	Whitehead and Brown (1985b)
TTT ↓ AAA	
GC ↓ TNAGC	Calléja *et al.* (1984)
	Roberts (1985)
GPu ↓ CGPyC	de Waard *et al.* (1978)
G ↓ G(A/T)CC	Whitehead and Brown (1985a)
C ↓ TTAAG	
A ↓ CPuPyGT	
TGC ↓ GCA	de Waard *et al.* (1979)
GPu ↓ CGPyC	
CC ↓ TNAGG	Roberts (1985)
GPu ↓ CGPyC	de Waard and Duyvesteyn (1980)
G ↓ GNCC	Hughes *et al.* (1980)
TT ↓ CGAA	de Waard and Duyvesteyn (1980)
GPu ↓ CGPyC	
C ↓ PyCGPuG	Murray *et al.* (1976)
G ↓ G(A/T)CC	Hughes and Murray (1980)
ATGCAT	Sutcliffe and Church (1978)
	Roizes *et al.* (1979)
CPyCGPuG	de Waard and Duyvesteyn (1980)
C ↓ CTAGG	
CCGCGG	Duyvesteyn *et al.* (1981)
TGCGCA	Roberts (1985)
TT ↓ CGAA	
G ↓ G(A/T)CC	van den Hondel *et al.* (1983)
TGC ↓ GCA	Roberts (1985)
TT ↓ CGAA	Duyvesteyn and de Waard (1980)
TGCGCA	Gingeras *et al.* (1978)
CC ↓ TNAGG	Roberts (1985)
TTCGAA	Duyvesteyn *et al.* (1983)
C(A/C)G ↓ C(T/G)G	
PuCATG ↓ Py	Reaston *et al.* (1982)
G(G/A/T)GC(C/A/T) ↓ C	
C ↓ PyCGPuG	
G ↓ GNCC	
TTCGAA	
PuCATG ↓ Py	Duyvesteyn *et al.* (1983)
GG(A/T)CC	
A ↓ GATCT	Lau *et al.* (1985)

Table 8.2   Cont.

Organism*	Enzyme	Isoschizomer of
Spirulina platensis	SplI	
	SplII	TthlIII
	SplIII	HaeIII
Tolypothrix tenuis	TtnI	HaeIII

* Names as in the original reference.
† Recognition sequence of only one strand of DNA, written from 5′ → 3′. The arrow indicates the point of cleavage. Bases in parentheses indicate that either base may occur in that position. Abbreviations are: Py (pyrimidine), Pu (purine), N (any base).

Several other shuttle vectors were later constructed, utilizing native cyanobacterial plasmids and the most commonly used *E. coli* cloning vectors (Sherman and van de Putte, 1982; Tandeau de Marsac *et al.*, 1982; Buzby *et al.*, 1983; Friedberg and Seijffers, 1983). However, these plasmids possessed only few unique restriction sites, thus reducing considerably the possibility of DNA insertion. To improve their usefulness as shuttle vectors, new biphasic plasmids have been constructed, introducing polylinkers carrying additional restriction sites (Gendel *et al.*, 1983a,b; Lau and Straus, 1985), or additional selectable markers (Kuhlemeier *et al.*, 1983; Lau and Straus, 1985).

All these experiments utilized as recipient strains naturally-transformable cyanobacteria such as *Synechococcus* 7942, 6311 and 7002, that bear endogenous plasmids from which the shuttle vectors were derived. In this type of experiment, recombination between the incoming plasmid and the resident one could take place (Kuhlemeier *et al.*, 1981). Indeed, although the two biphasic plasmids, pUC104 and pUC105, could transform *Synechococcus* 7942, only pUC104 yielded transformants bearing the intact plasmid independently of the drug used for the selection (ampicillin or chloramphenicol), whereas pUC105 was recovered only when chloramphenicol was used. Under ampicillin selection, 90% of the transformants contained deletion derivatives with the pACYC184 part removed, and identical to pUCl. The authors postulated that recombinational events occurred between pUC105 and the small resident plasmid pUH24, whereas the large one (32.6 Md) did not interfere. To confirm this hypothesis, transformation experiments utilizing plasmid-cured recipient strains of *Synechococcus* 7942 were performed (Chauvat *et al.*, 1983; Kuhlemeier *et al.*, 1983). In this type of experiment, the problem of deletion derivatives was eliminated, but the transformation efficiency was dramatically decreased, indicating that recombination with the resident plasmid is required to stabilize the shuttle vector. In fact, subsequent experiments demonstrated that recombination between the incoming vector and the endogenous plasmid is essential to obtain a powerful host–vector system for gene cloning in *Synechocystis* 6803 (Chauvat *et al.*, 1986).

All efforts to improve the cloning potential of the first shuttle vectors,

Recognition sequence†	Reference
C↓GTACG	Kawamura *et al.* (1986)
GACNNNGTC	
GGCC	
GGCC	Roberts (1985)

to optimize transformation conditions and to develop suitable strains, were aiming at the production of stable merodiploids. Once merodiploids were efficiently obtained, the problem of recombination between the cloned gene and its chromosomal counterpart arose (Tandeau de Marsac *et al.*, 1982; Kuhlemeier *et al.*, 1984a,b). This problem became evident for the first time from the analysis of transformants obtained when transposition-induced mutants were challenged with recombinant plasmids carrying the wild-type gene. Although restoration of the wild-type phenotype was observed, the recombinant plasmid was never recovered, thus suggesting that homologous recombination had occurred leading to the loss of the vector.

To obtain stable merodiploids care had, therefore, to be paid to minimize the homology between chromosomal and cloned DNA or to apply a stringent selective pressure (Kuhlemeier *et al.*, 1985). Stable maintenance of merodiploids in *Synechococcus* 7942 was demonstrated for the genes of the rRNA operon, from a closely related organism, by selecting for the chloramphenicol resistance gene present on the biphasic vector (Golden and Sherman, 1983). Although the authors could recover an intact plasmid from transformed cells, recombination between the rRNA operon on the plasmid and that on the genome, followed by the establishment of an equilibrium between autonomous and integrated state, could not be excluded. This possibility was confirmed by transforming *Synechococcus* 7002 with a plasmid carrying a fragment of genomic DNA that was capable of complementing an auxotrophic mutation in *E. coli*. The gene was subsequently inactivated by mutagenesis *in vitro* and became unable to transform the bacterium to prototrophy. However, when the same plasmid was recovered from the transformed cyanobacterial cells, the insert had regained the ability to complement the *E. coli* mutation. It is thus conceivable that recombination had occurred in the cyanobacterial host (Porter, 1986).

Experiments have also been performed to develop a method for stable integration of foreign DNA into the genome of cyanobacteria. The first experiment of this type was carried out in *Synechococcus* 7942 by Williams and Szalay (1983). A nonreplicating donor plasmid was constructed utilizing fragments of cyanobacterial genomic DNA, interrupted by insertion of the gene for chloramphenicol acetyltransferase (CAT), and cloned in pBR322. After transformation of *Synechococcus* 7942, three different types of transformants were isolated, all showing homologous recombination between the cyanobacterial sequence present on the plasmid and the sequence on the genome of the recipient cell. In the so-called type I transformants, resistant to chloramphenicol only, the sequence of the recipient DNA was replaced by the homologous portion, interrupted by the CAT gene present on the donor plasmid, whereas the pBR322 portion was lost. These transformants were very stable and could be obtained using either circular or linear, non-replicating plasmid DNA. Further investigations demonstrated that it was possible to obtain and maintain stable type I transformants with integration of foreign DNA as large as 20 kb (Kolowsky *et al.*, 1984). Type II transformants, resistant to ampicillin only, were generated by elimination of the CAT gene followed by the integration of the pBR322 sequence into the genome between two identical copies of the cyanobacterial fragments. Finally, type III transformants, resistant to both antibiotics, involved the integration of the entire donor plasmid between one interrupted and one uninterrupted cyanobacterial fragment.

Type II and III transformants were found to be unstable, probably for the subsequent homologous recombination between duplicate cyanobacterial sequences flanking the integrated foreign DNA. Thus it seems that foreign DNA fragments can be stably integrated in the genome only when these fragments are inserted into the cyanobacterial sequence present on the donor plasmid. A similar approach was utilized to genetically manipulate *Synechococcus* 7002, generating specific mutants by a gene-replacement mechanism (Porter *et al.*, 1986).

In addition, transformation with genomic DNA coding for specific genes of the photosystem II reaction centre, interrupted by heterologous insertions or modified by site-directed mutagenesis, has been performed in *Synechocystis* 6803 (Vermaas *et al.*, 1987; Williams, 1987), possibly the most suitable organism for these studies since it can also grow photoheterotrophically in the presence of glucose.

The role of *psb*A gene product in the activity of photosystem II was also investigated in *Synechococcus* 7942. The protein corresponding to this gene (Q_B protein) is essential for oxygenic photosynthetic electron transport and is the target of several herbicides. Transformation of wild-type cells with a nonreplicating plasmid, carrying an internal fragment of the *psb*A gene from a herbicide-resistant strain, demonstrated that a mutation in this gene was responsible for resistance to these herbicides, as already reported for higher plants (Golden and Haselkorn, 1985). It was also found that the organism contains three genes for the Q_B protein (*psb*AI,II,III), and that replacement of the *psb*AI gene only with the resistant counterpart converted wild-type, sensitive cells to herbicide resistance. By inactivating on the chromosome each of these genes, singly or in pairs, it was found that any single gene encodes a functional Q_B protein and that any single gene could provide enough protein to support normal photoautotrophic growth (Golden *et al.*, 1986).

All the plasmid-mediated transformations so far mentioned have been performed in the naturally-transformable, unicellular cyanobacteria of the *Synechococcus* and *Synechocystis* genera. At present, no reproducible transformation system has been developed for the filamentous forms which include most of the nitrogen-fixing or chromatically-adapting cyanobacteria.

For these organisms, conjugation appears to be an alternative approach for genetic manipulation. This system, reported for the first time by Wolk *et al.* (1984), requires the presence of a suitable conjugative plasmid such as RP-4, that has been shown to mediate the transfer of pBR322 derivatives from *E. coli* to a wide range of Gram-negative bacteria (Taylor *et al.*, 1983; van Haute *et al.*, 1983), of a helper plasmid coding for a DNA-nicking protein (*mob* gene) that specifically recognizes the *bom* (basis of mobility) site, and of a hybrid plasmid that has to be transferred with an intact *bom* region.

The conjugal transfer was performed utilizing triparental matings with *Anabaena* as recipient strain, *E. coli* J53 containing the conjugative plasmid RP-4 and *E. coli* HB101 containing the shuttle vector and the helper plasmids. The shuttle vectors contained a portion of pBR322 required for replication and mobilization, with a deletion of the sites for AvaI and AvaII (the restriction endonucleases present in *Anabaena*), the cyanobacterial replicon from an endogenous plasmid of *Nostoc*, and genes coding for the resistance to chloramphenicol, streptomycin, neomycin and erythromycin as selectable markers. Exconjugants able to grow on the selective media, and thus containing the shuttle vector, were found only when the three parents were present. To confirm the transfer of the shuttle vector to the recipient cells, the plasmids present in the presumptive exconjugants were recovered and used to transform *E. coli*. Restriction analysis of these plasmids demonstrated that the shuttle vectors could be transferred from *E. coli* to *Anabaena* and back without any modification. In addition, utilizing a mobilizable hybrid plasmid containing a *Synechococcus* replicon, it was possible to demonstrate that a conjugative transfer also occurs in this unicellular species. The possibility of DNA transfer from *E. coli* to cyanobacteria, via conjugation, was then tested in a number of strains of *Nostoc* and in *Fischerella* UTEX 1829 (Flores and Wolk, 1985). The results demonstrated that this shuttle vector was transferred and stably maintained in at least three strains of *Nostoc*. The three strains were able to fix nitrogen and to grow heterotrophically, thus representing good candidates for genetic analysis of aerobic nitrogen fixation and heterocyst formation. Further, two of these strains also displayed chromatic adaptation so they may be suitable for studying this phenomenon at the gene level.

Starting from the shuttle vectors that had been shown to replicate both in *E. coli* and in some *Anabaena* strains, hybrid plasmids have been developed carrying the genes for bacterial luciferase, *lux*A and *lux*B (Schmetterer *et al.*, 1986). Since the insert containing the fully functional *lux* genes appeared to have little or no promoter activity, the promoter for the structural genes of ribulose-bisphosphate carboxylase or nitrogenase from *Anabaena* was inserted upstream from the luciferase genes. After conjugation, by monitoring the production of light in the exconjugants, it was possible to demonstrate that the genes for luciferase were expressed in *Anabaena*. Further, the level of expression was greatly enhanced when the strong promoter of ribulose-bisphosphate carboxylase was inserted upstream from these genes in the correct orientation. A shuttle vector has been constructed in which it is possible to clone DNA upstream from the promoter-less genes encoding bacterial luciferase (Elhai, personal communication). Several promoters have been cloned into this vector and it is now possible to monitor the level of the expression of these promoters in a single, normal cell or in one differentiating into a heterocyst.

Cloning, sequencing and expression of cyanobacterial genes

The cyanobacterial genes so far cloned, sequenced and/or expressed are reported in Table 8.3. Different strategies were utilized to identify such genes and their products.

The most obvious and direct strategy is the one that relies on the utilization of probes derived from the same organism, such as the ribosomal RNAs that were utilized to isolate the corresponding genes in *Synechococcus* 6301 (Tomioka *et al.*, 1981), or from the same gene cloned from another cyanobacterium. The latter approach was followed to isolate the gene for glutamine synthetase of *Spirulina*, employing a probe derived from the cloned *Anabaena* gene (Riccardi *et al.*, 1985). However, so far such probes are rarely available and more often advantage is taken of the similarity in the nucleotide sequence of bacterial and cyanobacterial genes and probes from bacteria have been successfully utilized to identify the corresponding genes on cyanobacterial DNA. This approach was used, for instance, for isolating the glutamine synthetase gene of *Anabaena* (Fisher *et al.*, 1981), a *rec*A gene of *Synechococcus* 7002 (Murphy *et al.*, 1987), the genes for the elongation factors and three ribosomal proteins of *Spirulina* (Tiboni and Di Pasquale, 1987; Tiboni, unpublished results) and that for thioredoxin of *Anabaena* (Lim *et al.*, 1986). Similarly, the cloned *nif* genes from *Klebsiella pneumoniae* have been extensively utilized to identify some of the genes involved in nitrogen fixation in many cyanobacteria, e.g. *Anabaena* (Mazur *et al.*, 1980; Ruvkun and Ausubel, 1980; Hirschberg *et al.*, 1985).

Advantage has been taken also of the similarities existing in the genetic organization of chloroplasts and cyanobacteria by using probes from the former to identify genes in the latter. Thus, for instance, the genes for the apoproteins of the phycobiliproteins cloned from the chloroplast (cyanelle) DNA of *Cyanophora paradoxa* allowed the identification of the genes for the phycocyanins in *Calothrix* (Conley *et al.*, 1985) and the allophycocyanins in *Synechococcus* 6301 (Houmard *et al.*, 1986). Similarly, the gene for the large subunit of ribulose-bisphosphate carboxylase from spinach or *Chlamydomonas reinhardtii* chloroplasts was used for isolating the corresponding gene in *Synechococcus* 6301 (Shinozaki *et al.*, 1983) and in *Spirulina* (Tiboni *et al.*, 1984) and that for the 32 kd protein from spinach and maize for *Anabaena* (Curtis and Haselkorn, 1984), *Calotrix* 7601, *Nostoc* and *Synechococcus* 7942 (Mulligan *et al.*, 1984). In the case of the gene for the small subunit of the carboxylase that, in eukaryotes, is coded in the nuclear DNA, probes from the nuclear genes from pea and soybean allowed the identification of this gene in *Synechococcus* 6301 (Shinozaki and Sugiura, 1983) and *Spirulina* (Tiboni *et al.*, 1984), respectively.

When the amino acid sequence of a protein, or a portion of it, is known, a different strategy may be followed. Oligonucleotides, corresponding to a part of the protein's gene, are synthesized and utilized as probes to identify the DNA fragment carrying the gene for such a protein. This approach was followed for the phycoerythrin genes in *Calothrix* 7601 (Mazel *et al.*, 1986), those for phycocyanins in *Synechococcus* 7942 (Lau *et al.*, 1987b) and *Synechococcus* 7002 (de Lorimier *et al.*, 1984) and for the so-called linker polypeptides in *Calothrix* (Lomax *et al.*, 1987). Similarly, synthetic probes allowed the identification of the genes for the gas vesicle protein in *Calothrix* (Tandeau de Marsac *et al.*, 1985), and that for ferredoxin I in *Anabaena* (Alam *et al.*, 1986) and *Synechococcus* 7942 (Reith *et al.*, 1986).

Finally, another approach is that of complementing well-defined mutations in bacteria by a 'shotgun' type of experiment. In this approach, cyanobacterial DNA, fragmented by digestion with restriction endonucleases, is ligated to a linearized *E. coli* plasmid and the ligation mixture, or a mixture of hybrid plasmids, propagated in a bacterial host, used to transform to prototrophy, auxotrophic mutants of bacteria. This approach has identified the glutamine synthetase gene of *Anabaena* (Fisher *et al.*, 1981) and *Spirulina* (Riccardi *et al.*, 1985), the *rec*A gene of *Synechocystis* 6308 (Geoghegan and Houghton, 1987) and some genes of the leucine, threonine and proline pathways in *Synechococcus* 7002 (Porter *et al.*, 1986). However, in the latter investigation it was shown that not all bacterial mutations could be complemented in this way. Thus, for instance, only one mutation in the leucine pathway was successfully complemented by plasmids carrying portions of the cyanobacterial DNA, whereas mutants with lesions in other genes of the same pathway could not be transformed to prototrophy. The explanation for such failures has not been found, but it is possible that the pathways for the biosynthesis of amino acids differ to a certain extent in the two organisms and only certain enzymes may function in the heterologous host. Alternatively, it is possible that a cyanobacterial protein cannot interact in the bacterial host with the bacterial proteins to give the multienzyme complexes required for the activity of a biosynthetic pathway. However, at least in one case, a cyanobacterial protein appeared to be capable of integrating correctly in a multiprotein/RNA complex since a streptomycin-resistant mutant of *E. coli* was transformed to streptomycin sensitivity by a plasmid carrying, among others, the gene for the ribosomal protein S12 of *Spirulina* (Tiboni and Di Pasquale, 1987). This demonstrated that the S12 ribosomal protein of the cyanobacterium became an integral part of the *E. coli* ribosome and, due to the dominance of streptomycin sensitivity over streptomycin resistance, conferred to the bacterial cells the sensitive phenotype.

Apart from the above-reported complementation of well-defined bacterial mutations, expression of cloned cyanobacterial genes was demonstrated in other ways. Detection of an enzymatic activity, absent in the bacterial cells, but present after transformation with plasmids carrying

Table 8.3 Cloned cyanobacterial genes sequenced and/or expressed

Gene	Coding for	Organism
Genes for photosynthesis and CO$_2$ fixation		
*rbc*L.	Large subunit of ribulose-bisphosphate carboxylase	*Anabaena* 7120 *Chlorogloeopsis* 6912 *Spirulina* *Synechococcus* 6301
*rbc*S	Small subunit of ribulose-bisphosphate carboxylase	*Anabaena* 7120 *Chlorogloeopsis* 6912 *Spirulina* *Synechococcus* 6301
*psa*A,B	P700 chlorophyll *a* apoproteins	*Synechococcus* 7002
*psb*A	32kd (or Q$_B$) protein	*Anabaena* 7120 *Calothrix* 7601 *Synechococcus* 7942
*psb*B	CP-47 protein	*Synechocystis* 6803
*wox*A	Extrinsic 33kd protein	*Synechococcus* 7942
*pet*F	Ferredoxin I	*Anabaena* *Anabaena* 7120 *Synechococcus* 7942
ppc	Phosphoenolpyruvate carboxylase	*Synechococcus* 6301
*trx*A	Thioredoxin	*Anabaena* 7119
atp	Proteins of the ATP-synthase complex	*Synechococcus* 6301
*atp*B	β-Subunit of ATPase	*Anabaena* 7120
*atp*E	ε-Subunit of ATPase	*Anabaena* 7120
*cpc*A	α-Subunit of C-phycocyanin	*Anabaena* 7120 *Synechococcus* 7002 *Calothrix* 7601 *Synechococcus* 7942
*cpc*B	β-Subunit of C-phycocyanin	*Anabaena* 7120 *Synechococcus* 7002 *Calothrix* 7601 *Synechococcus* 6301 *Synechococcus* 7942
*cpe*A	α-Subunit of phycoerythrin	*Calothrix* 7601
*cpe*B	β-Subunit of phycoerythrin	*Calothrix* 7601
*apc*A	α-Subunit of allophycocyanin	*Synechococcus* 6301
*apc*B	β-Subunit of allophycocyanin	*Synechococcus* 6301
*cpc*C, D, E	Linker polypeptides of phycobilisomes	*Anabaena* 7120
*lpc*A, B, C	Linker polypeptides of phycobilisomes	*Calothrix* 7601

Sequenced	Expressed
Curtis and Haselkorn (1983)	Gurevitz et al. (1985)
	Vakeria et al. (1986)
	Tiboni et al. (1984)
Shinozaki et al. (1983)	Reichelt and Delaney (1983)
Reichelt and Delaney (1983)	Gatenby et al. (1985)
Shinozaki and Sugiura (1985)	Christeller et al. (1985)
	Tabita and Small (1985)
	van der Vies et al. (1986)
Nierzwicki-Bauer et al. (1984)	Gurevitz et al. (1985)
	Vakeria et al. (1986)
	Tiboni et al. (1984)
Shinozaki and Sugiura (1983)	Christeller et al. (1985)
Shinozaki and Sugiura (1985)	Gatenby et al. (1985)
	Tabita and Small (1985)
	van der Vies et al. (1986)
Cantrell and Bryant (1987)	
Curtis and Haselkorn (1984)	Curtis and Haselkorn (1984)
Mulligan et al. (1984)	
Golden and Haselkorn (1985)	Golden and Haselkorn (1985)
Golden et al. (1986)	Golden et al. (1986)
Vermaas et al. (1987)	
Kuwabara et al. (1987)	
van der Plas et al. (1986a)	
Alam et al. (1986)	
van der Plas et al. (1986b)	
Reith et al. (1986)	
Katagiri et al. (1985)	Kodaki et al. (1985)
	Lim et al. (1986)
Cozens and Walker (1987)	
Curtis (1987)	
Curtis (1987)	
Belknap and Haselkorn (1987)	Belknap and Haselkorn (1987)
Pilot and Fox (1984)	Bryant et al. (1985)
de Lorimier et al. (1984)	
Coenly et al. (1985)	Conley et al. (1985)
Lau et al. (1987b)	Lau et al. (1987b)
Belknap and Haselkorn (1987)	Belknap and Haselkorn (1987)
Pilot and Fox (1984)	Bryant et al. (1985)
de Lorimier et al. (1984)	
Conley et al. (1985)	Conley et al. (1985)
Lind et al. (1985)	
Lau et al. (1987a)	
Mazel et al. (1986)	Mazel et al. (1986)
Mazel et al. (1986)	Mazel et al (1986)
Houmard et al. (1986)	
Houmard et al. (1986)	
Belknap and Haselkorn (1987)	Belknap and Haselkorn (1987)
Lomax et al. (1987)	Lomax et al. (1987)

Table 8.3 Cont.

Gene	Coding for	Organism
*apc*C	Linker polypeptides of phycobilisomes	*Synechococcus* 6301
Genes for nitrogen fixation or metabolism		
*nif*D	α-Subunit of dinitrogenase	*Anabaena* 7120
*nif*K	β-Subunit of dinitrogenase	*Anabaena* 7120
*nif*H	Nitrogenase reductase	*Anabaena* 7120
*xis*A	Excision enzyme	*Anabaena* 7120
*gln*A	Glutamine synthetase	*Anabaena* 7120
		Spirulina
nar	Nitrate reductase	*Synechococcus* 7942
Genes for the protein synthetic apparatus		
*rrn*A,B	23S rRNA	*Synechococcus* 6301
	16S rRNA	*Synechococcus* 6301
	5S rRNA	*Synechococcus* 6715
		Synechococcus 6301
*trn*I	tRNA$_{Ile}$	*Synechococcus* 6301
*trn*A	tRNA$_{Ala}$	*Synechococcus* 6301
*tuf*A	Elongation factor Tu	*Spirulina*
fus	Elongation factor G	*Spirulina*
tsf	Elongation factor Ts	*Spirulina*
*rps*12	Ribosomal protein S12	*Spirulina*
*rps*7	Ribosomal protein S7	*Spirulina*
*rps*2	Ribosomal protein S2	*Spirulina*
Other genes		
*gvp*A B, C	Gas vesicle proteins	*Calothrix* 7601
*rec*A	*rec*A protein	*Anabaena*
		Synechococcus 7002
		Synechocystis 6308
arg		*Synechococcus* 7002
leu		*Synechococcus* 7002
met		*Synechococcus* 7942
*thi*1		*Synechococcus* 7942

fragments of cyanobacterial DNA, has been achieved in the case of the genes for ribulose-bisphosphate carboxylase (Christeller *et al.*, 1985; Gatenby *et al.*, 1985; Gurevitz *et al.*, 1985; Tabita and Small, 1985; van der Vies *et al.*, 1986) and such results were often substantiated by immunological assays. In the case of the gene for thioredoxin, the protein produced in the bacterial host was purified and shown to have characteristics identical to those of the protein present in the cyanobacterium, from which the gene

Sequenced	Expressed
Houmard et al. (1986)	
Lammers and Haselkorn (1983)	
Mazur and Chui (1982)	
Mevarech et al. (1980)	
Hirschberg et al. (1985)	
Lammers et al. (1986)	Lammers et al. (1986)
Tumer et al. (1983)	Fisher et al. (1981)
	Riccardi et al. (1985)
	Kuhlemeier et al. (1984a)
	Kuhlemeier et al. (1984b)
Kumano et al. (1983)	
Douglas and Doolittle (1984b)	
Tomioka and Sugiura (1983)	
Delihas et al. (1982)	
Delihas et al. (1985)	
Douglas and Doolittle (1984a)	
Sugiura (pers. comm.)	
Williamson and Doolittle (1983)	
Tomioka and Sugiura (1984)	
Williamson and Doolittle (1983)	
Tomioka and Sugiura (1984)	
Tiboni, unpublished	Tiboni and Di Pasquale (1987)
Tiboni, unpublished	
Tiboni, unpublished	
Tiboni, unpublished	Tiboni and Di Pasquale (1987)
Tiboni, unpublished	
Tiboni, unpublished	
Tandeau de Marsac et al. (1985)	
Damerval et al. (1987)	
	Owttrim and Coleman (1987)
Murphy et al. (1987)	Murphy et al. (1987)
	Geoghegan and Houghton (1987)
	Porter et al. (1986)
	Porter et al. (1986)
	Tandeau de Marsac et al. (1982)
	Williams and Szalay (1983)

was isolated, and different from those of the protein present in the untransformed bacterium (Lim et al., 1986). Other approaches have relied on the expression of the cyanobacterial genes in E. coli 'maxicells' (Fisher et al., 1981; Bryant et al., 1985; Murphy et al., 1987) or 'minicells' (Tiboni et al., 1984; Tiboni and Di Pasquale, 1987) followed by confirmation of the authenticity of the protein produced either by immunology (Fisher et al., 1981; Bryant et al., 1985) or by comparison of the patterns obtained on partial proteolytic digestion of the protein isolated from the transformed bacterial cells and that isolated from the cyanobacteria (Tiboni *et al.*, 1984). Finally, in a few cases, plasmids carrying cyanobacterial genes were transcribed and translated in *in vitro* systems resulting in the production of proteins with the same electrophoretic mobility of the authentic cyanobacterial proteins (Riccardi *et al.*, 1985; Owttrim and Coleman, 1987).

Potential uses of genetically-manipulated cyanobacteria and of cyanobacterial genes

As previously discussed in this chapter, genetic transformation of cyano-bacteria is just beginning; thus it is not surprising that, so far, only a few genes from other organisms have been inserted and expressed in cyanobacteria.

The gene coding for the entomocidal toxin of *Bacillus sphaericus* has been successfully expressed in *Synechococcus* 7942 (Tandeau de Marsac *et al.*, 1987). Utilizing a bifunctional vector, the gene was introduced into the cyanobacterium where it directed the synthesis of substantial amounts of the fully active toxin. Since cyanobacteria are present in all aquatic ecosystems, including those of disease-carrying insects, one can envisage the utilization of genetically-manipulated cyanobacteria for biological control of insect larvae. Experiments along this line are, indeed, under way in the case of the gene coding for the larvicidal toxin of *Bacillus thuringiensis* (van Montagu, personal communication).

As already mentioned, the *lux* genes, coding for the luciferase of the bacteria *Vibrio fischeri* and *V. harvey*, were cloned in a shuttle vector capable of replicating in *E. coli* and *Anabaena* and used to transform the latter (Schmetterer *et al.*, 1986). If a strong promoter was inserted upstream from the bacterial genes, the genes were found to be expressed in the cyanobacterium to such an extent that enough luminescence was produced to allow its detection in a single cell of *Anabaena*. Thus, at least in principle, expression of the luciferase genes in *Anabaena* could be employed to assess the utilization of different promoters by this organism. A somewhat similar approach was followed to ascertain the ability of *Synechococcus* 7942 to utilize promoters from chloroplasts (Dzelzkalns *et al.*, 1984). The putative promoter sequence of the *psb*A gene from the chloroplast of *Amaranthus hybridus* was inserted in a shuttle vector that contained the gene for the bacterial chloramphenicol acetyltransferase lacking its own promoter. Expression of the bacterial gene in the transformed cyanobacterial cells demonstrated that the chloroplast promoter was recognized and correctly transcribed in *Synechococcus* 7942. The latter results indicate the possibility of utilizing cyanobacteria for expressing chloroplast genes. Indeed, at least in one case, the genes for the allophycocyanins present in the cyanellar DNA of the flagellate *Cyanophora paradoxa* were correctly expressed in *Synechococcus* 7002 (de Lorimier *et al.*, 1987). Given the similarities that exist in the structure and the function of the genetic apparatus of chloroplasts (and other plastids such as the cyanelles) and cyanobacteria, such a finding is not too unexpected.

Cyanobacterial genes are often correctly expressed in bacteria and, indeed, some of the cyanobacterial restriction enzymes, such as AvaI and AvaII, are commercially produced in bacterial hosts. However, there are reports indicating that some cyanobacterial genes are not expressed in bacterial hosts. For instance, two *Anabaena* genes for nitrogen fixation, *nif*H and *nif*D, were found not to be expressed when introduced in point mutants of *Klebsiella pneumoniae* (Hirschberg *et al.*, 1985). The authors conclude that it is possible that at least some cyanobacterial promoters are not recognized by the bacterial transcription factors since no RNA transcript corresponding to the two cyanobacterial *nif* genes were found in the bacteria transformed with a plasmid carrying such genes. Such a possibility would explain why, as already mentioned, it has been impossible to transform to prototrophy some bacterial mutants by transformation with chimeric plasmids bearing cyanobacterial DNA that was capable of complementing other mutations in the same pathway.

Genetic manipulation of cyanobacteria is also of potential interest for studying phenomena of considerable biological significance, some of which occur solely in these organisms such as aerobic nitrogen fixation and heterocyst differentiation. For investigations in these areas, as well as in those of photosynthesis and cell differentiation, cyanobacteria appear to represent unique model organisms of unsurpassed simplicity.

The study of the structure and the expression of cyanobacterial genes may also be of importance in attempts to genetically manipulate the genomes of higher plants. A case in point is that of the genes for ribulose-bisphosphate carboxylase, the enzyme that catalyses the first reaction for carbon dioxide fixation and photorespiration in chloroplasts. In all photosynthetic eukaryotes, any investigation of the genes for this enzyme, including those attempting its modification to improve photosynthetic yields (e.g. by eliminating or lowering photorespiration), is rendered particularly difficult by the fact that the two protein subunits which make up the active hexadecamer are coded in different cell compartments (that for the large, or L, subunit being coded in the chloroplast DNA and that for the small, or S, subunit in the nuclear DNA). Further, since the enzyme is assembled in the chloroplast where it expresses its catalytic activity, the nuclear coded S subunit is produced in the cytoplasm as a larger precursor that is cleaved into the mature protein inside the chloroplast. Cyanobacteria possess a ribulose-bisphosphate carboxylase similar to that present in chloroplasts (for a review see McFadden and Purohit, 1978) but the genetic organization of the genes is much simpler. In these organisms, the genes for the L and S subunit are contiguous (Shinozaki and Sugiura, 1983) and are transcribed as a single operon (Shinozaki and Sugiura, 1985). In addition, no post-translational process for maturation of the S subunit exists in cyanobacteria since, in these organisms, the protein is synthesized without a transit peptide. Thus the cyanobacterial genes for ribulose-bisphosphate carboxylase are ideally suited for investigations on the mechanisms for the regulation of the expression of these genes and the assembly of the protein as well as for evaluating techniques, such as site directed mutagenesis, designed to improve the efficiency of the protein. Not only, as reported in Table 8.3, have the cyanobacterial genes for the L and S subunit been isolated, cloned and sequenced but they have also been expressed in *E. coli* to give the complete enzyme, indistinguishable from that present in the cyanobacterial cells and endowed with enzymatic activity (Christeller *et al.*, 1985; Gatenby *et al.*, 1985; Gurevitz *et al.*, 1985; Tabita and Small, 1985). Further, a hybrid enzyme, endowed with a reduced but still significant catalytic activity, was produced in the bacterial host transformed with a plasmid containing the L subunit gene of *Synechococcus* 6301 and infected with a phage carrying the gene for the wheat S subunit from which the portion coding for the transit peptide was removed (van der Vies *et al.*, 1986). These results show that the study of the genes for ribulose-bisphosphate carboxylase in cyanobacteria may also be of great help in studies on the manipulation of higher plants' genes. One can envisage, for instance, the production in bacterial hosts of novel oligomers of ribulose-bisphosphate carboxylase originating from higher plant genes (for either the L or the S subunit), together with the wild-type gene for the cognate subunit from a cyanobacterium. Such hybrid oligomers could be utilized for screening mutant genes that could give rise to enzymes endowed with better properties. Similarly, controlled manipulation of other components of the photosynthetic apparatus, including the production of mutants resistant to chemicals of agronomic importance, such as herbicides, could be evaluated by expressing mutagenized higher plant genes in cyanobacterial hosts. Indeed, as already mentioned, resistance to the herbicide diuron in *Synechococcus* 7942 was found, just as in higher plants, to be the result of a mutation in the *psb*A gene (Golden and Haselkorn, 1985).

Acknowledgement

The work performed in the authors' laboratory was supported by grants from Consiglio Nazionale delle Ricerche and Ministero della Pubblica Istruzione. During the preparation of this manuscript, one of us (O.C.) was visiting professor in the Department of Biochemistry, University of Ottawa and he gratefully acknowledges the help and support given to him, especially by Dr I. Altosaar.

References

Alam J, Whitaker R A, Krogmann D W, Curtis S E 1986 Isolation and sequence of the gene for ferredoxin I from the cyanobacterium *Anabaena* sp. strain PCC 7120. *J. Bact.* **168**: 1265–71

Anderson L K, Eiserling F A 1985 Plasmids of the cyanobacterium *Synechocystis* 6701. *FEMS Microbiol. Lett.* **29**: 193–5

Asato Y, Ginoza H S 1973 Separation of small circular DNA molecules from the blue–green alga *Anacystis nidulans*. *Nature New Biology* **244**: 132–3

Astier C, Espardellier F 1976 Mise en evidence d'un système de transfert génétique chez une cyanophycée du gendre *Aphanocapsa. C. R. Acad. Sci. Paris* **282**: 795–7

Bazin M J 1968 Sexuality in a blue–green alga: genetic recombination in *Anacystis nidulans. Nature (Lond.)* **218**: 282–3

Belknap W R, Haselkorn R 1987 Cloning and light regulation of expression of the phycocyanin operon of the cyanobacterium *Anabaena. EMBO J.* **6**: 871–84

Bryant D A, Dubbs J M, Fields P I, Porter R D, de Lorimier R 1985 Expression of phycobiliprotein genes in *Escherichia coli. FEMS Microbiol. Lett.* **29**: 343–9

Buzby J S, Porter R D, Stevens Jr S E 1983 Plasmid transformation in *Agmenellum quadruplicatum* PR-6: construction of biphasic plasmids and characterization of their transformation properties. *J. Bact.* **154**: 1446–50

Calléja F, Dekker B M M, Coursin T, de Waard A 1984 A new sequence specific endonuclease, *EspI,* of cyanobacterial origin. *FEBS Lett.* **178**: 69–72

Calléja F, Tandeau de Marsac N, Coursin T, van Ormondt H, de Waard A 1985 A new endonuclease recognizing the deoxynucleotide sequence C ↓ CNNGG from the cyanobacterium *Synechocystis* 6701. *Nucl. Acids Res.* **13**: 6745–51

Cantrell A, Bryant D A 1987 Molecular cloning and nucleotide sequence of the *psa*A and *psa*B genes of the cyanobacterium *Synechococcus* sp. PCC 7002. *Pl. Mol. Biol.* **9**: 453–68

Castets A-M, Houmard J, Tandeau de Marsac N 1986 Is cell motility a plasmid-encoded function in the cyanobacterium *Synechocystis* 6803? *FEMS Microbiol. Lett.* **37**: 277–81

Chauvat F, Astier C, Vedel F, Joset-Espardellier F 1983 Transformation in the cyanobacterium *Synechococcus* R2: improvement of efficiency; role of the pUH24 plasmid. *Mol. Gen. Genet.* **191**: 39–45

Chauvat F, de Vries L, van der Ende A, van Arkel G A 1986 A host–vector system for gene cloning in the cyanobacterium *Synechocystis* PCC 6803. *Mol. Gen. Genet.* **204**: 185–91

Christeller J T, Terzaghi B E, Hill D F, Laing W A 1985 Activity expressed from cloned *Anacystis nidulans* large and small subunit ribulose bisphosphate carboxy-lase genes. *Pl. Mol. Biol.* **5**: 257–63

Conley P B, Lemaux P G, Grossman A R 1985 Cyanobacterial light-harvesting complex subunits encoded in two red light-induced transcripts. *Science* **230**: 550–3

Cozens A L, Walker J E 1987 The organization and sequence of the genes for ATP synthase subunits in the cyanobacterium *Synechococcus* 6301. Support for an endosymbiotic origin of chloroplasts. *J. Mol. Biol.* **194**: 359–83

Curtis S E 1987 Genes encoding the beta and epsilon subunits of the proton-trans-locating ATPase from *Anabaena* sp. strain PCC 7120. *J. Bact.* **169**: 80–6

Curtis S E, Haselkorn R 1983 Isolation and sequence of the gene for the large subunit of ribulose-1,5-bisphosphate carboxylase from the cyanobacterium *Anabaena* 7120. *Proc. Nat. Acad. Sci. USA* **80**: 1835–9

Curtis S E, Haselkorn R 1984 Isolation, sequence and expression of two members of the 32kd thylakoid membrane protein gene family from the cyanobacterium *Anabaena* 7120. *Pl. Mol. Biol.* **3**: 249–58

Damerval T, Houmard J, Guglielmi G, Csiszar K, Tandeau de Marsac N 1987 A developmentally regulated *gvp*ABC operon is involved in the formation of gas vesicles in the cyanobacterium *Calothrix* 7601. *Gene* **54**: 83–92

Daniell H, Sarojini G, McFadden B A 1986 Transformation of the cyanobacterium *Anacystis nidulans* 6301 with the *Escherichia coli* plasmid pBR322. *Proc. Nat. Acad. Sci. USA* **83**: 2546–50

Daniell H, Sarojini G, McFadden B A 1987 Cyanobacterial transformation: expression of Col El plasmids in *Anacystis nidulans* 6301. In Biggins J (ed) *Prog-*

ress in Photosynthesis Research, Vol 4, Martinus Nijhoff Publishers, Dordrecht, pp 837–40

Delihas N, Andresini W, Andersen J, Berns D 1982 Structural features unique to the 5S ribosomal RNAs of the thermophilic cyanobacterium *Synechococcus lividus* III and the green plant chloroplasts. *J. Mol. Biol.* **162**: 721–7

Delihas N, Andersen J, Berns D 1985 Phylogeny of the 5S ribosomal RNA from *Synechococcus lividus* II: the cyanobacterial/chloroplast 5S RNAs form a common structural class. *J. Mol. Evol.* **21**: 334–7

de Lorimier R, Bryant D A, Porter R D, Liu W-Y, Jay E, Stevens Jr S E 1984 Genes for the α and β subunits of phycocyanin. *Proc. Nat. Acad. Sci. USA* **81**: 7946–50

de Lorimier R, Guglielmi G, Bryant D A, Stevens Jr S E 1987 Functional expression of plastid allophycocyanin genes in a cyanobacterium. *J. Bact.* **169**: 1830–5

Devilly C I, Houghton J A 1977 A study of genetic transformation in *Gloeocapsa alpicola*. *J. Gen. Microbiol.* **98**: 277–80

de Waard A, Duyvesteyn M 1980 Are sequence-specific deoxyribonucleases of value as taxonomic markers of cyanobacterial species? *Arch. Microbiol.* **128**: 242–7

de Waard A, Korsuize J, van Beveren C P, Maat J 1978 A new sequence-specific endonuclease from *Anabaena cylindrica*. *FEBS Lett.* **96**: 106–10

de Waard A, van Beveren C P, Duyvesteyn M, van Ormondt H 1979 Two sequence-specific endonucleases from *Anabaena oscillarioides*. *FEBS Lett.* **101**: 71–6

Douglas S E, Doolittle W F 1984a Nucleotide sequence of the 5S rRNA gene and flanking regions in the cyanobacterium, *Anacystis nidulans*. *FEBS Lett.* **166**: 307–10

Douglas S E, Doolittle W F 1984b Complete nucleotide sequence of the 23S rRNA gene of the cyanobacterium, *Anacystis nidulans*. *Nucl. Acids Res.* **12**: 3373–86

Duyvesteyn M, de Waard A 1980 A new sequence-specific endonuclease from a thermophilic cyanobacterium, *Mastigocladus laminosus*. *FEBS Lett.* **111**: 423–6

Duyvesteyn M G C, Korsuize J. de Waard A 1981 Isolation and characterization of a sequence-specific deoxyriboendonuclease from *Calothrix scopulorum*. *Pl. Mol. Biol.* **1**: 75–9

Duyvesteyn M G C, Korsuize J, de Waard A, Vonshak A, Wolk C P 1983 Sequence-specific endonucleases in strains of *Anabaena* and *Nostoc*. *Arch. Microbiol.* **134**: 276–81

Dzelzkalns V A, Bogorad L 1986 Stable transformation of the cyanobacterium *Synechocystis* sp. PCC 6803 induced by UV irradiation. *J. Bact.* **165**: 964–71

Dzelzkalns V A, Owens G C, Bogorad L 1984 Chloroplast promoter driven expression of the chloramphenicol acetyl transferase gene in a cyanobacterium. *Nucl. Acids Res.* **12**: 8917–25

Engwall K S, Gendel S M, 1985 A rapid method for identifying plasmids in cyanobacteria. *FEMS Microbiol. Lett.* **26**: 337–40

Fisher R, Tuli R, Haselkorn R 1981 A cloned cyanobacterial gene for glutamine synthetase functions in *Escherichia coli*, but the enzyme is not adenylated. *Proc. Nat. Acad. Sci. USA* **78** 3393–7

Flores E, Wolk C P 1985 Identification of facultatively heterotrophic, N_2-fixing cyanobacteria able to receive plasmid vectors from *Escherichia coli* by conjugation. *J. Bact.* **162**: 1339–41

Friedberg D, Seijffers J 1979 Plasmids in two cyanobacterial strains. *FEBS Lett.* **107**: 165–8

Friedberg D, Seijffers J 1983 A new hybrid plasmid capable of transforming *Escherichia coli* and *Anacystis nidulans*. *Gene* **22**: 267–75

Gallagher M L, Burke Jr W F 1985 Sequence-specific endonuclease from the transformable cyanobacterium *Anacystis nidulans* R2. *FEMS Microbiol. Lett.* **26**: 317–21

Gatenby A A, van der Vies S M, Bradley D 1985 Assembly in *E. coli* of a functional multi-subunit ribulose bisphosphate carboxylase from a blue–green alga. *Nature (Lond.)* **314**: 617–20

Gendel S, Straus N, Pulleyblank D, Williams J 1983a A novel shuttle cloning vector for the cyanobacterium *Anacystis nidulans*. *FEMS Microbiol. Lett.* **19**: 291–4

Gendel S, Straus N, Pulleyblank D, Williams J 1983b Shuttle cloning vector for the cyanobacterium *Anacystis nidulans*. *J. Bact.* **156**: 148–54

Geoghegan C M, Houghton J A 1987 Molecular cloning and isolation of a cyanobacterial gene which increases the UV and methyl methanesulphonate survival of *rec*A strains of *Escherichia coli* K12. *J. Gen. Microbiol.* **133**: 119–26

Gingeras T R, Milazzo J P, Roberts R J 1978 A computer assisted method for the determination of restriction enzyme recognition sites. *Nucl. Acids Res.* **5**: 4105–27

Golden S S, Haselkorn R 1985 Mutation to herbicide resistance maps within the *psb*A gene of *Anacystis nidulans* R2. *Science* **229**: 1104–7

Golden S S, Sherman L A 1983 A hybrid plasmid is a stable cloning vector for the cyanobacterium *Anacystis nidulans* R2. *J. Bact.* **155**: 966–72

Golden S S, Sherman L A 1984 Optimal conditions for genetic transformation of the cyanobacterium *Anacystis nidulans* R2. *J. Bact.* **158**: 36–42

Golden S S, Brusslan J, Haselkorn R 1986 Expression of a family of *psb*A genes encoding a photosystem II polypeptide in the cyanobacterium *Anacystis nidulans* R2. *EMBO J.* **5**: 2789–98

Grigorieva G, Shestakov S 1982 Transformation in the cyanobacterium *Synechocystis* sp. 6803. *FEMS Microbiol. Lett.* **13**: 367–70

Gurevitz M, Somerville C R, McIntosh L 1985 Pathway of assembly of ribulosebisphosphate carboxylase/oxygenase from *Anabaena* 7120 expressed in *Escherichia coli*. *Proc. Nat. Acad. Sci. USA* **82**: 6546–50

Herdman M 1973a Transformation in the blue–green alga *Anacystis nidulans* and an associated phenomenon of mutation. In Archer L J (ed) *Bacterial Transformation*, Academic Press, New York, pp 369–86

Herdman M 1973b Mutations arising during transformation in the blue–green alga *Anacystis nidulans*. *Mol. Gen. Genet.* **120**: 369–78

Herdman M 1982 Evolution and genetic properties of cyanobacterial genomes. In Carr N G, Whitton B A (eds) *Biology of Cyanobacteria*, Blackwell, Oxford, pp 263–305

Herdman M, Carr N G 1971 Recombination in *Anacystis nidulans* mediated by an extracellular DNA/RNA complex. *J. Gen. Microbiol.* **68**: XIV

Herdman M, Carr N G 1972 The isolation and characterization of mutant strains of the blue–green alga *Anacystis nidulans*. *J. Gen. Microbiol.* **70**: 213–20

Herdman M, Janvier M, Waterbury J B, Rippka R, Stanier R Y, Mandel M 1979 Deoxyribonucleic acid base composition of cyanobacteria. *J. Gen. Microbiol.* **111**: 63–71

Hirschberg R, Samson S M, Kimmel B E, Page K A, Collins J J, Myers J A, Yarbrough L R 1985 Cloning and characterization of nitrogenase genes from *Anabaena variabilis*. *J. Biotechnol.* **2**: 23–37

Houmard J, Mazel D, Moguet C, Bryant D A, Tandeau de Marsac N 1986 Organization and nucleotide sequence of genes encoding core components of the phycobilisomes from *Synechococcus* 6301. *Mol. Gen. Genet.* **205**: 404–10

Hughes S G, Murray K 1980 The nucleotide sequences recognized by endonucleases *Ava*I and *Ava*II from *Anabaena variabilis*. *Biochem. J.* **185**: 65–75

Hughes S G, Bruce T, Murray K 1980 The isolation and characterization of a sequence-specific endonuclease from *Anabaena subcylindrica*. *Biochem. J.* **185**: 59–63

Karreman C, Tandeau de Marsac N, de Waard A 1986 Isolation of a deoxycytidylate methyl transferase capable of protecting DNA uniquely against cleavage by endonuclease R.*Aqu* I (isoschizomer of *Ava*I). *Nucl. Acids Res.* **14**: 5199–205

Katagiri F, Kodaki T, Fujita N, Izui K, Katsuki H 1985 Nucleotide sequence of the phosphoenolpyruvate carboxylase gene of the cyanobacterium *Anacystis nidulans*. *Gene* **38**: 265–9

Kawamura M, Sakakibara M, Watanabe T, Kita K, Hiraoka N, Obayashi A, Takagi T, Yano K 1986 A new restriction endonuclease from *Spirulina platensis*. *Nucl. Acids Res.* **14**: 1985–9

Kodaki T, Katagiri F, Asano M, Izui K, Katsuki H 1985 Cloning of phosphoenol-pyruvate carboxylase gene from a cyanobacterium, *Anacystis nidulans*, in *Escherichia coli. J. Biochem.* **97**: 533–9

Kolowsky K S, Williams J G K, Szalay A A 1984 Length of foreign DNA in chimeric plasmids determines the efficiency of its integration into the chromosome of the cyanobacterium *Synechococcus* R2. *Gene* **27**: 289–99

Kuhlemeier C J, Borrias W E, van den Hondel C A M J J, van Arkel G A 1981 Vectors for cloning in cyanobacteria: construction and characterization of two recombinant plasmids capable of transformation to *Escherichia coli* K12 and *Anacystis nidulans* R2. *Mol. Gen. Genet.* **184**: 249–54

Kuhlemeier C J, Hardon E M, van Arkel G A, van de Vate C 1985 Self-cloning in the cyanobacterium *Anacystis nidulans* R2: fate of a cloned gene after reintro-duction. *Plasmid* **14**: 200–8

Kuhlemeier C J, Logtenberg T, Stoorvogel W, van Heugten H A A, Borrias W E, van Arkel G A 1984a Cloning of nitrate reductase genes from the cyanobacterium *Anacystis nidulans. J. Bact.* **159**: 36–41

Kuhlemeier C J, Teeuwsen V J P, Janssen M J T, van Arkel G A 1984b Cloning of a third nitrate reductase gene from the cyanobacterium *Anacystis nidulans* R2 using a shuttle cosmid library. *Gene* **31**: 109–16

Kuhlemeier C J, Thomas A A M, van der Ende A, van Leen R W, Borrias W E, van den Hondel C A M J J, van Arkel G A 1983 A host–vector system for gene cloning in the cyanobacterium *Anacystis nidulans* R2. *Plasmid* **10**: 156–63

Kumano M, Tomioka N, Sugiura M 1983 The complete nucleotide sequence of a 23S rRNA gene from a blue–green alga, *Anacystis nidulans. Gene* **24**: 219–25

Kumar H D, 1962 Apparent genetic recombination in a blue–green alga. *Nature (Lond.)* **196**: 1121–2

Kuwabara T, Reddy K J, Sherman L A 1987 Nucleotide sequence of the gene from the cyanobacterium *Anacystis nidulans* R2 encoding the Mn-stabilizing protein involved in photosystem-II water oxidation. *Proc. Nat. Acad. Sci. USA* **84**: 8230–4

Lambert G R, Carr N G 1982 Rapid small-scale plasmid isolation by several methods from filamentous cyanobacteria. *Arch. Microbiol.* **133**: 122–5

Lambert G R, Scott J G, Carr N G 1984 Characterization and cloning of extra-chromosomal DNA from filamentous cyanobacteria. *FEMS Microbiol. Lett.* **21**: 225–31

Lammers P J, Haselkorn R 1983 Sequence of the *nif*D gene coding for the subunit of dinitrogenase from the cyanobacterium *Anabaena. Proc. Nat. Acad. Sci. USA* **80**: 4723–7

Lammers P J, Golden J W, Haselkorn R 1986 Identification and sequence of a gene required for a developmentally regulated DNA excision in *Anabaena. Cell* **44**: 905–11

Lau R H, Doolittle W F 1979 Covalently closed circular DNAs in closely related unicellular cyanobacteria. *J. Bact.* **137**: 648–52

Lau R H, Doolittle W F 1980 *Aqu*I: a more easily purified isoschizomer of *Ava*I. *FEBS Lett.* **121**: 200–2

Lau R H, Straus N A 1985 Versatile shuttle cloning vectors for the unicellular cyanobacterium *Anacystis nidulans* R2. *FEMS Microbiol. Lett.* **27**: 253–6

Lau R H, Alvarado-Urbina G, Lau P C K 1987b Phycocyanin α-subunit gene of *Anacystis nidulans* R2: cloning, nucleotide sequencing and expression in *Esche-richia coli. Gene* **52**: 21–9

Lau P C K, Condie J A, Alvarado-Urbina G, Lau R H 1987a Nucleotide sequence of phycocyanin β-subunit gene of cyanobacterium *Anacystis nidulans* strain R2. *Nucl. Acids Res.* **15**: 2394

Lau R H, Sapienza C, Doolittle W F 1980 Cyanobacterial plasmids: their widespread occurrence, and the existence of regions of homology between plasmids in the same and different species. *Mol. Gen. Genet.* **178**: 203–11

Lau R H, Visentin L P, Martin S M, Hofman J D, Doolittle W F 1985 Site-specific restriction endonuclease from the filamentous cyanobacterium *Nostoc* sp. MAC PCC 8009. *FEBS Lett.* **179**: 129–32

Laudenbach D E, Straus N A, Gendel S, Williams J P 1983 The large endogenous plasmid of *Anacystis nidulans*: mapping, cloning and localization of the origin of replication. *Mol. Gen. Genet.* **192**: 402–7

Lim C-J, Gleason F K, Fuchs J A 1986 Cloning, expression, and characterization of the *Anabaena* thioredoxin gene in *Escherichia coli. J. Bact.* **168**: 1258–64

Lind L K, Kalla S R, Lönneborg A, Oquist G, Gustafsson P 1985 Cloning of the β-phycocyanin gene from *Anacystis nidulans. FEBS Lett.* **188**: 27–32

Lomax T L, Conley P B, Schilling J, Grossman A R 1987 Isolation and characterization of light-regulated phycobilisome linker polypeptide genes and their transcription as a polycistronic mRNA. *J. Bact.* **169**: 2675–84

Mazel D, Guglielmi G, Houmard J, Sidler W, Bryant D A, Tandeau de Marsac N 1986 Green light induces transcription of the phycoerythrin operon in the cyanobacterium *Calothrix* 7601. *Nucl. Acids Res.* **14**: 8279–90

Mazur B J, Chui C-F 1982 Sequence of the gene coding for the β-subunit of dinitrogenase from the blue–green alga *Anabaena. Proc. Nat. Acad. Sci. USA* **79**: 6782–6

Mazur B J, Rice D, Haselkorn R 1980 Identification of blue–green algal nitrogen fixation genes by using heterologous DNA hybridization probes. *Proc. Nat. Acad. Sci. USA* **77**: 186–90

McFadden B A, Purohit K 1978 Chemosynthetic, photosynthetic, and cyanobacterial ribulose bisphosphate carboxylase. In Siegelman H W, Hind G (eds) *Photosynthetic Carbon Assimilation*, Plenum Press, New York, pp 179–207

Mevarech M, Rice D, Haselkorn R 1980 Nucleotide sequence of a cyanobacterial *nif*H gene coding for nitrogenase reductase. *Proc. Nat. Acad. Sci. USA* **77**: 6476–80

Mulligan B, Schultes N, Chen L, Bogorad L 1984 Nucleotide sequence of a multiple-copy gene for the B protein of photosystem II of a cyanobacterium. *Proc. Nat. Acad. Sci. USA* **81**: 2693–7

Murphy R C, Bryant D A, Porter R D, Tandeau de Marsac N 1987 Molecular cloning and characterization of the *rec*A gene from the cyanobacterium *Synechococcus* sp. strain PCC 7002. *J. Bact.* **169**: 2739–47

Murray K, Hughes S G, Brown J S, Bruce S A 1976 Isolation and characterization of two sequence-specific endonucleases from *Anabaena variabilis. Biochem. J.* **159**: 317–22

Nierzwicki-Bauer S A, Curtis S E, Haselkorn R 1984 Cotranscription of genes encoding the small and large subunits of ribulose-1,5-bisphosphate carboxylase in the cyanobacterium *Anabaena* 7120. *Proc. Nat. Acad. Sci. USA* **81**: 5961–5

Orkwiszewski K G, Kaney A R 1974 Genetic transformation of the blue–green bacterium, *Anacystis nidulans. Arch. Microbiol.* **98**: 31–7

Owttrim G W, Coleman J R 1987 Molecular cloning of a *rec*A-like gene from the cyanobacterium *Anabaena variabilis. J. Bact.* **169**: 1824–9

Pikalek P 1967 Attempt to find genetic recombination in *Anacystis nidulans. Nature (Lond.)* **215**: 866–7

Pilot T J, Fox J L 1984 Cloning and sequencing of the genes encoding the α and β subunits of C-phycocyanin from the cyanobacterium *Agmenellum quadruplicatum. Proc. Nat. Acad. Sci. USA* **81**: 6983–7

Porter R D 1986 Transformation in cyanobacteria. *Crit. Rev. Microbiol.* **13**: 111–32

Porter R D, Buzby J S, Pilon A, Fields P I, Dubbs J M, Stevens Jr. S E 1986 Genes from the cyanobacterium *Agmenellum quadruplicatum* isolated by complementation: characterization and production of merodiploids. *Gene* **41**: 249–60

Potts M 1984 Distribution of plasmids in cyanobacteria of the LPP group. *FEMS Microbiol. Lett.* **24**: 351–4

Prasad A B, Vaishampayan A 1984 Genetic recombination in *Nostoc muscorum* mutant strains. In Manna G K, Sinha U (eds) *Perspectives in Cytology and Genetics*, **4**: 179–84

Reaston J, Duyvesteyn M G C, de Waard A 1982 *Nostoc* PCC 7524, a cyanobacterium which contains five sequence-specific deoxyribonucleases. *Gene* **20**: 103–10

Reaston J, van den Hondel C A M J J, van der Ende A, van Arkel G A, Stewart W D P, Herdman M 1980 Comparison of plasmids from the cyanobacterium *Nostoc* PCC 7524 with two mutant strains unable to form heterocysts. *FEMS Microbiol. Lett.* **9**: 185–8

Rebière M-C, Castets A-M, Houmard J, Tandeau de Marsac N 1986 Plasmid distribution among unicellular and filamentous cyanobacteria: occurrence of large and mega-plasmids. *FEMS Microbiol. Lett.* **37**: 269–75

Reichelt B Y, Delaney S F 1983 The nucleotide sequence of the large subunit of ribulose-1,5-bisphosphate carboxylase from a unicellular cyanobacterium, *Synechococcus* PCC 6301. *DNA* **2**: 121–9

Reith M E, Laudenbach D E, Straus N A 1986 Isolation and nucleotide sequence analysis of the ferredoxin I gene from the cyanobacterium *Anacystis nidulans* R2. *J. Bact.* **168**: 1319–24

Riccardi G, De Rossi E, Della Valle G, Ciferri O 1985 Cloning of the glutamine synthetase gene from *Spirulina platensis. Pl. Mol. Biol.* **4**: 133–6

Rippka R, Deruelles J, Waterbury J B, Herdman M, Stanier R Y 1979 Generic assignments, strain histories and properties of pure cultures of cyanobacteria. *J. Gen. Microbiol.* **111**: 1–61

Roberts R J 1985 Restriction and modification enzymes and their recognition sequences. *Nucl. Acids Res.* **13**: 2165–200

Roberts T M, Koths K E 1976 The blue–green alga *Agmenellum quadruplicatum* contains covalently closed DNA circles. *Cell* **9**: 551–7

Roizes G, Nardeux P-C, Monier R 1979 A new specific endonuclease from *Anabaena variabilis. FEBS Lett.* **104**: 39–44

Ruvkun G B, Ausubel F M 1980 Interspecies homology of nitrogenase genes. *Proc. Nat. Acad. Sci. USA* **77**: 191–5

Schmetterer G, Wolk C P, Elhai J 1986 Expression of luciferases from *Vibrio harveyi* and *Vibrio fischeri* in filamentous cyanobacteria. *J. Bact.* **167**: 411–14

Sherman L A, van de Putte P 1982 Construction of a hybrid plasmid capable of replication in the bacterium *Escherichia coli* and the cyanobacterium *Anacystis nidulans. J. Bact.* **150**: 410–13

Shestakov S V, Khyen N T 1970 Evidence for genetic transformation in blue–green alga *Anacystis nidulans. Mol. Gen. Genet.* **107**: 372–5

Shinozaki K, Sugiura M 1983 The gene for the small subunit of ribulose-1,5-bisphosphate carboxylase/oxygenase is located close to the gene for the large subunit in the cyanobacterium *Anacystis nidulans* 6301. *Nucl. Acids Res.* **11**: 6957–64

Shinozaki K, Sugiura M 1985 Genes for the large and small subunits of ribulose-1,5-bisphosphate carboxylase/oxygenase constitute a single operon in a cyanobacterium *Anacystis nidulans* 6301. *Mol. Gen. Genet.* **200**: 27–32

Shinozaki K, Yamada C, Takahata N, Sugiura M 1983 Molecular cloning and sequence analysis of the cyanobacterial gene for the large subunit of ribulose-1,5-bisphosphate carboxylase/oxygenase. *Proc. Nat. Acad. Sci. USA* **80**: 4050–4

Simon R D 1978 Survey of extrachromosomal DNA found in the filamentous cyanobacteria. *J. Bact.* **136**: 414–18

Singh H N 1967 Genetic control of sporulation in the blue–green alga *Anabaena doliolum* Bharadwaja. *Planta* **75**: 33–8

Singh R N, Sinha R 1965 Genetic recombination in a blue–green alga, *Cylindrospermum majus* Kuetz. *Nature (Lond.)* **207**: 782–3

Singh D T, Nirmala K, Modi D R, Katiyar S, Singh H N 1987 Genetic transfer of herbicide resistance gene(s) from *Gloeocapsa* spp. to *Nostoc muscorum. Mol. Gen. Genet.* **208**: 436–8

Stassi D L, Lopez P, Espinosa M, Lacks S A 1981 Cloning of chromosomal genes in *Streptococcus pneumoniae. Proc. Nat. Acad. Sci. USA* **78**: 7028–32

Stevens Jr S E, Porter R D 1980 Transformation in *Agmenellum quadruplicatum. Proc. Nat. Acad. Sci. USA* **77**: 6052–6

Stevens Jr S E, Porter R D 1986 Heterospecific transformation among cyanobacteria. *J. Bact.* **167**: 1074–6

Stewart W D P, Singh H N 1975 Transfer of nitrogen-fixing (*nif*) genes in the blue–green alga *Nostoc muscorum. Biochem. Biophys. Res. Commun.* **62**: 62–9

Sutcliffe J G, Church G M 1978 The cleavage site of the restriction endonuclease AvaII. *Nucl. Acids Res.* **5**: 2313–19

Tabita F R, Small C L 1985 Expression and assembly of active cyanobacterial ribulose-1,5-bisphosphate carboxylase/oxygenase in *Escherichia coli* containing stoichiometric amounts of large and small subunits. *Proc. Nat. Acad. Sci. USA* **82**: 6100–3

Tandeau de Marsac N, Borrias W E, Kuhlemeier C J, Castets A M, van Arkel G A, van den Hondel C A M J J 1982 A new approach for molecular cloning in cyanobacteria: cloning of an *Anacystis nidulans met* gene using a Tn901-induced mutant. *Gene* **20**: 111–19

Tandeau de Marsac N, de la Torre F, Szulmajster J 1987 Expression of the larvicidal gene of *Bacillus sphaericus* 1593M in the cyanobacterium *Anacystis nidulans* R2. *Mol. Gen. Genet.* **209**: 396–8

Tandeau de Marsac N, Mazel D, Bryant D A, Houmard J 1985 Molecular cloning and nucleotide sequence of a developmentally regulated gene from the cyanobacterium *Calothrix* PCC 7601: a gas vesicle protein gene. *Nucl. Acids Res.* **13**: 7223–36

Taylor D P, Cohen S N, Clark W G, Marrs B L 1983 Alignment of genetic and restriction maps of the photosynthesis region of the *Rhodopseudomonas capsulata* chromosome by a conjugation-mediated marker rescue technique. *J. Bact.* **154**: 580–90

Tiboni O, Di Pasquale G 1987 Organization of genes for ribosomal proteins S7 and S12, elongation factors EF-Tu and EF-G in the cyanobacterium *Spirulina platensis. Biochim. Biophys. Acta* **908**: 113–22

Tiboni O, Di Pasquale G, Ciferri O 1984 Cloning and expression of the genes for ribulose-1,5-bisphosphate carboxylase from *Spirulina platensis. Biochim. Biophys. Acta* **783**: 258–64

Tomioka N, Sugiura M 1983 The complete nucleotide sequence of a 16S ribosomal RNA gene from a blue–green alga, *Anacystis nidulans. Mol. Gen. Genet.* **191**: 46–50

Tomioka N, Sugiura M 1984 Nucleotide sequence of the 16S–23S spacer region in the *rrn*A operon from a blue–green alga, *Anacystis nidulans. Mol. Gen. Genet.* **193**: 427–30

Tomioka N, Shinozaki K, Sugiura M 1981 Molecular cloning and characterization of ribosomal RNA genes from a blue–green alga, *Anacystis nidulans. Mol. Gen. Genet.* **184**: 359–63

Trehan K, Sinha U 1981 Genetic transfer in a nitrogen-fixing filamentous cyanobacterium. *J. Gen. Microbiol.* **124**: 349–52

Trehan K, Sinha U 1982 DNA-mediated transformation in *Nostoc muscorum*, a nitrogen-fixing cyanobacterium. *Aust. J. Biol. Sci.* **35**: 573–7

Tumer N E, Robinson S J, Haselkorn R 1983 Different promoters for the *Anabaena* glutamine synthetase gene during growth using molecular or fixed nitrogen. *Nature (Lond.)* **306**: 337–42

Vaishampayan A, Prasad A B 1984 Inter-strain transfer of a pesticide-resistant marker in the N_2-fixing cyanobacterium *Nostoc muscorum. Mol. Gen. Genet.* **193**: 195–7

Vakeria D, Codd G A, Hawthornthwaite A M, Stewart W D P 1986 Construction of a gene bank and identification of the ribulose bisphosphate carboxylase/oxygenase genes from the nutritionally versatile cyanobacterium *Chlorogloeopsis fritschii. Arch. Microbiol.* **145**: 228–33

van den Hondel C A M J J, van Arkel G A 1980 Development of a cloning system in cyanobacteria. *Antonie van Leeuwenhoek* **46**: 228–9

van den Hondel C A M J J, Keegstra W, Borrias W E, van Arkel G A 1979 Homology of plasmids in strains of unicellular cyanobacteria. *Plasmid* **2**: 323–33

van den Hondel C A M J J, van Leen R W, van Arkel G A, Duyvesteyn M, de Waard A 1983 Sequence-specific nucleases from the cyanobacterium *Fremyella diplosiphon*, and a peculiar resistance of its chromosomal DNA towards cleavage by other restriction enzymes. *FEMS Microbiol. Lett.* **16**: 7–12

van den Hondel C A M J J, Verbeek S, van der Ende A, Weisbeek P J, Borrias W E, van Arkel G A 1980 Introduction of transposon Tn901 into a plasmid of *Anacystis nidulans*: preparation for cloning in cyanobacteria. *Proc. Nat. Acad. Sci. USA* **77**:1570–4

van der Plas J, de Groot R P, Weisbeek P J, van Arkel G A 1986a Coding sequence of a ferredoxin gene from *Anabaena variabilis* ATCC 29413. *Nucl. Acids Res.* **14**: 7803

van der Plas J, de Groot R P, Woortman M R, Weisbeek P J, van Arkel G A 1986b Coding sequence of a ferredoxin gene from *Anacystis nidulans* R2 (*Synechococcus* PCC7942). *Nucl. Acids Res.* **14**:7804

van der Vies S M, Bradley D, Gatenby A A 1986 Assembly of cyanobacterial and higher plant ribulose bisphosphate carboxylase subunits into functional homologous and heterologous enzyme molecules in *Escherichia coli. EMBO J.* **5**: 2439–44

van Haute E, Joos H, Maes M, Warren G, van Montagu M, Schell J 1983 Intergeneric transfer and exchange recombination of restriction fragments cloned in pBR322: a novel strategy for the reversed genetics of the Ti plasmids of *Agrobacterium tumefaciens. EMBO J.* **2**: 411–17

Vermaas W F J, Williams J G K, Arntzen C J 1987 Sequencing and modification of *psb*B, the gene encoding the CP-47 protein of photosystem II, in the cyanobacterium *Synechocystis* 6803. *Pl. Mol. Biol.* **8**: 317–26

Whitehead P R, Brown N L 1982 AhaIII: a restriction endonuclease with a recognition sequence containing only A:T basepairs. *FEBS Lett.* **143**: 296–300

Whitehead P R, Brown N L 1985a Three restriction endonucleases from *Anabaena flos-aquae. J. Gen. Microbiol.* **131**: 951–8

Whitehead P R, Brown N L 1985b A simple and rapid method for screening bacteria for type II restriction endonucleases: enzymes in *Aphanothece halophytica. Arch. Microbiol.* **141**:70–4

Williams J G K 1987 Construction of specific mutations in the photosystem II photosynthetic reaction center by genetic engineering methods in the cyanobacterium *Synechocystis* 6803. In Povoker L, Glazer A N (eds) *Methods in Enzymology* Academic Press, New York, Vol **167**: 766–78

Williams J G K, Szalay A A 1983 Stable integration of foreign DNA into the chromosome of the cyanobacterium *Synechococcus* R2. *Gene* **24**: 37–51

Williamson S E, Doolittle W F 1983 Genes for tRNA[Ile] and tRNA[Ala] in the spacer between the 16S and 23S rRNA genes of a blue–green alga: strong homology to chloroplast tRNA genes and tRNA genes of the *E. coli rrn*D gene cluster. *Nucl. Acids Res.* **11**: 225–35

Wolk C P, Vonshak A, Kehoe P, Elhai J 1984 Construction of shuttle vectors capable of conjugative transfer from *Escherichia coli* to nitrogen-fixing filamentous cyanobacteria. *Proc. Nat. Acad. Sci. USA* **81**: 1561–5

Yoshikawa H 1966 Mutations resulting from the transformation of *Bacillus subtilis*. *Genetics* **54**: 1201–14

Note added in proof: While this chapter was in preparation, related topics were dealt with by Tandeau de Marsac and Houmard (1987).

Tandeau de Marsac N, Houmard J 1987 Advances in cyanobacterial molecular genetics. In Fay P, van Baalen C (eds) *The cyanobacteria*, Elsevier, Amsterdam, pp 251–302

9

Immobilized cells: An appraisal of the methods and applications of cell immobilization techniques

M. Brouers, H. deJong, D. J. Shi and D. O. Hall

Introduction

In the past two decades, the use of immobilized microbial and plant cell components as biocatalysts has become a rapidly advancing area of biotechnology; the interest in immobilization as a method for the production of metabolites accounts for a significant number of scientific reports that appear in biotechnological journals. Several reviews and books describe the characteristics, the preparation and applications of immobilized biocatalysts (Mosbach, 1982; Bucke, 1983, 1986; Atkinson, 1986; Brodelius and Mosbach, 1987; Gisby et al., 1987; Brouers et al., 1988).

Initially most of the research was centred around immobilization of enzymes. Subsequently, the fact that isolation and purification of intracellular enzymes was an expensive and often inefficient process and the difficulties encountered with immobilization of multienzymatic complexes, has led to the development of immobilized cell systems. In parallel, new simpler and easier ways of immobilization were developed. In this regard, it must be emphasized that many microorganisms normally exist naturally in an immobilized-like state, either as films on a surface (interaction of bacteria and surfaces) or entrapped within gels or slimes of their own synthesis. They can also be encapsulated as a part of a more complex system such as in symbiosis, e.g. symbiotic algae and cyanobacteria.

Until recently known commercial applications were few due to the difficulty of introducing new techniques into well established processes. However, examples of individual applications using immobilized cells are increasing significantly and showing that the prospects are fundamentally bright.

In comparison to batch or continuous culture fermentations where free cells are used, immobilized cells may offer certain specific advantages such as (1) accelerated reaction rates due to increased cell density per unit volume, (2) increased cell metabolism and cell wall permeability, (3) no wash-out of cells, (4) higher operational stability and better control of the catalytic processes, and (5) separation and reuse of catalyst. In addition, one of the main features of immobilization systems is the reduction of costs due to the easier separation of cells and excreted product.

Many methods are available for the immobilization of cells but several conditions must be fulfilled if the development of an industrial process is envisaged. The method must be safe and thus employ chemically inert materials that present no hazard to operators or cells; it must be simple and must lead to a long-lived process. This implies resistance to abrasion of the matrix, long-term maintenance of cell activity and avoiding extremes of heat and pH; it must be cheap in order to compete successfully with the alternative processes.

Cell immobilization techniques have developed from the experience gained with enzyme immobilization. There are four important procedures available for cell and enzyme immobilization, viz. adsorption, entrapment in gels or polymers, covalent coupling, and cross linking to insoluble matrices. A variety of matrices are now commercially available to the biotechnologist interested in cell immobilization and include polyacrylamide, agar, carrageenan, calcium alginate gels, glass, ceramic and silica beads, and polyurethane and polyvinyl foams. The choice of the matrix and the immobilization procedure will depend on the nature of the cell or organelle to be immobilized, the nature of the substrates and products formed, and on the reaction conditions.

Entrapment in porous gels and foams has become the most popular technique for the immobilization of chloroplasts, algae, cyanobacteria, plant cells, etc., as they often do not involve reagents which could modify or damage the organelle or cell. A special requirement for photosynthetic systems, where light is one of the substrates, is that the material should be translucent or transparent. Because of this limitation the most commonly used matrices are alginate and agar gels and polyurethane and polyvinyl foams. A disadvantage of the use of agar and alginate is their low mechanical stability for long-term use in bioreactors. Moreover, calcium alginate gels are disrupted by phosphate ions. Polyurethane and polyvinyl matrices offer better mechanical properties and are neutral to most commonly used ions. Some of the techniques that have been used for immobilization of photosynthetic organelles and cells are listed in Table 9.1.

Photosynthesis by plants, algae, cyanobacteria (blue–green algae) and photosynthetic bacteria converts large quantities of solar radiation into chemical energy in the form of carbohydrates, lipids (oils and fats), proteins, ammonia, hydrogen, ATP, pyridine nucleotides, etc. The importance of photosynthetic processes as energy converters lies in the fact that the substrates used such as water, CO_2 and N_2 are inexpensive and readily available.

To date there have been a small number of reports of the effects of immobilization on photosynthetic activities of immobilized microbes. The thermophilic cyanobacterium *Mastigocladus laminosus*, for example, has been successfully immobilized on an SnO_2 optically transparent electrode, which exhibited both photosystem I and photosystem II activities in its intact

Table 9.1 Immobilization of photosynthetic systems

Organelle or cells	Immobilization agent and conditions
Chloroplasts (Thylakoids)	0.05% Glutaraldehyde at 4 °C
Chloroplasts (Thylakoids)	Encapsulation in protamine + toluene diisocyanate
Chloroplasts (Thylakoids)	0.37% Glutaraldehyde + serum albumin at −30°C
Chloroplasts (Thylakoids)	3% Polyvinyl alcohol + serum albumin
Chloroplasts (Thylakoids)	Agar in hollow fibre reactor
Chloroplasts (Thylakoids)	2% Calcium alginate as films
Chloroplasts (Thylakoids)	0.063% Glutaraldehyde + 5% gelatin
Chloroplasts (Thylakoids)	Polyurethane + 5% serum albumin
Chloroplasts (Thylakoids)	2% Agar + 0.5% serum albumin
Mastigocladus laminosus	5% Calcium alginate deposited on SnO_2 electrode
Mastigocladus laminosus	2% Agar or 2% calcium alginate
Mastigocladus laminosus *Nostoc muscorum* *Chlorogloea fritschii*	Polyurethane foam pieces (cyanobacteria cultured in the presence of foam)
Anabaena cylindrica	Cells suspended in glass beads and cultured outdoors aerobically
Scenedesmus obliquus	(a) Polyurethane foam pieces and kept in nutrient flux
	(b) Calcium alginate
Porphyridium purpureum	(a) Polyurethane foam pieces and kept in nutrient flux
	(b) Calcium alginate
Anabaena 27893	Calcium alginate beads in a fluidized bed column
Rhodospirillum rubrum	Agar slabs
Rhodopseudomonas capsulata chromatophores	2% Barium alginate
Chromatium Miami PBS 1071	Agar gel matrix

From Cammack *et al.*, 1985.

state, and functioned as a 'living electrode' catalysing the photodecomposition of water and producing a steady electrical current for 20 days (Ochiai *et al.*, 1980).

After immobilization of *Chlorella vulgaris* and *Scenedesmus obliquus* in urethane prepolymer, photosynthetic oxygen evolution was virtually unchanged for 25 days (Brouers *et al.*, 1982). It has been reported recently that undiminished capacities for photosynthetic oxygen evolution and glycollate excretion were obtained over a six month period after the immobilization of *Chlorella emersonii* by entrapment in calcium alginate gel

Stability and photosynthetic activity after immobilization compared to free

Higher stability in storage. No improvement in light stability.
H_2 evolution with ferredoxin and hydrogenase.
PSI Activity retained. Loss of PSII activity.

70% O_2 evolution activity. 27% ATP synthesis.
Stable in light for 9 days.
100% PSI and 80% PSII activity retained. Only 20% loss of activity in 5 weeks storage.
NADP reduction for 2 h; 31% activity after one week at 5 °C.

100% O_2 exchange, up to 67% H_2 evolution. Light stability increased to 7 h compared to 4 h for non-immobilized.
45% of activity in ferricyanide reduction. 50% ferricyanide reduction activity; stability up to 400 h.
PSII and PSI stabilized without loss of activity. Better H_2 evolution rates in PSI-mediated H_2 evolution with other catalysts also immobilized.
70% PSII and 50% PSI activity. Photocurrent generated on continuous illumination for 20 days.
Continuous PSI-mediated H_2 production with methylviologen and hydrogenase for more than 10 days.
Both PSII and PSI activity retained. Higher stability of PSI in light. PSI mediated H_2 production with exogenous hydrogenase in continuous light lasted more than 10 days.
Continuous H_2 production for more than 20 days.

O_2 evolution maintained for one to three months. H_2 evolution after anaerobic adaptation.

O_2 evolution maintained for one to three months. H_2 evolution after anaerobic adaptation.
No loss in O_2 evolution rate even after 48 days. Sustained photosynthesis and N_2 fixation over a 130 h period.
Continuous photo H_2 production with malate for 150 h in a reactor.
70% of ATP synthesis activity.

Continuous H_2 production for several weeks from sulphide compared to free cells which lasted only for days.

(Day and Codd, 1985). An alginate matrix conserved the photosynthetic activity and fluorescence characteristics of the green alga *Scenedesmus obliquus* (Jeanfils and Collard, 1983), chloroplasts (Jeanfils *et al.*, 1986) and the cyanobacterium *Phormidium laminosum* (Brouers and Hall, 1985). On the contrary, the serum albumin glutaraldehyde matrix inhibited algal photosynthetic activity and modified the kinetics of the room temperature 690 nm fluorescence (Jeanfils and Collard, 1983). *Botryococcus braunii* immobilized in alginate gel has a higher photosynthetic capacity than free cells (Baillez *et al.*, 1983, 1986). It is unfortunate that the otherwise thorough study of Adlercreutz and Mattiasson (1982) on oxygen evolution by immobilized *Chlorella* made no comparison with that of free-living cells. The induction kinetics of variable fluorescence of immobilized *Porphyridium cruentum* cells in a urethane prepolymer showed the same characteristics as observed with free-living cells and photosynthetic oxygen evolution was the same for both free-living and immobilized cells (Thepenier *et al.*, 1985). For *Euglena gracilis* the immobilization of cells in a calcium alginate matrix maintained respiratory and photosynthetic activities and ultrastructural integrity. Moreover, immobilization did not prevent *Euglena* cells from greening inside the gel beads (Tamponnet *et al.*, 1985). *Chlorella emersonii* cells entrapped in a calcium alginate matrix had a higher chlorophyll content than free cells (Robinson *et al.*, 1985).

Other groups of workers have reported on the production of metabolites by immobilized algae, cyanobacteria and photosynthetic bacteria. Most-studied metabolites are hydrogen and NADPH (green algae and cyanobacteria), ammonia and amino acids (cyanobacteria), polysaccharides (green and red algae) and hydrocarbons (green algae, *Botryococcus*). A number of studies have been done on hydrogen photoproduction, using a wide variety of supports and substrates. Of special interest are systems capable of using solar energy in the biophotolytic decomposition of water to produce hydrogen and oxygen. Chloroplasts co-immobilized with *Clostridium butyricum* are capable of producing hydrogen for up to six hours (Karube *et al.*, 1981); however, whole cells of *Chlorella vulgaris* which are co-immobilized with *Clostridium butyricum* and supplemented with 0.8 mM NADP, are capable of continuously evolving hydrogen for six days (Kayano *et al.*, 1981). Various authors have observed hydrogen evolution by cultures of photosynthetic bacteria immobilized in agar or alginate (Vincenzini *et al.*, 1981; Weetall *et al.*, 1981). Usually they utilize organic carbon substrates or inorganic sulphides as electron donors. Cyanobacteria, however, are able to use water in the process of photolysis as the substrate to produce hydrogen through the utilization of light energy. Lambert *et al.*, (1979) reported that *Anabaena cylindrica* immobilized with glass beads catalysed hydrogen production at 30-fold greater rates than free-living cells for up to 30 days (Lambert and Smith, 1981).

Continuous hydrogen production mediated by PSI (Photosystem I) with ascorbate as the electron donor, in the presence of methylviologen and a bacterial hydrogenase as catalysts, was studied using several species of cyanobacteria both in suspension and in the immobilized state (Smith *et al.*, 1982; Muallem *et al.*, 1983). The H_2 photoproduction activity of the cyanobacteria was maintained under saturating illumination for at least nine days and the stability of H_2 evolution activity was extended further by immobilization of the cells in polyvinyl and polyurethane foams and also agar. The photoproduction of NADP by *Nostoc muscorum* cells immobilized in polyurethane foams was also demonstrated (Muallem *et al.*, 1983). Photoproduction of hydrogen by adapted (anaerobic dark treatment) *Scenedesmus* cells immobilized in polyurethane and alginate was demonstrated by Brouers *et al.* (1983). High rates of hydrogen photoproduction (up to 13 μl H_2 mg dry wt^{-1} h^{-1}) over prolonged periods by the cyanobacterium *Oscillatoria* sp.

(Miami BG7) immobilized in agar were reported by Philips and Mitsui (1986).

Photoproduction of ammonia by N_2-fixing cyanobacteria immobilized in a Ca-alginate matrix was demonstrated by Musgrave *et al.* (1982) and Kerby *et al.* (1986) using either inactivation of glutamine synthetase (GS) by L-methionine-D,L-sulphoximine (MSX) or using GS-deficient selected mutant strains. Continuous photoproduction of amino acids was also demonstrated for a period of 10 weeks in air-lift reactors from selected mutant cyanobacterial strains immobilized in Ca-alginate beads (Kerby *et al.*, 1987). Immobilization of the red alga *Porphyridium cruentum* in polyurethane foams was used by Thepenier *et al.* (1985) for the long-term production of sulphated polysaccharides. Production of hydrocarbons by *Botryococcus braunii* immobilized in polyurethane or alginate matrices was studied by Baillez (1984).

In addition to the various metabolites listed above, it must be emphasized that other compounds synthesized by algae and cyanobacteria such as antibiotics, vitamins, plant growth regulators, pigments, etc. (with particular interest in pharmaceutical, agricultural or agro-food industry) are envisaged as potential products from photobioreactors by immobilized photosynthetic cell systems.

The foregoing short review has shown that immobilization of cells is a new and valuable technology and that immobilized photosynthetic cells have unique advantages. However, up to now very little is known about the mechanisms which induce changes in photosynthetic cell function and structure that occur when cells are immobilized. One of the methods used in our laboratory to find at least a partial answer to this question was to compare the characteristics of cells in free-living state, in a natural immobilization state (symbiosis) and in an artificial immobilized system. The chosen model is the cyanobacterium *Anabaena azollae* either as part of the symbiotic association in the water fern *Azolla* or as the presumptive isolate grown in independent free-living culture and then immobilized in polyvinyl or polyurethane foam (Shi and Hall, 1988). One of the main features of the symbiotic *A. azollae* cells is their ability to release ammonia to the host. In experiments with symbiotic *A. azollae* recently isolated from the symbiotic association, it was reported that about 40% of the total nitrogen fixed by the cyanobacterium was found excreted into the suspension medium as ammonia (Meeks *et al.*, 1985). Another characteristic of symbiotic cells is the high heterocyst frequency as compared with free-living cyanobacterial cells. These characteristics of the symbiotic association are the main reason for its use in Asia and other parts of the world as a biofertilizer to enrich flooded rice paddies.

We have examined the morphology, compared the hydrogenase and nitrogenase activities, and determined the yield of ammonia excreted by *A. azollae* in the free-living state or immobilized in polyurethane, polyvinyl or alginate matrices. The practical aim of the research is to screen the potentialities for use of such immobilized systems for the production of nitrogen biofertilizers.

Methods used in our research

Cyanobacterium

Anabaena azollae, a presumptive isolate from *Azolla filiculoides* (Tel-Or *et al.*, 1983), obtained from Dr E. Tel-Or (Hebrew University, Rehovot, Israel), was grown in BG-11 medium (Stanier *et al.*, 1971) without combined nitrogen at 28 °C under 'cool-white' fluorescent lamps at a photon flux density of 75 μmol m^{-2} s^{-1}. The cells were grown in a 5% CO_2/air mixture in 250 ml Erlenmeyer flasks kept agitated on a rotary shaker at 125–140 rpm.

Immobilization of cyanobacterial cells

This was carried out in polyurethane (PU) foam (codes 3300A, 4300A and 74165A) and in hydrophilic polyvinyl (PV) foam (codes D, Jan. 84 and PR22/60; Caligen Foam Ltd, Accrington, Lancs., UK). The foams were cut into 5 mm cubes and washed 3–5 times in distilled water for a few days with the bubbles being removed from the foam after each wash by squeezing. One hundred pieces of foam were added to each 250 ml flask containing 140 ml growth medium; the flasks and contents were then autoclaved at 120 °C and 101.3 kPa. After cooling, the media were inoculated with the cyanobacteria and grown under the same conditions as free-living cells. After a few days, cell growth was observed on the surface and in the internal pores of the foam where the immobilized cells adhered. When the cultures were kept for a prolonged time the growth media were renewed aseptically every two weeks. Alternatively, immobilization of cyanobacteria was performed by mixing an aqueous suspension of cells with a urethane prepolymer. We have routinely used the commercially available TDI (tolylene diisocyanate)-based hydrophilic urethane prepolymer Hypol FHP2002 with low TDI content (W. R. Grace Ltd, UK). A typical procedure consists of mixing a concentrated suspension of cyanobacteria with an equal amount (w/w) of prepolymer. The mixture was first mixed for 1 min in an ice bath and polymerization was then continued at room temperature without mixing. A spongy blue–green mass forms in the vial. After complete polymerization (*c.* 15 min) the foam was cut into small pieces (*c.* 5 mm square), washed several times with nutrient medium and then suspended in the medium for continued growth or direct use in photobioreactors.

Immobilization of cells in alginate beads was carried out as follows: 5.5 g sodium alginate (Protonal 10/60; Protan A/S, Drammen, Norway) was dissolved in 150 ml growth medium at 80 °C. After cooling, the solution was mixed with 50 ml of sterile concentrated suspension of cyanobacteria. The sodium alginate–cyanobacterium mixture was then added dropwise from a separating funnel into a 0.1 M $CaCl_2$ solution at room temperature. The alginate beads formed by crosslinking via Ca^{2+} were harvested, washed in growth medium, resuspended in fresh growth medium and cultured in 2 L Erlenmeyer flasks. The growth and wash media were prepared so as not to contain phosphate in order to prevent the formation of calcium phosphate. Free-living control cultures were grown in parallel; for renewal of the medium they were centrifuged every two weeks and the pellets transferred to fresh, sterile growth medium.

Hydrogen production

Hydrogen production by nitrogenase was assayed by incubating the cell suspensions or immobilized cells in 7 ml glass vials fitted with Suba-seal stoppers in an argon/4% CO atmosphere. Incubates were shaken continuously at 28 °C and illuminated with white incandescent lamps at a photon flux density of 230–250 μmol m^{-2} s^{-1}. Hydrogen evolved was measured using a gas chromatograph (Taylor Servomex, Crowborough, UK). Hydrogen production by hydrogenase was measured, in the presence of nitrogenase, as hydrogen formation from reduced methylviologen (Daday et al., 1979). The assay was done in two replicates.

Nitrogen fixation

Nitrogenase activity was assayed using the acetylene-reduction technique. The method used was essentially similar to that developed by Stewart et al. (1968). The incubation conditions were the same as with H_2 production except C_2H_2 (about 10% of the gas phase in the vials) was added instead of CO. The ethylene production was measured using gas chromatography.

Ammonia production

In batch experiments, immobilized or free-living cyanobacteria were incubated in the light in the growth medium for defined periods (generally 24 h) in the presence or absence of MSX. The media were sampled at the end of each period for determination of ammonia concentration by the colorimetric method of Solorzano (1969). The algae were then resuspended in the growth media for the next experimental period. The scheme for the experimental device for continuous measurement of ammonia in the effluent of a continuous flow packed bed reactor is shown in Fig. 9.1. The reactor consisted of a column (2 cm internal diameter, 25 cm high) containing 50 ml liquid nutrient medium and packed with polyvinyl foam pieces (c. 200) with immobilized cyanobacteria (total chlorophyll, 1–4 mg). Nutrient medium (BG-11 with or without MSX) was added at the top of the column and collected from the bottom using peristaltic pumps; the dilution rate was 0.4 h^{-1} (flow 20 ml h^{-1}). The outflow effluent was directed towards a mixing cell (5 ml total volume) for continuous addition of 1 M NaOH + 0.1 M NaEDTA (10% v/v) in order to liberate ammonia. The resulting ammonia solution was pumped to an ammonia electrode (Kent Industrial Measurements Ltd, UK) fitted with a flow device for continuous monitoring of ammonia concentration by recording the electrode potential. Calibration was performed by pumping standard NH_4Cl solution in the mixing cell.

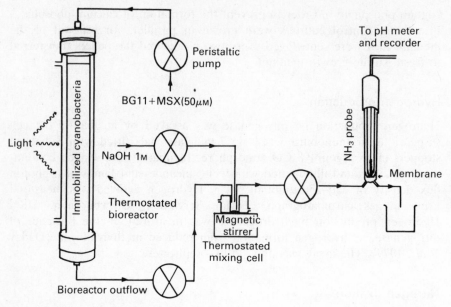

Fig. 9.1 Experimental device for continuous measurement of ammonia production

Samples of the effluent were collected from time to time for parallel measurement of ammonia by the colorimetric method of Solorzano (1969). The reactor column and mixing cell were thermostated at 28°C. Cool-white fluorescent lamps were used for illumination of both sides of the bioreactor (photon flux density 100 μmol m^{-2} s^{-1} at the surface of the column).

Chlorophyll determination

Free-living cells were pelleted by centrifugation (6000 g, 5 min). Cells were released from alginate beads containing immobilized cyanobacteria by dissolving the alginate in 0.1 M sodium citrate prior to centrifugation. Chlorophyll (Chl) a was extracted from these pellets, or foam particles containing cyanobacterial cells, by homogenization in a ground-glass homogenizer with 80% acetone followed by sonication (3 × 40 s, 60 W, 1.5 A; MSE, Crawley, Sussex, UK). Extraction of chlorophyll was continued overnight at 3°C in the dark. The suspensions were then centrifuged as above and chlorophyll a was determined in the supernatant by spectrophotometry at 663 nm (absorption coefficient Chl a = 89 × 10^{-3} l.mg^{-1}.cm^{-1}). An alternative method used was to extract the cells with 90% methanol overnight under the same conditions as above but without homogenization and to determine Chl a at 665 nm (absorption coefficient Chl a = 75 × 10^{-3} l.mg^{-1}.cm^{-1}).

Estimation of heterocysts

Heterocysts were observed under phase contrast at 400× magnification.

Material for microscopic observation was fresh or fixed in glutaraldehyde (2.5% final concentration). Estimation was done using a Hand Tally Counter and Standard Counting Chamber (Gelman Hawksley, Northampton, UK); triplicate counts of at least 1000 cells were done on each sample. Heterocyst frequency was expressed as a percentage of the total number of cells, the mean value being given.

Scanning electron microscopy (SEM)

Free-living or immobilized cyanobacterial cells were viewed either as fixed, critical-point-dried specimens or following low-temperature preparation. For low-temperature SEM, specimens were prepared and viewed essentially as described by Robins *et al.* (1986). Alternatively, cells were fixed in 2.5% glutaraldehyde in 0.1 M phosphate (pH 7.2) containing 6% sucrose (w/v) for 2 h at room temperature. The samples were washed in the same buffer and transferred to 1% osmium tetroxide in buffer under the same conditions. The specimens were washed with buffer, dehydrated in graded (50–100%) ethanol and dried in a Samdri-780 Critical Point Drying Apparatus (Tousimis Research Corp., Rockville, Md, USA) using liquid carbon dioxide. Dry specimens were coated with platinum in a Sputter Coater (EM-Scope, London, UK) and specimens examined in a model S-510 scanning electron microscope (Hitachi Scientific Instruments Co., Tokyo, Japan) at an accelerating voltage of 25 kV.

Discussion of research results

Structure and growth of immobilized and symbiotic cyanobacteria

Measurement of chlorophyll at various times following the immobilization process in polyvinyl foam or in alginate matrices demonstrated a progressive colonization of the matrix. In freshly prepared alginate beads with immobilized *A. azollae* the chlorophyll content was *c.* 60 μg per 100 beads: it increased to 350 μg per 100 beads during the first 40 days following immobilization and then remained approximately constant for up to four months. Forty days after immobilization the chlorophyll content per unit volume of the alginate matrix (proportional to the biomass loading) was 77 μg ml^{-1} which was five times greater than a 10-day-old free-living culture. Scanning electron microscope studies (Figs. 9.2, 9.3) showed that polyvinyl and polyurethane foam-immobilized cyanobacteria were imprisoned inside pores and closely connected with the matrix (Fig. 9.2c). One form of connection covered the whole surface of the cell interacting with the foam surface (Fig. 9.2a), whereas a second form of connection observed in polyurethane foam consisted of filaments (possibly composed of extracellular polysaccharides) connecting the cell to the matrix (Fig. 9.2b). A more detailed analysis of immobilized (Fig. 9.3c,d) and symbiotic *A. azollae* (Fig. 9.3a,b) using the frozen hydrated SEM technique (Shi *et al.*, 1987) indicated close similarities

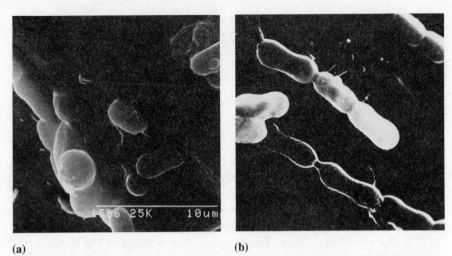

(a)   **(b)**

Fig. 9.2(a),(b)   Scanning electron microscope (SEM) pictures of *Anabaena azollae* immobilized in polyurethane foam. (Brouers and Hall, 1986.)

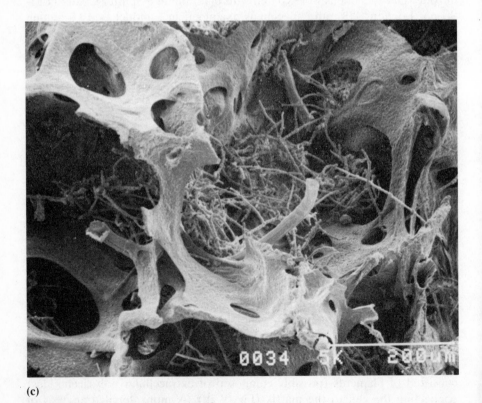

(c)

Fig. 9.2(c)   Scanning electron microscope picture of *A. azollae* growing in polyvinyl foam. (Brouers *et al.*, 1988.)

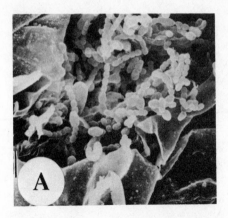
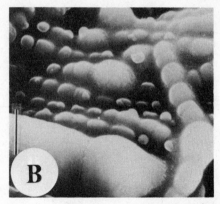

Fig. 9.3(a),(b) Scanning electron micrographs of the frond cavity of *Azolla* spp. showing the associated symbiotic cyanobacterium *A. azollae*: (a) critical-point dried specimens of *A. filiculoides*. Low-power view showing densely packed filaments within cavity. Bar = 20 µm, × 500; (b) lightly-etched frozen-hydrated specimens of *A. imbricata*. Filaments and transfer cell covered in a thick layer of mucilage. Bar = 10 µm, × 1680. (Shi *et al.*, 1987.)

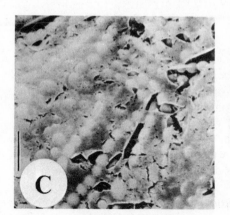
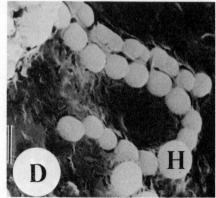

Fig. 9.3(c),(d) Scanning electron micrographs of frozen-hydrated *A. azollae* immobilized in polyvinyl foam PR22/60 (10 days old): (c) was lightly etched; (d) was sublimed for about 40 min. (c) Medium-power view of filaments. Note high heterocyst (H) frequency. Bar = 5 µm. (d) Filaments of cells underlying a thick layer of mucilage. Bar = 10 µm. (Shi *et al.*, 1987.)

between immobilized cells and cells living in the symbiotic association. In both cases a thick mucilage layer covered the cell surface. Although also present in free-living cells, this layer was then much thinner. Another interesting observation was that the heterocyst frequency of the immobilized *A. azollae* doubled relative to free-living cells and reached a level of 14–17% (Fig. 9.4). This again resembles the situation in the symbiotic *A. azollae*.

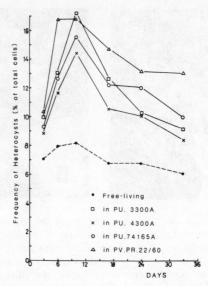

Fig. 9.4 Heterocyst frequency of free-living and immobilized *A. azollae* during incubation in BG-11 growth medium. PU. 3300A; PU. 4300A; PU. 74165A; PV. PR22/60 = *A. azollae* immobilized in polyurethane (PU) and polyvinyl (PV) foam; ●, free-living *A. azollae*. (Shi *et al.*, 1987.)

Hydrogen production

In an argon atmosphere free-living cells of *A. azollae* showed a net H_2 photoproduction up to 9 h after onset of illumination with a total of 8 μmol H_2 mg chl^{-1}. A decrease in the H_2 content was then observed. When CO (an inhibitor of hydrogenase activity) was present, the total amount accumulated after 9 h was 7 μmol mg chl^{-1}, which remained constant up to 20 h. This shows that most of the H_2 was produced via nitrogenase activity and that the decrease under argon was due to the development of an uptake hydrogenase activity. When *A. azollae* was immobilized in foam the rate of H_2 evolution under argon was doubled compared to free-living cells. Comparison with H_2 production under argon plus 4% CO showed that *c*. 50% of the H_2 evolution was hydrogenase mediated (instead of *c*. 20% in the free-living cells), indicating that the increased H_2 yield was mainly due to increased hydrogenase activity. The long-term stability of the hydrogenase activity (assayed as oxidation of reduced methylviologen) of both free-living and foam-immobilized *A. azollae* over a period of five months is shown in Fig. 9.5.

Nitrogenase activity

Nitrogenase activity of free-living or immobilized *A. azollae* measured as acetylene reduction (equivalent to N_2 fixation) is shown in Fig. 9.6. Nitrogenase activity was always higher in immobilized than in free-living samples.

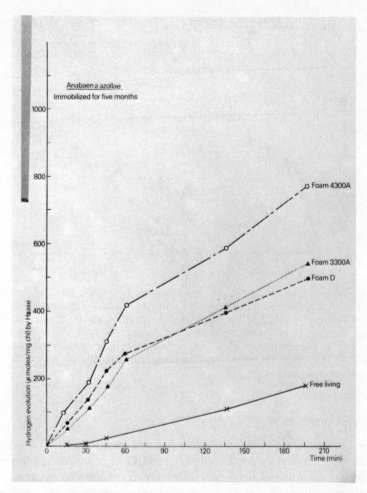

Fig. 9.5 H$_2$ production by *A. azollae* hydrogenase after growth for 5 months as free-living or immobilized cells. H$_2$ evolution was assayed using dithionite reduced methylviologen as electron donor. (Hall *et al.*, 1987.)

After 50 h continuous illumination, the total amount of acetylene reduced by freshly immobilized *A. azollae* was six times greater than the amount reduced by free-living controls. Forty days after immobilization in alginate and in polyvinyl foam, nitrogenase activity was still high whereas at that time no activity was measurable in the free-living controls. No acetylene reduction was observed in the dark. Introducing an intermediate light, aerobic period after 50 h led to a recovery of nitrogenase activity under argon + C$_2$H$_2$ (Fig. 9.6, right). The restored activity was still higher in immobilized than in free-living cells.

It must be stressed that the increase in nitrogenase activity observed as a result of immobilization is similar to that observed in the symbiotic association which showed nitrogen fixation activity 10 to 20 times higher than free-living cyanobacteria (Becking, 1976).

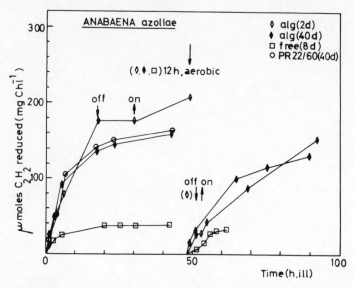

Fig. 9.6 Nitrogenase activity (C_2H_2 reduction) of free-living (□) and immobilized *A. azollae* in alginate beads, (◊) 2 days and (♦) 40 days after immobilization in polyvinyl foam; (○) PR22/60, 40 days after inoculation of the foam. The samples were incubated in an initial gas atmosphere of argon/10% C_2H_2. Light flux 250 μmol photons m^{-2} s^{-1}; temperature, 25 °C. off, light off; on, light on. 12 h aerobic: samples were put in the air in the light for 12 h and then reincubated under argon/10% C_2H_2. (Brouers & Hall, 1986.)

Ammonia production

In a first series of experiments, free-living and immobilized *A. azollae* were incubated for 33 days in combined N_2-free (BG-11) culture medium without MSX under continuous light and with shaking. Ammonia accumulation in the culture medium (Fig. 9.7) was assayed at various intervals. Low amounts of ammonia were produced by free-living cells throughout the time of the experiment. The same was true for immobilized cells up to day 10; however, the ammonia excreted into the medium then increased rapidly from day 10 to day 33. The accumulated ammonia after 33 days incubation varied from 18 to 52 μmol mg Chl^{-1} according to the matrix used. These results show that the ability to excrete ammonia (without MSX supply) observed in *A. azollae* recently isolated from the *Azolla–Anabaena* symbiotic association (Peters *et al.*, 1980; Meeks *et al.*, 1985) is lost after long-term subculture of isolated cyanobacteria in the free-living state, but is restored by immobilization in artificial matrices.

The effect of MSX supply on ammonia excretion was then studied in batch experiments by immobilized and free-living cultures of *A. azollae* after each of three successive 24 h incubation periods (in light and air) with or without MSX (Table 9.2). Low amounts of ammonia were produced by control free-living *A. azollae* even in the presence of MSX. High yields were, however, obtained in the presence of MSX from polyvinyl foam

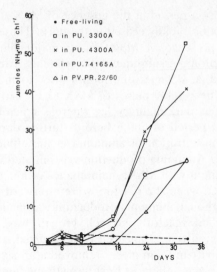

Fig. 9.7 Cumulative ammonia production by free-living and immobilized *A. azollae* during incubation in BG-11 growth medium without addition of MSX. Other conditions as in Fig. 9.4. (Shi *et al.*, 1987.)

Table 9.2 Ammonia production by free-living and immobilized *Anabaena azollae*

Sample	Ammonia production (µmol NH₃ per mg chlorophyll)			
	First 24 h period	Second 24 h period	Third 24 h period	Total
PR22/60−MSX	1	0	0	1
PR22/60+MSX	151	387	298	836
Free−MSX	9	0	0	9
Free + MSX	8	0	8	16

PR22/60: immobilization in polyvinyl foam PR22/60 (40 days after inoculation of foam).
Free: free-living control culture of immobilized filaments in PR22/60.
−MSX: no addition of methionine sulphoximine (MSX).
+MSX: in the presence of 50 μM MSX.
Photon flux density during incubation 70 μmol photons m^{-2} s^{-1}
Temperature: 30 °C.

immobilized cells (up to 390 μmol ammonia per mg chlorophyll per 24 h). Maximal production of ammonia by polyvinyl foam-immobilized cells was observed in the presence of MSX during the second 24 h period.

Sustained ammonia production was then studied in continuous flow packed bed reactors (Fig. 9.1). Initial experiments were designed to test the requirements for optimal continuous production in a flow-through reactor packed with foam-immobilized *M. laminosus*. It was found that in the continuous presence of MSX, ammonia production in the effluent ceased after about 70 h. The ability to produce ammonia could, however, be partly restored after a 24 h period in the absence of MSX. It was also shown that ammonia production rapidly ceased in the dark. These results indicated that the continuous presence of MSX was a major limitation for long-term ammonia production. In subsequent experiments, two packed bed reactors with immobilized *A. azollae* (one month after foam immobilization) were run in parallel with weekly pulses of MSX (5.10^{-4} M). Reactor I was operated under various dark–light cycles whereas reactor II was continuously illuminated for a period of 200 h before starting dark–light cycles. Figure 9.8 shows the concentration of ammonia in the effluent over time in both reactors I and II. Ammonia production was initiated in both reactors by an initial 10 h treatment with MSX; maintenance of a net ammonia production was then observed when the reactors were afterwards exposed to an MSX-free medium. Although ammonia production was maintained for a long time in the packed bed reactors, no cell growth was observed and finally ammonia production ceased and cell lysis occurred. It was, however, possible to revive activity in foam pieces removed after six weeks from a packed bed column, by placing them in fresh growth medium in conical flasks and shaking on an orbital incubator (still giving weekly MSX-pulses).

From the results of these experiments it was concluded that diffusion had been a major problem in the packed bed reactor, but also that the frequent and regular MSX-pulsing is too drastic a treatment. Experiments were thus designed in order to improve the reactor design. A fluidized bed reactor was used to obtain a continuous culture of immobilized *A. azollae*. Immobilization was initiated in the reactor itself. Instead of foam pieces that already contained *A. azollae*, sterilized polyvinyl foam pieces were added to the medium; these were found to circulate reasonably at random through the

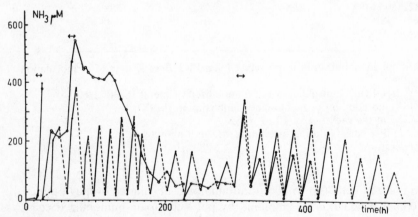

Fig. 9.8 Evolution of ammonia; concentration in the effluent outflow of bioreactors packed with polyvinyl foam-immobilized *A. azollae*. Dilution rate: $0.4 \, h^{-1}$ (BG-11 nutrient medium). Initial chlorophyll; 4.7 mg chl *a*. ●——●, reactor I; ○——○, reactor II; ——, light periods (flux 100 μmol photons $m^{-2} s^{-1}$); - - -, dark periods; ←→, periods with MSX supply (final concentration in effluent 5.10^{-4} M). At t = 24 h and 228 h, liquid reactor media were discarded, foam pieces were rinsed twice with fresh nutrient medium and the column was again supplied with 50 ml fresh nutrient medium.

reactor. Immobilization was then initiated by adding a small amount of free-living *A. azollae*. The reasoning behind this was that if cell growth was possible in such a reactor, then diffusion problems such as found in the packed bed column were not likely to occur and future problems would be more likely linked to the addition of MSX. Indeed, it was found that immobilization occurred within hours and that cell growth was very fast. Because of the continuous tumbling and collisions, however, cell loading of the foams remained low. Also, 'foam loading' is relatively low in a fluidized bed column, and therefore ammonia production would only result in low concentrations being produced in the effluent of the column; for this reason the design was abandoned. Other known designs all posed similar or their own particular problems. We have come up with a seemingly new solution. The new design is a reactor aerated from the bottom and containing strips of foam with immobilized cyanobacteria. The reactor design enabled rapid immobilization (even faster than in a fluidized bed reactor), very high cell loadings and good diffusion properties. Whereas in a packed bed column, the highest production rates were in the order of 300 μmol ammonia l^{-1} h^{-1}, in this new column a production rate of 1000 μmol l^{-1} h^{-1} was found, and the viability of cells was maintained. Different configurations are currently being tried out.

Conclusions

The results obtained in the particular application of immobilization of cyanobacterial cells described above, i.e. production of ammonia, illustrate some of the features of immobilization and of the main criteria that must be investigated during research for improving such systems. The similarities between the properties of artificially immobilized cyanobacteria and of the natural immobilized state in symbiosis illustrate how the confinement of cells in a particular microenvironment can induce the cells (here *A. azollae*) to acquire particular physiological, metabolic and morphological properties which can be useful for biotechnological applications; in this case, an increase in nitrogenase activity, long-term stabilization of other enzymatic activities, increased rates of ammonia production, etc. The research described for improving the long-term productivity and the yield of ammonia production also illustrates some of the prerequisites to be investigated with a view to industrial application:

– Use of matrices with good mechanical properties such as polyvinyl and polyurethane foams.
– Effective mass transfer and diffusion of nutrients, products and cofactors (choice of pore size, surface properties of the matrix, bioreactor design).
– Use of mild immobilization processes which preserve the viability of the cells and the long-term maintenance of the biological activities.
– Optimization of working conditions (light regime, light intensity and quality, flow rate, temperature etc.).

The example, from our work on the problem of ammonia production by immobilized cyanobacteria, shows that the use of the inhibitor methionine-D,L-sulphoximine (MSX) for inhibition of glutamine synthetase (GS) activity is a limiting factor for stable long-term ammonia production due to the difficulty of regulating nitrogenase activity and product yield and to inhibitory side-effects of MSX. Selection of GS-deficient natural strains or derepression through genetic manipulation can be the answer to that problem (Subramanian and Shanmugasundarum, 1986; Kerby *et al.*, 1986). However, in the case of genetic manipulation, growth of revertants and contaminating organisms may be a problem unless excreted ammonia concentrations are kept low, which is economically not desirable.

Future work requires pilot scale, field experiments in order to test the possibility for practical applications (biofertilization) using immobilized cyanobacteria. One possibility is the scaling-up of a non-sterile bioreactor incorporating foam-immobilized cyanobacteria which will provide ammonia in solutions directly to the field. Another is the suspension of foam pieces with immobilized cyanobacteria in irrigation ponds or hydroponic cultures. Such field experiments are necessary for making rational choices for meaningful bioengineering and economic analysis.

For the production of higher value products than ammonia, such as pigments, pharmaceutical compounds, growth regulators, polysaccharides, pesticides, etc., the use of immobilized cyanobacterial or algal cells in photobioreactors is very attractive. In this case the nutrient solution is circulated in the system and not the cells (except when they are removed) so obviating the problem of cell rupture and other mechanical factors. The use of chemically and mechanically stable immobilization matrices such as foams undoubtedly has distinct advantages for the algal biotechnologist.

Acknowledgements

M.B. was the recipient of a training contract from the Commission of the European Communities (Biomolecular Engineering Programme). We thank G. Morgan, Anatomy Department, King's College and A. J. Turner, Food Research Institute, Norwich, UK for scanning electron microscope work and collaboration.

References

Adlercreutz P, Mattiasson B 1982 Oxygen supply to immobilized cells. 1. Oxygen production by immobilized *Chlorella pyrenoidosa*. *Enzyme Microbiol. Technol.* **4**: 332–40

Atkinson B 1986 Immobilised cells, their applications and potential. In Webb C, Black G M, Atkinson B (eds) *Process Engineering Aspects of Immobilised cell Systems*, Pergamon Press, pp 3–19

Baillez C 1984 Immobilisation de l'algue *Botryococcus braunii* en gel d'alginate et en mousse de polyurethane, influence sur la capacite photosynthetique et sur la production d'hydrocarbures. These de doctorat, Université Pierre et Marie Curie, Paris, France

Baillez C, Casadevall E, Largeau C 1983 Effect of immobilization on the hydrocarbon rich alga *Botryococcus braunii*. In Strub A, Chartier P, Schleser G (eds) *Energy from Biomass*, Applied Science Pub., London, pp 286–93

Baillez C, Largeau C, Berkaloff C, Casadevall E 1986 Immobilization of *Botryococcus braunii* in alginate: influence on chlorophyll content, photosynthetic activity and degeneration during batch cultures. *Appl. Microbiol. Biotechnol.* **23**: 361–6

Becking J H 1976 Contributions of plant algal associations. In Newton W E, Nyman C J (eds) *Proceedings of 1st International Symposium on Nitrogen Fixation*, Vol 2, Washington State University Press, Pullman, pp 556–80

Brodelius P, Mosbach K (eds) 1987 Immobilised enzymes and cells. *Methods in Enzymology*, Vol 135, Academic Press, New York

Brouers M, Collard F, Jeanfils J, Jeanson A, Sironval C 1982 Immobilization and stabilization of green and blue–green algae in cross-linked serum albumin glutaraldehyde and in polyurethane matrices. In Hall D O, Palz W (eds) *Photochemical, Photoelectrochemical and Photobiological Processes* Vol. I, Reidel Publ., Dordrecht, pp 134–9

Brouers M, Collard F, Jeanfils J, Loudeche R 1983 Long term stabilization of photobiological activities of immobilized algae; photoproduction of hydrogen by immobilized adapted *Scenedesmus* cells. In Hall D O, Palz W, Pirrwitz D (eds) *Photochemical, Photoelectrochemical and Photobiological Processes*, Vol. 2, Reidel Publ., Dordrecht, pp 170–8

Brouers M, Hall D O 1985 Drying pretreatment enhances the photosynthetic stability of alginate immobilized *Phormidium laminosum*, *Biotechnol. Lett.* **7**: 567–72

Brouers M, Hall D O, 1986. Ammonia and hydrogen production by immobilized cyanobacteria. *J. Biotechnol.* **3**: 307–21

Brouers M, Shi D J, Hall D O, 1988 Immobilization methods for cyanobacteria in solid matrices. In Packer L, Glazer A N (eds) *Cyanobacteria. Methods in Enzymology*, Vol. 167, Academic Press, New York, pp. 629–36

Bucke C 1983 Immobilised cells. *Phil. Trans. R. Soc. Lond. B.* **300**: 369–89

Bucke C 1986 Methods of immobilising cells. In Webb C, Black G M, Atkinson B (eds) *Process Engineering Aspects of Immobilised Cell Systems*, Pergamon Press, pp 20–34

Cammack R, Hall D O, Rao K K 1985 Hydrogenase: Structure and application in hydrogen production. In Poole R K, Dow C (eds) *Microbial Gas Metabolism: Mechanistic, Metabolic and Biotechnological Aspects*, Academic Press, London, pp 75–102

Daday A, Lambert G R, Smith G C 1979 Measurement *in vivo* of hydrogenase-catalysed hydrogen evolution in the presence of nitrogenase enzyme in cyanobacteria. *Biochem. J.* **177**: 139–44

Day J C, Codd G A 1985 Photosynthesis and glycollate excretion by immobilized *Chlorella emersonii*. *Biotechnol. Lett.* **7**: 573–6

Gisby P E, Rao K K, Hall D O 1987 Entrapment techniques for chloroplasts, cyanobacteria and hydrogenases. In Brodelius P, Mosbach K (eds) *Methods in Enzymology*, Vol. 135, Academic Press, New York, pp 440–54

Hall D O, Brouers M, de Jong H, De la Rosa M A, Rao K K, Shi D J, Yang L W 1987 Immobilized photosynthetic systems for the production of fuels and chemicals. *Photobiochem. Photobiophys. Suppl.* 167–80

Jeanfils J, Collard F 1983 Effect of immobilizing *Scenedesmus obliquus* cells in a matrix on oxygen evolution and fluorescence properties. *Eur. J. Appl. Microbiol. Biotechnol.* **17**: 254–7

Jeanfils J, Cocquempot M F, Thomas D 1986 Study of fluorescence kinetics and PSII activity in fixed or immobilized chloroplast membranes. *Enz. Microbiol. Technol.* **8**: 157–60

Karube I, Matsunaga T, Otsuka T, Kayano H, Suzuki S 1981 Hydrogen evolution by co-immobilized chloroplasts and *Clostridium butyricum*. *Biochem. Biophys. Acta* **637**: 490–8

Kayano H, Matsunaga T, Karube I, Suzuki S 1981 Hydrogen evolution by co-immobilized *Chlorella vulgaris* and *Clostridium butyricum* cells. *Biochem. Biophys. Acta* **638**: 80–5

Kerby N W, Musgrave S C, Rowell P, Shestakov S V, Stewart W D P 1986 Photo-production of ammonium by immobilized mutant strains of *Anabaena variabilis*. *Appl. Microbiol. Biotechnol.* **24**: 42–6

Kerby N W, Niven G W, Rowell P, Stewart W D P 1987 Photoproduction of amino acids by mutant strains of N_2-fixing cyanobacteria. *Appl. Microbiol. Biotechnol.* **25**: 547–52

Lambert G R, Daday A, Smith G D 1979 Hydrogen evolution from immobilized cultures of the cyanobacterium *Anabaena cylindrica* B629. *FEBS Lett.* **101**: 125–8

Lambert G R, Smith G D 1981 The hydrogen metabolism of cyanobacteria (blue–green algae). *Biol. Rev.* **56**: 589–660

Meeks J C, Steinberg N, Joseph C M, Enderlin C S, Jorgensen P A, Peters G A 1985 Assimilation of exogenous dinitrogen derived [13]NH_4 by *Anabaena azollae* separated from *Azolla caroliniana* wild. *Arch. Microbiol.* **142**: 229–33

Mosbach K 1982 Use of immobilized cells with special emphasis on the formation of products formed by multistep enzyme systems and coenzyme. *J. Chem. Technol. Biotechnol.* **32**: 179–88

Muallem A, Bruce D, Hall D O 1983 Photoproduction of hydrogen and $NADPH_2$ by blue–green algae (cyanobacteria) immobilized in polyurethane foam. *Biotechnol. Lett.* **5**: 365–8

Musgrave S C, Kerby N W, Codd G A, Rowell P, Stewart W D P 1982 Sustained ammonia production by immobilized filaments of the nitrogen-fixing cyanobacterium *Anabaena* 27893. *Biotechnol. Lett.* **4**: 647–52

Ochiai H, Shibata H, Yoshihiro S, Katoh T 1980 'Living electrode' as a long-lived photoconverter for biophotolysis of water. *Proc. Nat. Acad. Sci. USA* **77**: 2442–4

Peters G A, Ray T B, Mayne B C, Toia R E 1980 *Azolla–Anabaena* association: morphological and physiological studies. In Newton W E Orme-Johnson W H (eds) *Nitrogen Fixation*, Vol. II, University Park Press, Baltimore, pp 293–309

Philips E J, Mitsui A 1986 Characterization and optimization of hydrogen production by a salt water blue–green alga *Oscillatoria* sp. Miami BG7. II. Use of immobilization for enhancement of hydrogen production. *Int. J. Hyd. Energy* **11**: 83–9

Robins R J, Hall D O, Shi D J, Turner R J, Rhodes M S C 1986 Mucilage acts to adhere cyanobacteria and cultured plant cells to biological and inert surfaces. *FEMS Microbiol. Lett.* **34**: 155–60

Robinson P K, Dainty A L, Goulding K H, Simpkins I, Trevan M D 1985 Physiology of alginate-immobilized *Chlorella*. *Enzyme Microbiol. Technol.* **7**: 212–16

Shi D J, Brouers M, Hall D O, Robins R J 1987 The effects of immobilization on the biochemical, physiological and morphological features of *Anabaena azollae*. *Planta* **172**: 298–308

Shi D J, Hall D O 1988 The *Azolla–Anabaena* association: historical perspective, symbiosis and energy metabolism. *Bot. Rev.* **54**: 353–86

Smith G D, Muallem A, Hall D O 1982 Hydrogenase catalyzed photoproduction of hydrogen by photosystem I of the thermophilic blue–green algae *Mastigocladus laminosus & Phormidium laminosum. Photobiochem. Photobiophys.* **4**: 307–19

Solorzano L 1969 Determination of ammonia in natural waters by the phenol–hypochlorite method. *Limnol. Oceanogr.* **14**: 799–801

Stanier R Y, Kunisawa R, Mandel M, Cohen Bazire G 1971 Purification and properties of unicellular blue–green algae (order Chroococcales). *Bact. Rev.* **35**: 171–205

Stewart W D P, Fitzgerald G P, Burris R H 1968 Acetylene reduction by the nitrogen fixing blue–green algae. *Arch. Microbiol.* **62**: 336–48

Subramanian G, Shanmugasundarum S 1986 Uninduced ammonia release by the nitrogen fixing cyanobacterium *Anabaena. FEMS Microbiol. Lett.* **37**: 151–4

Tamponnet C, Costantino F, Barbotin J N, Calvayrac R 1985 Cytological and physiological behaviour of *Euglena gracilis* cells entrapped in a calcium alginate gel. *Physiol. Plant.* **63**: 277–83

Tel-Or E, Sandovsky T, Kobiler D, Arad C, Weinberg R 1983 The unique symbiotic properties of *Anabaena* in the water fern *Azolla.* In Papageogiou G C, Packer L (eds) *Photosynthetic Prokaryotes: Cell Differentiation and Function,* Elsevier Biomedicine, New York, pp 303–14

Thepenier C, Gudin C, Thomas D 1985 Immobilization of *Porphyridium cruentum* in polyurethane foams for the production of polysaccharide. *Biomass* **7**: 225–40

Vincenzini M, Balloni N, Manelli D, Florenzano G 1981 A bioreactor for continuous treatment of waste waters with immobilized cells of photosynthetic bacteria. *Experientia* **37**: 710–12

Weetall H H, Sharma B P, Detar C C 1981 Photometabolic production of hydrogen from organic substrates by free and immobilized mixed cultures of *Rhodospirillum rubrum* and *Klebsiella pneumoniae. Biotechnol. Bioeng.* **23**: 605–14

10

Industrial production: methods and economics

L. J. Borowitzka and M. A. Borowitzka

Introduction

The ultimate aim of algal biotechnology is to take an idea from the laboratory to the production plant and the market place. This all-important transition requires the interaction of many specialists, some time, usually much money, and has to be carried out with a continued awareness of commercial realities.

In this chapter we propose to outline the steps taken in selecting a product or process, the selection of a cultivation and harvesting system, and the scale-up from the laboratory via the pilot plant to the production plant.

Product selection

It is the interaction between *ideas* on products and processes which are generated in the laboratory, and *market information* which leads to the concept for a potentially commercial process.

The first step is to define which products or processes are done better using algae compared with other organisms such as bacteria and yeasts, or with chemical synthesis. In recent years there have been several papers concerned with these potential products (i.e. Borowitzka, 1986, 1988a,b; Cohen, 1986). These products then have to be evaluated in the light of existing or potential market demand, market size and the market price which could be attained. The market determines:

1. Which products and/or processes are worth investigating beyond laboratory studies.
2. What formulation or concentration is required.
3. What price is attainable.

The considerations listed above act also to influence the decision on which alga to use (if there is a choice of more than one), and the appropriate culturing, harvesting and subsequent processing methods to be used.

Further evaluation then requires the estimation of:

1. the likely total value of sales over a five year period; and,
2. the possible profitability (i.e. sales minus costs of goods and other overheads and expenses).

With these considerations in mind, we feel that in the next decade algal biotechnology should concentrate on products with a value of about $US100 kg^{-1} or greater. The reasoning behind this is as follows.

At present, algal biomass costs between $US1 and $US20 kg^{-1} to produce (approximate costs: *Spirulina* $US7 kg^{-1}; *Dunaliella salina* $US10–20 kg^{-1}; see also Table 10.2). Cell components other than bulk protein, polysaccharides or lipids are generally present at concentrations of less than 10%, and often less than 1% of cell dry weight (i.e. β-carotene is 10% of dry wt only in *D. salina*, with related pigments usually less than 1% of dry wt in most algal species). Therefore, the price for a minor component (1–10% of dry wt) of the biomass must be around $US100 kg^{-1} excluding downstream processing costs. Products which sell for such relatively high prices include fine chemicals and vitamins, biologically active compounds such as pharmaceuticals and antibiotics, and some agrochemicals. Suitable revenue may also be derived from particular and/or specialized mineral leaching or waste-treatment systems (Oswald, 1988a) or possibly in the area of specialized feed for high value aquaculture species. The latter may be attractive due to reduced downstream processing costs.

Products unlikely to command sufficient prices to support algal biotechnology processes in the next few years include single-cell protein and lipids as fuels and oils. Bulk chemicals and common chemicals such as glycerol, sugars and amino acids constitute high volume but low unit price ($US1–10) products and are also unlikely, in our view, to sustain the technology at present.

The product range for consideration is further restricted to those of the above which can be produced more cost-effectively by algal rather than other biological systems or chemical synthesis. The latter restriction suggests concentrating on some uniquely algal pigments, oils and sterols, as well as a range of pigments, vitamins, oils, sterols and chemical intermediates in which algae are particularly rich.

The advantage of algal biotechnology is usually perceived to be the relatively cheap production of biomass for extraction of the desired product. Harvesting costs are generally greater than for fermentation systems, as the cell density achieved is lower in algal cultures, increasing the volume of culture to be harvested and therefore increasing costs. Downstream processing costs, excluding harvesting costs, are similar to those for more conventional biomass production systems. Thus, although interferon is a very valuable product, there is no reason to believe it could be cheaper to produce in genetically engineered algae, compared to suitably engineered bacterial or eukaryotic cell cultures. In the case of interferon, most of the production cost comes from downstream processing and purification, and the cost saving in cheaper biomass production is irrelevant.

Once a suitably priced product, from a process which appears technically viable, is selected, comparison of the production cost with costs from alternative biological and chemical processes is necessary. For example, in the case of β-carotene, chemical synthesis provides most of the world's β-carotene, in an annual market worth approximately $US50–100 million. The cost of chemical synthesis is a matter of speculation, but selling prices on the basis of β-carotene content are generally above $US500 kg^{-1} depending on final formulation. Algal-derived natural β-carotene sells for up to $US1000 kg^{-1}, to specialist customers who market the 'natural' aspects to the food and health food industries. The higher selling price reflects higher production costs and smaller volumes. In the future, other biological systems including genetically engineered fungi, yeasts and bacteria grown in fermenters, could compete for the 'natural' β-carotene market, and the relative production costs should determine the eventual winner.

Once there is an awareness of the types of products which may find markets at suitable prices, it can combine with, or even generate, ideas on specific products and processes from the laboratory, or from observations of nature. For example, a knowledge of the commercial value of the pigment β-carotene coupled with laboratory results on environmental induction of β-carotene in *D. salina*, and also with observations of immense natural blooms of β-carotene-containing algae in salt lakes and solar salt works. This melding of diverse information led to the development of several different commercial processes and a viable industry.

It is of interest to note that the laboratory and salt lake observations were products of basic, undirected research and yet were the critical factors in establishing the industry. Furthermore, the integration of market and technical information to develop a commercial process can take considerable time. The early observations on *D. salina* in the laboratory and in salt lakes date back to 1905 (cf. Borowitzka and Borowitzka, 1988a), and synthetic β-carotene market development to the 1950s. First attempts at developing commercial production of β-carotene from *D. salina* were made in the USSR and the Ukraine in the 1960s, then in Israel and Australia in the 1970s. In Australia, the USA and Israel, in the late 1980s, the industry is now finally established, stimulated by an increasing market demand for 'natural' products.

The establishment of a viable *Spirulina* industry is also the result of melding a market demand for new health foods with the original observations by Léonard and others in Africa in the early 1960s. Production commenced in Mexico, Thailand, Taiwan, Japan, Israel and the USA in the 1970s (Richmond, 1988).

Cultivation systems

Industrial cultivation of microalgae may be either extensive or intensive, and may be either in open systems or in closed, non-axenic or axenic, systems. No one culture method can be said to be better than another and the choice of culture system will depend, in large part, on the alga to be used and the product. Much of the technology of microalgal mass-culture has been reviewed by Terry and Raymond (1985) who also give a good historical overview, and by Dodd (1986) and Oswald (1988b).

Extensive cultivation in large outdoor open ponds is the oldest of the industrial systems for algal cultivation and has been used for the cultivation of algae such as *Chlorella, Scenedesmus* and *Spirulina* for single-cell protein and health· food (cf. Kawaguchi, 1980; Soong, 1980; Venkatamaran and Becker, 1985), and for the cultivation of *Dunaliella salina* for β-carotene (Borowitzka and Borowitzka, 1988a). This cultivation method is inherently the simplest and cheapest method, but is generally only applicable if the alga to be cultured grows in a selective and specialized environment. Thus, for example, *D. salina* grows in brines with NaCl concentrations > 20% w/v, and *Spirulina platensis* grows in alkaline waters with a pH > 9.2. That is not to say, however, that monocultures of algae growing in less selective environments (i.e. seawater or freshwater) cannot be maintained for long periods with judicious pond management.

The simplest type of extensive cultivation is used by Western Biotechnology in their *Dunaliella* plant at Hutt Lagoon, WA where the unlined 5 ha production ponds are approximately 500 m long and 100 m wide, constructed on the lake bed of Hutt Lagoon with earthen berms (Fig. 10.1). The cultures are approximately 15–20 cm deep and are not mixed other than by wind and convection.

The most common extensive open-air pond design, however, is the 'raceway' design which consists of long channels arranged in a single or in multiple loops and mixed by a paddle-wheel. These shallow mixed ponds were originally introduced in the 1950s and 1960s by Oswald *et al.* (1959). Many of the early designs used a configuration consisting of relatively narrow channels with many 180° bends and using propeller pumps to produce a channel velocity of about 30 cm s^{-1}. In the 1970s paddle-wheel mixers of various design were introduced and found to be more effective, with less energy requirements and reduced stress forces on the algal cells. Aside from propeller pumps and paddle-wheels, air-lifts have also been proposed for the mixing of 'raceway' ponds. These have, however, not been tested at a large scale to any extent and at this stage paddle-wheels are the preferred mixing device (Persoone *et al.*, 1980; Weissman and Goebel, 1987). The design and efficiency of air-lifts is considered in detail by Augenstein (1987) and Parker and Suttle (1987).

The numerous bends in the channels of the older designs also led to hydraulic losses and problems with solids deposition. These were minimized by using a single-loop (racetrack) configuration, with suitable baffles being

Fig. 10.1 The production plant for the mass culture of *Dunaliella salina* for β-carotene operated by Western Biotechnology Ltd at Hutt Lagoon, Western Australia, showing the 5 ha production ponds (left) and the pilot plant ponds (centre) constructed directly on the lake bed of Hutt Lagoon.

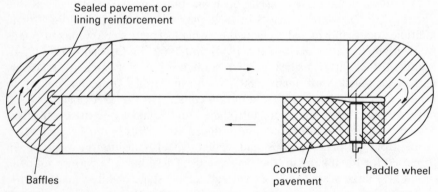

Fig. 10.2 Typical design of a 'raceway' -type of open-air algal culture facility.

incorporated. Figure 10.2 shows the design of what is presently considered the optimal configuration. The optimal pond size is a compromise between the hydraulics of the pond and the ability to actually construct the pond (Dodd, 1986). Simple geometric optimization shows that a large pond with a low length to width (L/W) ratio gives the largest pond area for the least wall length, and is therefore cheaper to construct. The actual pond length needs then to be evaluated to take account of head loss (which depends on velocity and roughness) relative to the mixing system to be used.

Several factors need to be taken into account when designing the optimally sized pond. These include:

1. Optimal pond depth, taking into account the degree of light penetration.
2. Mixing velocity. This relates to the need to keep the algae in suspension, avoiding any dead spaces and the effects of turbulence on the pond materials.
3. Energy requirement for mixing.
4. Materials from which the pond is constructed.

Experience has shown that velocities generally greater than 10 cm s^{-1} are necessary to avoid settling of cells. However, in order to cope with unavoidable variations in velocity, especially near the pond ends, a minimum design velocity of 20 cm s^{-1} appears to be necessary (Dodd, 1986). In order to maximize productivity by ensuring the maximum exposure to light of each algal cell, velocities of up to 50 cm s^{-1} have been suggested (e.g. Richmond and Vonshak, 1978); however such velocities are generally uneconomical (Oswald, 1988b).

Since hydraulic energy is low due to friction and because friction increases as the square of velocity, the depth of flowing water in the channel decreases as a complex function of channel length, i.e.

$$\frac{\delta d}{\delta L} = f\left(V^2, R, f\right) \tag{10.1}$$

where d is the channel depth (m), L the length of the channel (m), V the velocity (m s^{-1}), R the hydraulic radius (m) and f is the friction.

Now R, the hydraulic radius, is equal to the area of flow (A) divided by the perimeter in contact with water (P) for a unit length of channel, i.e.

$$R = A/P \tag{10.2a}$$

and

$$A = dw$$

where w is the width of the channel and

$$P = w + 2d$$

so that

$$R = dw/(w + 2d) \tag{10.2b}$$

and for very wide, shallow channels R is approximately equal to d.

The Manning equation, which is an empirical solution for the complex functions of flow in conduits (i.e. streams and canals) can be used to determine the finite length L that corresponds to an assumed change in depth for a given friction factor, hydraulic radius and velocity.

The Manning equation is:

$$V = \frac{1}{n} R^{2/3} s^{1/2} \tag{10.3}$$

where V is the mean channel velocity (m s^{-1}), n the Manning friction factor (s m$^{-1/3}$) which represents friction due to channel roughness, R the hydraulic radius defined above, and s the rate of loss of energy in the channel per unit length, that is $\Delta d/L$ (dimensionless).

We can solve this equation for s by squaring both sides:

$$s = \frac{V^2 n^2}{R^{4/3}} \tag{10.4}$$

and, since $s = \Delta d/L$ and $R = dw/(w + 2d)$

$$\Delta d = \frac{L V^2 n^2}{(dw/(w + d))^{4/3}} \tag{10.5}$$

and

$$L = \frac{\Delta d(dw/(w + 2d))^{4/3}}{V^2 n^2} \tag{10.6}$$

Using these equations and estimates for n, the Manning friction factor determined empirically by several workers (Table 10.1), we can estimate the permissable length L for a channel for a desired maximum change in depth, d, at a given velocity.

A minimum mixing velocity of about 10 cm s^{-1} is generally considered sufficient to avoid settling of cells; however, most designs use a minimum design value of 20 cm s^{-1} based on an average over the channel cross-section to cope with unavoidable variations in velocity, especially near the pond ends. Many ponds operating in the velocity range of 10–30 cm s^{-1}

Table 10.1 Estimated mean values for Manning's n in open channels (from Oswald, 1988b)

Material for channel liner	Manning's n
Smooth plastic on smooth concrete	0.008
Plastic with 'skrim' on smooth earth	0.010
Smooth plastic on granular earth	0.012
Smooth Portland cement concrete	0.013
Smooth asphalt concrete	0.015
Coarse trowelled concrete, rolled asphalt	0.016
Gunnite or sprayed membranes	0.020
Compacted smooth earth	0.020
Rolled coarse gravel, coarse asphalt	0.025
Rough earth	0.030

have, however, encountered difficulty with solids settling in stagnant areas which can only be partially overcome with baffles. Two solutions to the above have been developed. The first, which has been tested with two 1230 m^2 ponds in Singapore (Dodd, 1986) consisted of an eccentrically placed curved wall and baffles at the end of the pond without the paddle-wheel. This creates a curved zone of accelerating flow followed by a flow expansion zone after the directional change has been made. The rate of constriction of the curved zone is sufficient to avoid eddies or velocities below that causing deposition on the back side of the centre wall and baffles. Similar baffles were not necessary at the paddle-wheel end due to the head drop created by the paddle-wheel.

An alternative has been developed using closely spaced turning vanes along the diagonals from the centre wall to the corners in *Spirulina* ponds. These vanes which are arranged less than 50 cm apart produce two right angle changes in direction (Shimamatsu, 1987).

The power requirement for mixing is given by:

$$P = \frac{9810 \, A V^3 n^2}{d^{0.3} e} \tag{10.7}$$

where P is the power (watts), A the pond area (m^2), d the depth (m) and e the overall mixing system efficiency (approximately 30–40% for paddle-wheels). From eqn (10.7) it can be seen that the power varies as the cube of the mixing speed, and actually increases slightly at lower depths; therefore, the velocity should be minimized wherever energy is a major cost factor. Velocities greater than 30 cm s^{-1} will result in large values of Δd in long channels and many require high channel walls and higher divider walls, thus adding to cost. Scour may also be a problem at higher velocities and may necessitate more expensive pond construction.

In order to optimize productivity, pond depth should also be optimal. Bush (quoted in Burlew, 1953) derived a theoretical equation to estimate the fraction, f_m, of the maximum possible photosynthetic efficiency that can be attained by a culture. Bush's equation states:

$$f_m = (S_s/S_o)[\ln(S_o/S_s) + 1] \tag{10.8}$$

where S_s is the photon flux density at which photosynthesis is saturated and S_o is the photon flux density at any instant. This predicted efficiency cannot, however, be attained in real cultures since they are also limited by the quantum efficiency of various wavelengths. Furthermore, one must also take account of night-time respiration before a net efficiency is obtained. The actual light utilization efficiency of mixed cultures of algae ranges from less than 1% to a (rarely achieved) maximum of 5%.

Assuming that the major light absorbing components in a shallow algal culture are the algae, Oswald (1988b) has empirically determined the following relationship (for green algae) between the concentration of algae, C_c (mg l^{-1}) and the light penetration depth d_p (cm):

$$d_p = 6\,000/C_c \tag{10.9}$$

Field observations of large-scale cultures further indicate that the culture concentration in light-limited, continuously mixed cultures approaches that which permits light to penetrate two-thirds of the actual culture depth, i.e.

$$d_p = (2/3)\ d \tag{10.10}$$

and

$$C_c = 9\ 000/d \tag{Oswald, 1988b).}$$

A continuously mixed outdoor culture of green algae at 30 cm depth will therefore achieve an average maximum light-limited algal concentration of about 300 mg l^{-1}. Higher photon flux densities will increase this only slightly, since the penetration of light is proportional to the log of its intensity. High concentrations of algae can therefore only be attained at shallow culture depth. Since shallow cultures are more prone to major temperature fluctuations, greater changes in medium composition due to evaporation and are more difficult and expensive to mix (see Oswald, 1988b), the actual culture depth used must be a compromise.

Recent studies by Laws and coworkers (1984) using a system that introduces ordered vertical mixing into the culture by suspending small foils, similar to airplane wings, at regular intervals in the flowstream at a relatively high angle of attack and so generating regular vortices (Fig. 10.3), have shown that culture yield can be increased by 50–100% for *Phaeodactylum tricornutum*. Whether this approach will be cost effective has yet to be determined.

So far we have only considered open-air systems for large-scale algal culture. Such systems are, however, only suited to algal species which grow in a relatively extreme environment (i.e. *Dunaliella salina*, high salinity; *Spirulina*, high pH) where competition by other species and problems with predators and pathogens are reduced. Open systems may also work with fast-growing algae under optimum conditions (i.e. *Chlorella, Scenedesmus, Phaeodactylum*) which can outgrow most of their competitors.

In order to overcome these problems several large-scale closed bioreactor systems are being developed based on earlier designs (e.g. Jüttner, 1977; Krüger and Eloff, 1978; Pirt *et al.*, 1983), and some of these are at a pilot stage of development. The most successful of these to date are the tubular

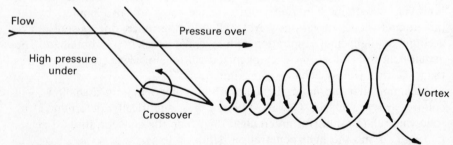

Fig. 10.3 Design of a single foil showing mechanism of vortex production by the foil in an algal 'raceway' (redrawn from Laws, 1984).

photobioreactors developed in the UK and France. Basically, these systems consist of a tubular solar receptor, a carbonation tower or similar system for CO_2 supply and a pump to circulate the culture between these two parts. For some algae and at some sites, these reactors appear to be superior (at least in biomass production) to open systems (cf. Torzillo *et al.*, 1986).

The photobioreactor at the Association pour la Recherche en Bioénergie Solaire at Cadarache, France, has been scaled up since 1981 from a 3 m² (70 l of culture) culture unit with the tubular solar receptors made of glass tubes, to a 100 m² (7000 l) pilot unit (Gudin and Thepenier, 1986; Chaumont *et al.*, 1988) with the solar receptor composed of five identical 20 m² units constructed of 25 cm diameter polyethylene tubes floating on or in a large pool of water (Figs 10.4, 10.5). Since tubular reactors accumulate heat similarly to solar hot water systems, they require cooling. In cool temperate climates heating may also be required in winter. The Cadarache unit controls the culture temperature by either floating the solar receptors on the surface of the pool of water (to heat) or by immersing them in the water (to cool). This is achieved by an air flotation system attached to the tubular units. The water also provides a convenient support for the long tubes of the solar receptors.

Pilot 'biocoil' facilities at Luton in the UK and in Australia have solar receptors arranged as a coil of approximately 30 mm diameter low density polyethylene tubing arranged around an open circular framework (Fig. 10.6).

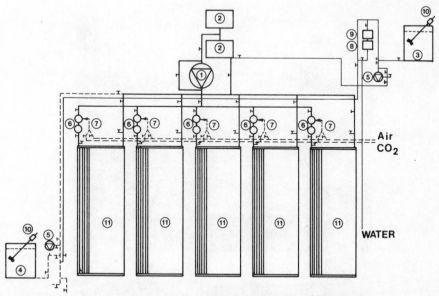

Fig. 10.4 Schematic diagram of the 100 m² tubular algal photobioreactor at Cadarache. France (redrawn from Chaumont *et al*,. 1988). (1) pump for culture circulation; (2) 200 l buffer tanks; (3) culture medium tank; (4) harvested culture tank; (5) pump; (6) carbonation towers; (7) flowmeter for air/CO_2 mixture; (8) 40 μm water filter; (9) 1 μm water filter; (10) stirrers; (11) solar receptor units (see Fig. 10.5)

Fig. 10.5 Photograph of the 100 m² tubular algal photobioreactor at Cadarache, France showing the five solar receptor units floating in a large pool (photograph courtesy of D. Chaumont and C. Gudin).

Fig. 10.6 The pilot tubular photobioreactor at Luton, UK operated by Biotechna Ltd.

Temperature control is achieved by a heat exchange unit installed between the reactor and the pump (Robinson, 1987). A 1300 l unit would have an approximate photosynthetic surface area of 100 m².

Although these systems are uni-algal, they are, at this stage at least, not axenic and the Cadarache group have reported bacterial and other contamination of their *Porphyridium* cultures. For some algal species and/or products some means of sterilizing the system and maintaining sterility may be required and this has not yet been developed. The choice of pump used for circulating the culture is also critical to avoid damage to the algal cells. Another potential problem which has appeared during scale-up is the build-up of O_2 in the system as a result of photosynthesis by the algae in the long tubes. Since high levels of O_2 inhibit algal photosynthesis this may limit the productivity of the system. CO_2 addition is also more of a problem at the larger scale, as it is for all bioreactors (cf. Ho and Shanahan, 1986). CO_2 is quite expensive and more efficient carbonation towers are required to reduce operating costs.

Irrespective of the problems, the tubular photobioreactors represent an exciting new way to culture algae on a large scale and the pilot reactor of Cadarache has already achieved a steady state of operation using *Porphyridium cruentum* for periods in excess of 50 days, with a maximum production of 20 g m^{-2} d^{-1} (Chaumont *et al.*, 1988). Similarly, the Luton units have achieved good long-term production of *Spirulina* (Robinson, personal communication). A recent economic analysis (Tapie and Bernard, 1988) gives an estimated production cost of FF 26 kg^{-1} (approx. \$US4.70) of dry biomass for a plant size of 10 ha surface area and a productivity of 60 t ha^{-1} yr^{-1} and this compares well with other estimates for the cost of large-scale algal production (Table 10.2). At this stage the major question is whether other algal species, such as *Chlamydomonas* and *Haematococcus*, can also be scaled-up in such a reactor and whether such systems are economically viable.

A related development to the tubular photobioreactors are immobilized cell bioreactors. Such systems are being tested and developed for various processes such as ammonia production using blue–green algae (cyanobacteria) and polysaccharide production (Gudin and Thepenier, 1986; Robinson *et al.*, 1986). At this stage these systems must be considered experimental only and much work needs to be done before such systems become a commercial reality.

One final point which must be considered is whether or not to add CO_2 to increase productivity. In open ponds CO_2 addition is not necessary, but if added will significantly increase productivity. In closed systems, however, CO_2 must be added. In these systems CO_2 addition can also be used to strip the system of O_2 and to control pH. Various economic analyses of algal mass-culture systems have shown that CO_2 addition is a major cost factor in algal production (e.g. Weissman and Goebel, 1987) and this cost must therefore be weighed against the gain in algal productivity and product yield. At present almost no large-scale commercial algal production systems use CO_2 addition.

Table 10.2 Theoretical economic analyses of a range of large-scale algal production systems (adapted from Tapie and Bernard, 1988)

Product	Cuture method	Culture area (ha)	Productivity (t ha^{-1} yr^{-1})	Cost ($US kg^{-1} dry wt)	Reference
Scenedesmus (protein)	Raceway	4	70	3.09	1
Spirulina (protein, vitamins)	Ponds	20	45	2.7	2
Spirulina	Raceway	5	45	7.1	2
Spirulina	Raceway	10	50	2.69	3
Liquid fuel	Ponds	800	62.5	0.16	4
Liquid fuel	Ponds	10 000	73	0.15	5
Liquid fuel	Ponds	405	112	0.26	6
Liquid fuel	Ponds	1 000	33	0.39	7
Porphyridium (polysaccharide)	Solar water heater	34	112	6.95 (polysaccharide only)	8
Dunaliella (glycerol)	Ponds	530	66 (glycerol only)	0.92 (glycerol only)	9
Microalgae	Tubular photobio-reactor	10	60	4.7	10

1 Becker and Venkatamaran, (1980)
2 Rebeller, (1982)
3 Richmond, (1983)
4 Benneman et al., (1982)
5 Regan and Gartside, (1983)
6 Weissman and Goebel, (1987)
7 Neenan et al., (1986)
8 Anderson and Eakin, (1984)
9 Chen and Chi, (1981)
10 Tapie and Bernard, (1988)

Harvesting

One of the most difficult and, at this stage, expensive aspects of large-scale algal culture is harvesting of the algal biomass prior to processing. This is because the algae achieve only a concentration of 100–600 mg l^{-1} or 0.01–0.06% (w/v) water-free solids under practical outdoor culture conditions. Even in closed systems with CO_2 addition values of only up to 1300 mg l^{-1} have been reported (Chaumont et al., 1988).

Depending on the application, the algae must be concentrated to a water-free solids concentration of 1–20% (i.e. 'milk-like' to 'cheese-like' consistency) before further processing or drying (cf. Tables 10.3, 10.4). This represents a 20- to 1000-fold concentration. Filamentous algae such as Spirulina may be concentrated effectively through the use of vibrating screen filters; however, smaller unicellular algae require centrifugation, filtration, micro-straining, flocculation or some similar process.

Table 10.3 Levels of algal concentration comparing consistency, % dry solids (DS) and the concentration factor

Consistency	% DS	Concentration factor
Culture	0.2–0.6	–
Milk-like	1–2	17–100
Cream-like	10–12	160–600
Cheese-like	15–20	250–1000

Table 10.4 Potential use of algal biomass and % dry solids required in harvested biomass

Purpose	Form	% DS
Feed		
Zooplankton	Suspension	
Fish larvae	(not harvested)	0.02–0.06
Fish pellets	Spray-dried, drum-dried	12–22
Ruminants	Wet preserved	
Pigs & Poultry	(ruminants only)	8–12
	Storable powder	8–22
	Pellets	8–30
Fine chemicals, human food etc.		
Protein food	Noodles, etc.	8–30
Chemicals	For extraction	8–>22
Health food	Tablets	8–15

The various options available and their advantages and disadvantages have been well reviewed by Mohn (1980, 1988) and will not be considered further here. Suffice it to say that harvesting is usually a major cost in any algal process and an optimum, cost-effective method must be developed for each alga and product. Some idea of the relative costs of the various processes is presented in Table 10.5.

Scale-up

The decision on whether to extend the technical study, including scaling-up, must depend on the total value of the market for the product. Unless there are additional, strategic influences on the decision, the total market value should also determine the budget appropriate for R & D scale-up and piloting. Ideally, the 'market-pull' should also determine the timescale and scheduling of the scale-up.

The scale-up sequence

We have found scale-up by a series of steps, each representing an approxi-

Table 10.5 Cost and efficiency of harvesting methods based on data of Mohn (1988) for *Coelastrum, Scenedesmus* and *Spirulina* (approximate costs in Australian dollars)

Method	Cost of apparatus (15 m³ h⁻¹)	Cost m⁻³	% WFS
Centrifugation			
Multichamber centrifuge	$100 000	0.44	20
Self-cleaning plate centrifuge	$ 95 000	0.40	12
Nozzle centrifuge	$ 85 000	0.36	<15
Sedimentation Tank & Flocculant			
With potato starch	$ 25 000	0.10	>1.5
With Al³⁺	$ 25 000	0.16	>1.5
Flotation & Flocculant			
With potato starch	$ 90 000	0.35	7
With Al³⁺	$ 90 000	0.42	7
Filtration			
Filter press (14 m²)	$ 35 000	0.16	20–22
Membrane press (11 m²)	$ 46 000	0.19	25
Vacuum band-filter	$110 000	0.52	22
Microstrainer (25 μm)	$ 35 000	0.13	1.5
Vibrating screen	$ 37 000	0.15	6
Sand filter	$?	?	6

mate factor of ten increase, is most manageable in biological and engineering terms, and yields fewer surprises.

Phase 1: Growth of first cultures, 10 ml to 10 l

These cultures are used to:

1. screen for the desired product and determine its concentration in the cells;
2. start to develop processes for growth (e.g. axenic vs mixed cultures, photoautotrophic vs heterotrophic or mixotrophic conditions) and to establish broad parameters for growth medium constituents;
3. develop product analysis methods;
4. assess the suitability of various harvesting techniques for typical cultures; and
5. assess extraction or other processing methods. (This is usually necessary, anyway, in order to develop sampling and analytical methods.)

At this stage, none of these processes should be optimized, as most parameters will change significantly before piloting. For example, knowledge of bulk water and chemical availability on site, and the type of mixing possible in particular ponds and vessels will alter concepts on medium constituents and vessel or pond design.

For simplicity, the process should be batch at this stage, but the advantages of a continuous or semi-continuous process should be anticipated.

A rough costing is needed, even at the 20 l culture volume, to confirm that the process is in the right cost range. For example, the need for a particularly costly medium component may eliminate a particular process at this stage.

Phase 2: Scale-up to 100 l

Cultures of 100–500 l can be grown in bags, mini-ponds, or fermenters; they can be indoors or outside. They should still be batch cultures, to simplify the engineering.

At this scale, scale-up problems and advantages can be studied prior to the next, more expensive and rigorous stage. The importance of mixing, gas distribution, cell adhesion to vessels and culture medium sterility can all be determined at this stage.

This stage may also become part of a routine inoculum sequence for a commercial process, and its technology may have later value.

Phase 3: Scale-up to 1000 l

At this stage, a growth process close to the final envisaged commercial process should be run. A decision should have been made and implemented on such factors as axenic or mixed cultures, agitation, vessel shape and illumination.

Detailed and rigorous trials should now be undertaken, optimizing conditions for growth and product accumulation, optimizing the selected harvesting method(s), processing, product formulation, stability and shelf-life trials and applications trials.

Cultures of this scale can model expected representative production parameters, and provide the product needed for:

1. application trials, e.g. animal feeding trials, incorporation in typical food products, etc.;
2. samples to gauge customer reaction;
3. extensive analysis and testing for contaminants and pollutants (especially important if the final product is for the food or feed industry); and
4. preliminary toxicity studies, if required.

Most of the steps in the sequence should now be optimized and a detailed costing of the process, projected to commercial scale, should be undertaken. Financial modelling is used to examine the influence of size, volume and other parameters on costing. Sensitivity analysis using the model then pinpoints areas warranting experimental attention as major contributors to cost or to process inefficiency.

The production costs should then be compared with costs known or estimated for alternative production systems.

It is worthwhile to recheck the market research data at this step since time has elapsed since the first decision to scale-up. In particular, market information may suggest a change of product concentration or formulation as, for example, addition of a stabilizer. Further process development and recosting may also be necessary. Up-to-date market value and price information is also critically important before the piloting stage, which can cost 10% to 50% of a commercial plant.

Piloting

In general, a process should be fully piloted at about one-tenth commercial scale.

Example of a scale-up. β-Carotene production from *Dunaliella salina* at Hutt Lagoon, Western Australia

Some of the basic research of particular importance to generating ideas leading to commercial process development of β-carotene production using the halophilic green alga *D. salina* included studies of the mechanism and extent of salt tolerance (Loeblich, 1972; Ben-Amotz and Avron, 1973; Borowitzka and Brown, 1974; Borowitzka *et al.*, 1977) and studies on β-carotene accumulation (Mil'ko, 1963; Aasen *et al.*, 1969; Loeblich, 1972; Ben-Amotz and Avron, 1973). This was supplemented by detailed studies in our own laboratory. A large number of reports on open pond trials, with extensive scientific supporting studies in Russia and the Ukraine (see Borowitzka and Borowitzka, 1988a for references) regrettably had limited application in the early stages of commercialization because these papers were published in Russian and Ukranian (e.g. Massyuk and Abdula, 1969; Massyuk, 1973).

The basic research suggested processes for growing open monocultures of *D. salina* using particular environmental conditions to maintain the monoculture, and to produce cells with up to 14% β-carotene on a dry weight basis. Concurrent market research indicated a world β-carotene market of approximately 50 t, with selling prices ranging between $A400 and $A1000 kg^{-1}. The market was growing at approximately 5% per annum, and was being filled with chemically synthesized β-carotene. The potential for a niche market of 1–5 t 'natural' β-carotene was identified, based on a trend towards 'natural' foods and additives.

Assuming a price of at least $A600 kg^{-1} for natural β-carotene, the niche market value was estimated at $A600 000 to $A3 million (for 1–5 t). It was decided that production costs of half the selling price, say $A300 kg^{-1}, were an achievable goal, and scale-up commenced.

The scale-up from laboratory studies by factors of 10 to 'pilot scale' at

Hutt Lagoon is described in Borowitzka *et al.* (1984; 1985). Laboratory cultures of 10 ml increased to 20 l bottles were used to develop growth media.

The first open vessel experiments were in 10 l microcosms (Borowitzka and Borowitzka, 1988b) and modelled the effect of intrinsic and external contaminants of Hutt Lagoon derived *D. salina* cultures under controlled temperature and light conditions. These were extended to temperature controlled, open plastic ponds of several hundred litres in Sydney (Australia), and led to the development of a management system, controlling the stability of the monoculture using inoculum regimes and medium composition.

Open plastic ponds of several hundred litres were then established at Hutt Lagoon on the west coast of Australia, the site chosen for meeting predetermined criteria relating to climate, water and salt availability and logistics.

Scale-up to earth-walled ponds constructed on the bed of Hutt Lagoon started with ponds of 100 m² area (approx. 20 000 l). Smaller ponds were tried briefly, but the large wall area to volume ratio resulted in unacceptably high rates of leakage. In the larger ponds this leakage became relatively insignificant. Other disadvantages of high perimeter to volume ratios in shallow ponds were identified as:

– the loss of pond holding capacity due to erosion of banks;
– loss of algal cells through windrowing on the banks;
– runoff of rainwater from the sloping walls of the ponds reducing the salinity, eroding the banks and increasing turbidity.

All of these disadvantages were reduced in scale-up. A series of earth pond scale-ups at Hutt Lagoon, 100 m² (20 000 l), 250 m² (50 000 l) and 600 m² (120 000 l) were used to explore the influences of nutrient additions, pond depth (in relation to light penetration), salinity, use of blends of high and low salinity brines, maximum and minimum daily temperatures, rainfall patterns and cloud cover, on growth and β-carotene production rate. The roles of endemic bacteria, algae and predatory protozoa were also studied and processes for controlling them developed. Several pond management regimes were trialled and operational protocols developed.

Before final design of the production plant, trials were undertaken in 0.5 ha ponds (i.e. one-tenth of the proposed production pond size) to provide data for accurate economic analysis and to evaluate further certain features of pond design and operation.

Compared with laboratory cultures under similar conditions of salinity, nutrient addition and temperature, pond cultures grew slower and had lower maximum concentrations of cells and β-carotene (Moulton *et al.*, 1987b). This phenomenon has been observed with other large-scale operations (Goldman, 1979; Richmond and Grobbelaar, 1986) and is probably due to a combination of factors including reduced light penetration, gas exchange and mixing.

Transfers of inoculum culture between ponds and to the harvester required scale-up of pumps and plumbing. Pumps and narrow pipes causing high shear forces were found to impair cell viability and lead to oxidation loss of β-carotene. Low-shear pumps suitable for the fragile cells and the high salinity brines were selected and tested.

The challenge of gaining representative samples to enable monitoring of growth, product concentration and nutrient status is greatest in large, unmixed open ponds. At each scale-up above 100 l, to the commercial 5 ha (10^7 l) ponds the sampling method was rechecked for statistical accuracy. The appropriate minimum number of samples and selection of their positions in the ponds was determined from tests with many samples taken from different positions in the ponds under different prevailing wind and climatic conditions.

The cost-effectiveness of recycling culture medium following cell harvesting became apparent in the scale-up series. All harvesting methods were re-evaluated to ensure that no chemical modifications were made to the growth medium which inhibited algal regrowth. This excluded some types of harvesting by flocculation.

Once culture volumes reach thousands of litres, commercial scale continuous centrifuges and fitration equipment are required if direct scale-up of laboratory techniques is applied. Both processes are capital intensive, and in the case of *D. salina*, equipment is subject to extensive salt corrosion. At this scale, in any algal system, cheaper harvesting alternatives must be explored. Observations from smaller cultures should suggest alternative methods, for example, for *D. salina*, gyrotaxis (Kessler, 1982), hydrophobic binding (Curtain and Snook, 1983) and migration in a density gradient (Bloch *et al.*, 1982) have all been patented as harvesting methods derived from observations of small cultures.

Sufficient biomass was available from thousand litre cultures for trials with pilot-scale drying equipment (freeze-driers, spray driers, drum or belt driers) to produce β-carotene-rich dried algal powder. Systems for extracting, concentrating and purifying the β-carotene were also significantly modified from laboratory and analytical techniques.

Bioeconomic modelling was used to predict the cost of production of β-carotene and to highlight the expensive components of the system (Moulton *et al.*, 1987a).

Conclusions

Large-scale cultures often differ significantly in their behaviour from small laboratory cultures; however, laboratory studies will often be necessary in parallel with the large-scale pilot studies to establish the cause of observed phenomena and to suggest a possible strategy for implementation at the pilot scale. Successful scale-up and long-term management of microalgal mass cultures therefore requires good laboratory backup.

The scale-up of a laboratory microalgal process to commercial scale must proceed in a series of logical steps and requires the collaboration of biol-

ogists, ecologists, engineers, economists and marketing experts. For this
scale-up process to lead to a commercially viable process in the shortest
time, and for the least development cost, the economics of the process must
be critically evaluated at each step and the critical components identified so
that available resources can be effectively allocated. In any algal process
there is much room for improvement, especially since we are still in the
early phase of the development of the technology of commercial algal
biotechnology. Critical analysis of the process under development should
indicate where improvement in the process will lead to significant
production cost savings, rather than just marginal improvement in the economics of the process.

References

Aasen A J, Eimhjellen K E, Liaaen-Jensen S 1969 An extreme source of β-carotene.
  Acta Chim. Scand. **23**: 2544–5
Anderson D B, Eakin D E 1984 A process for the production of polysaccharides from
  microalgae. *Batelle Lab Reports* (Quoted in Tapie and Bernard, 1988)
Augenstein D C 1987 Gaslift pumps for combined pumping and gas contacting in
  algal production systems. *SERI Subcontractors Report STR-231-2840*: 154–73

Becker E W, VenkataMaran L V 1980 Production and processing of algae in pilot
  plant scale. Experiences of the Indo–German project. In Shelef G, Soeder C J
  (eds) *Algae Biomass*, Elsevier/North Holland Biomedical Press, Amsterdam,
  pp 35–50
Ben-Amotz A, Avron M 1973 The role of glycerol in the osmotic regulation in the
  halophilic alga, *Dunaliella parva. Pl. Physiol.* **51**: 875–8
Benemann J R, Goebel R P, Weissman J C, Augenstein D C 1982 Microalgae as a
  source of liquid fuels. *Report to DOE Office of Energy Research, May 1982*, 17 pp
Bloch M R, Sasson J, Ginzburg M E, Goldman Z, Ginzburg B Z 1982 Oil products
  from algae. US Patent No 4 341 038
Borowitzka L J, Brown A D 1974 The salt relations of the marine and halophilic
  strains of the unicellular green alga, *Dunaliella. Arch. Microbiol.* **96**: 37–52
Borowitzka L J, Borowitzka M A, Moulton T P 1984 The mass culture of *Dunaliella
  salina* for fine chemicals: from laboratory to pilot plant. *Hydrobiologia*
  116/117: 115–21
Borowitzka L J, Kessley D S, Brown A D 1977 The salt relations of *Dunaliella*.
  Further observations on glycerol production and its regulation. *Arch. Microbiol.*
  113: 131–8
Borowitzka L J, Moulton T P, Borowitzka M A 1985 Salinity and the commercial
  production of beta-carotene from *Dunaliella salina*. In Barclay W J, McIntosh R
  (eds) *Algal Biomass: An Interdisciplinary Perspective*, J. Cramer Verlag, Vaduz,
  pp 217–22
Borowitzka M A 1986 Microalgae as a source of fine chemicals. *Microbiol. Sci.*
  3: 372–5
Borowitzka M A 1988a Vitamins and fine chemicals from micro-algae. In Borowitzka
  M A, Borowitzka L J (eds) *Micro-Algal Biotechnology*, Cambridge University
  Press, Cambridge, pp 153–96
Borowitzka M A 1988b Fats, oils and hydrocarbons. In Borowitzka M A, Borowitzka
  L J (eds) *Micro-Algal Biotechnology*, Cambridge University Press, Cambridge,
  pp 257–87
Borowitzka M A, Borowitzka L J 1988a *Dunaliella*. In Borowitzka M A, Borowitzka

L J (eds) *Micro-Algal Biotechnology*, Cambridge University Press, Cambridge. pp 27–58

Borowitzka M A, Borowitzka L J 1988b Limits to growth and carotenogenesis in laboratory and large-scale outdoor cultures of *Dunaliella salina*. In Stadler T, Mollion J, Verdus M-C, Karamanos Y, Morvan H, Christiaen D (eds) *Algal Biotechnology*, Elsevier Applied Science, London, pp 371–81

Burlew J S 1953 Current status of large scale cultures of algae. In Burlew J S (ed) *Algal Cultures from Laboratory to Pilot Plant*, Carnegie Institution, Washington, DC, pp 3–23

Chaumont D, Thepenier C, Gudin C 1988 Scaling-up a tubular photoreactor for continuous culture of *Porphyridium cruentum* from laboratory to pilot plant (1981–1987). In Stadler T, Mollion J, Verdus M-C, Karamanos Y, Morvan H, Christiaen D (eds) *Algal Biotechnology*, Elsevier Applied Science, London, pp 199–208

Chen B H, Chi C H 1981 Process development and evaluation for glycerol production. *Biotechnol. Bioeng.* 23: 1267–87

Cohen Z 1986 Products from microalgae. In Richmond A (ed) *Handbook of Microalgal Mass Culture*, CRC Press, Boca Raton, pp 421–54

Curtain C C, Snook H 1983 Method for harvesting algae. US Patent No 511 135

Dodd J C 1986 Elements of pond design and construction. In Richmond A (ed) *Handbook of Microalgal Mass Culture*, CRC Press. Boca Raton, pp 265–83

Goldman J C 1979 Outdoor algal mass cultures I. Applications. *Wat. Res.* 13: 1–19

Gudin C, Thepenier C 1986 Bioconversion of solar energy into organic chemicals by microalgae. *Adv. Biotechnol. Processes* 6: 73–110

Ho C S, Shanahan J F 1986 Carbon dioxide transfer in bioreactors. *CRC Crit. Rev. Biotechnol.* 4: 185–252

Jüttner F 1977 Thirty liter tower-type pilot plant for the mass cultivation of light and motion sensitive planktonic algae. *Biotechnol. Bioeng.* 19: 1679–88

Kawaguchi K 1980 Microalgae production systems in Asia. In Shelef G. Soeder C J (eds) *Algae Biomass*, Elsevier/North Holland Biomedical Press. Amsterdam, pp 25–33

Kessler J O 1982 Algal cell harvesting. US Patent No 4 324 067

Krüger C H J, Eloff J N 1978 Mass culture of *Microcystis* under sterile conditions. *J. Limnol. Soc. S. Afr.* 4: 119–24

Laws E A 1984 Research, development and demonstration of algal production raceway (APR) systems for the production of hydrocarbon resources. *SERI Subcontractors Report STR–231–2206*

Loeblich L A 1972 Studies on the brine flagellate *Dunaliella salina*. Unpublished PhD dissertation (thesis), University of California, San Diego

Massyuk N P 1973 *Morphology, Taxonomy, Ecology and Geographic Distribution of the Genus Dunaliella Teod. and Prospects for its Potential Utilisation*, Naukova Dumka, Kiev (in Russian)

Massyuk N P, Abdula Y H 1969 First experiments of growing carotene-containing algae under semi-industrial conditions. *Ukranskva Botanichnya Zhournal* 26: 21–7 (In Ukranian)

Mil'ko E S 1963 Effect of various environmental factors on pigment production in the alga *Dunaliella salina*. *Mikrobiologya* 32: 299–307 (In Ukranian)

Mohn F H 1980 Experience and strategies in the recovery of biomass from mass cultures of microalgae. In Shelef G, Soeder C J (eds) *Algae Biomass*, Elsevier/North-Holland Biomedical Press, Amsterdam, pp 547–71

Mohn F H 1988 Harvesting of micro-algal biomass. In Borowitzka, M A, Borowitzka L J (eds) *Micro-Algal Biotechnology*, Cambridge University Press, Cambridge, pp 395–414

Moulton T P, Borowitzka L J, Vincent D J 1987a The mass culture of *Dunaliella salina* for β-carotene: from pilot plant to production plant. *Hydrobiologia* **151/152**: 99–105

Moulton T P, Sommer T R, Burford M A, Borowitzka L J 1987b Competition between *Dunaliella* species at high salinity. *Hydrobiologia* **151/152**: 107–16

Neenan B, Feinberg D, Hill A, McIntosh R, Terry K 1986 Fuels from microalgae: Technology status, potential, and research requirements. *SERI Report SERI/SP-231-2550*: 1–149

Oswald W J 1988a Micro-algae and waste-water treatment. In Borowitzka M A, Borowitzka L J (eds) *Micro-Algal Biotechnology*, Cambridge University Press, Cambridge, pp 305–28

Oswald, W J 1988b Large-scale algal culture systems (engineering aspects). In Borowitzka M A, Borowitzka L J (eds) *Micro-Algal Biotechnology*, Cambridge University Press, Cambridge, pp 357–94

Oswald W J, Golucke C G, Gee H K 1959 Waste water reclamation through the production of algae. *Water Research Center Contribution 22*, Sanitary Engineering Research Laboratory, University of California, Berkeley

Parker N C, Suttle M A 1987 Design of airlift pumps for water circulation and aeration in aquaculture. *Aqua. Eng.* **6**: 97–110

Persoone G, Morales J, Verlet H, De Pauw N 1980 Air-lift pumps and the effect of mixing on algal growth. In Shelef G, Soeder C J (eds) *Algae Biomass*, Elsevier/North-Holland Biomedical Press, Amsterdam, pp 505–22

Pirt S J, Lee Y K, Walach M R, Pirt M W, Bayuzi H H M, Bazin M J 1983 A tubular bioreactor for photosynthetic production of biomass from CO_2: design and performance. *J. Chem. Technol. Biotechol.* **33b**: 35–58

Rebeller M 1982 Techniques de culture et de récolte des algues spirulines. *Doc. IFP*, 14 pp

Regan D L, Gartside G 1983 Liquid fuels from micro-algae in Australia. CSIRO, Melbourne, 55 pp

Richmond A 1983 Phototrophic microalgae. In Rehm H J, Reed G, Dallweg H (eds) *Biotechnology*, Vol 3, Verlag Chemie, Weinheim, pp 109–44

Richmond A 1988 *Spirulina*. In Borowitzka M A, Borowitzka L J (eds) *Micro-Algal Biotechnology*, Cambridge University Press, Cambridge, pp 85–121

Richmond A, Grobbelaar J U 1986 Factors affecting the output rates of *Spirulina platensis* with reference to mass cultivation. *Biomass* **10**: 253–64

Richmond A, Vonshak A 1978 *Spirulina* culture in Israel. *Arch. Hydrobiol. Beih. Ergeb. Limnol.* **11**: 247–80

Robinson, L F 1987 Improvements relating to biomass production. *European Patent* 0 239 272

Robinson P K, Mak A L, Trevan M D 1986 Immobilized algae: A review. *Process Biochem.* **21**: 122–7

Shimamatsu H 1987 A pond for edible *Spirulina* production and its hydraulic studies. *Hydrobiologia* **151/152**: 83–9

Soong P 1980 Production and development of *Chlorella* and *Spirulina* in Taiwan. In Shelef G, Soeder C J (eds) *Algae Biomass*, Elsevier/North-Holland Biomedical Press, Amsterdam, pp 97–113

Tapie P, Bernard A 1988 Microalgae production: technical and economic evaluations. *Biotechnol. Bioeng.* In Press

Terry K L, Raymond L P 1985 System design for the autotrophic production of microalgae. *Enz. Microbial Technol.* **7**: 474–87

Torzillo G, Pushparat B, Bocci F, Balloni W, Materassi R, Florenzano G 1986 Production of *Spirulina* biomass in closed photobioreactor. *Biomass* **11**: 75–9

Venkatamaran L V, Becker E W 1985. *Biotechnology and Utilization of Algae – The Indian Experience*, Department of Science and Technology, New Delhi

Weissman J C, Goebel R P 1987 Design and analysis of microalgal open pond systems for the purpose of producing fuels. *SERI Subcontractors Report STR-231-2840*: 1–214

11

The future of microalgal biotechnology

J. R. Benemann

Introduction

The beginning of the field of algal biotechnology can be dated to the work carried out in the late 1940s by several groups in the US and other countries, summarized in a book, edited by Burlew (1953), *Algae Culture: from Laboratory to Pilot Plant*. Progress over the past three decades has moved microalgal culture from the pilot plant to a commercial reality. A large number of experimental and pilot plant scale facilities have been operated in several developed and developing countries over the past few decades. In Japan and Taiwan a number of commercial micoalgal production plants have operated for about 20 years, and several new plants were established in the US and other countries in recent years. An example of a commercial microalgal production facility is shown in Fig. 11.1

Microalgae are being commercially produced in these plants on a large scale (over one hectare in size) for food pigments, vitamin supplements (β-carotene) and health foods. Microalgal cultures are also used in waste-water treatment and in the aquacultural production of shellfish. In the US and Europe major R & D programmes, with cumulative budgets exceeding $25 million, have been underway for over a decade to develop algal biomass production as a future source of fuels. In addition to the products already commercialized, a large number of potential microalgal products are in various stages of development (Table 11.1).

Thus it would appear that microalgal mass culture is one of the successes of modern biotechnology and that the future promises rapid expansion and new applications in the field of algal biotechnology.

There are, however, some cautionary signs: most commercial enterprises in this field are only in their initial stages and still struggling to achieve a positive cash-flow (let alone profits), with a number having already failed. The commercial successes of the first enterprises for algal mass culture, the *Chlorella* production systems in Japan and Taiwan and the *Spirulina* production systems more recently established in the US and elsewhere, were

Fig. 11.1 Microalgal production plant of Cyanotech, Inc., Kona Hawaii. Each pond is about one acre in size; the two ponds on the left produce *Spirulina*, the two on the right *Dunaliella*. (Courtesy, Cyanotech, Inc.)

based on unsupported claims that algal pills and powders promote health and well-being. The current costs of production of microalgae by existing commercial facilities can be estimated at over $10 000 per ton of organic matter, with the plants in Japan and Taiwan greatly exceeding even this high cost. Current production costs of β-carotene by *Dunaliella* exceed those of the industrial synthesis process by a factor of over ten-fold. The use of microalgae in wastewater treatment and aquacultural production of shellfish is based on low productivity algal ponds, which lack control over the algal populations. The production of low cost commodities (foods, feeds, fuels, fertilizers) with microalgal cultures appear even further in the future today than 40 years ago.

It is the objective of this chapter to predict the likely future developments in this field, both in the short range (the next ten years) and the long range (the next 25 years). Any such gazing into the crystal ball is fraught with danger. It is amusing to go back to books and articles from the early 1960s with titles such as '25 Years From Now . .'. However, the freedom that such speculation allows outweighs prudence. Therefore I make specific predictions about the future of this technology, to stimulate thought, discussion, speculation, and even controversy. That, rather than the predictions themselves, I hope to be the contribution of this chapter. I make no attempt to review the literature, which is covered in the preceding chapters and in other recent books (Shelef and Soeder, 1980; Becker, 1985; Richmond, 1986; Borowitzka and Borowitzka, 1988; Lembi and Waaland, 1988). I begin with the history of microalgal biotechnology, as an instructive starting point.

Historical background

Microalgae (colloquially known as 'pond scum') have historically been considered a nuisance or at least worthless. Virgil is quoted as writing: '*Nihil Vilior Algae*' – 'there is nothing more worthless than algae'. Most practical interest in microalgae has been in their eradication. They foul pipes and contaminate water supplies (some cyanobacteria are toxic). Algal blooms, resulting from over-fertilization due to run-off from farms and towns, cause oxygen depletion and eutrophication in rivers and lakes. The amount of money expended on controlling algal fouling and blooms exceed those spent on their purposeful cultivation by several orders of magnitude.

There have, however, been a few traditional practical uses of algae, particularly as a food source. The use of *Spirulina* by the Aztecs in Mexico and the natives of Chad are well documented, and the consumption of *Nostoc* 'balls' growing in rivers appears to be a worldwide (though not widespread) phenomenon.

The first, laboratory scale, work on the culture of microalgae for practical applications was carried out in England on the growth of algae for the rearing of oyster larvae (Bruce *et al.*, 1940). During the Second World War in Germany, Harder and von Witsch (1942) proposed diatoms as a source of β-carotene (to improve the night-time vision of flyers) and vegetable oils.

After the war programmes were initiated for the development of microalgal mass cultures in the US at the Carnegie Institution of Washington, and (with US funding) in Japan and Germany. The impetus for this research came from the prospect that growing world populations would soon outstrip conventional food supplies, and the perception that microalgal cultures were inordinately productive. (This perception, resulting from Otto Warburg's early work still persists, as is discussed below.) Burlew's book brought together articles by most of the pioneers in this field and describes the pilot plant work carried out by the A. D. Little Company, representing the state of the art in the early 1950s.

Burlew's book discussed the key technical problems to the development of a successful microalgal biotechnology, many of which remain to be solved today. How to maintain one desired species of microalgae and prevent contamination of the cultures by competing algal 'weeds', infections (of the algae) by bacteria, fungi, or viruses, or grazing by rotifers, copepods, and

Table 11.1 Products from microalgae

Algal products	Uses and examples	Approx. value, $/kg	Approx. market*
Isotopic compounds	Medicine Research	>1000	Small
Phycobiliproteins	Diagnostics Food colours	>10 000 >100	Small Medium
Pharmaceuticals	Anticancer Antibiotics	Unknown (very high)	Large Large
β-Carotene	Food supplement Food colour	>500 300	Small Medium
Xanthophylls	Chicken feeds Fish feeds	200–500 1000	Medium Medium
Vitamins C & E	Natural vitamins	10–50	Medium
Health foods	Supplements	10–20	Medium to large
Polysaccharides	Viscocifiers, gums Ion exchangers	5–10	Medium to large Large
Bivalve feeds	Seed production Aquaculture	20–100 1–10	Small Large
Soil inocula	Conditioner, Fertilizers	>100	Unknown Unknown
Amino acids	Proline Arginine, aspartate	5–50 5–50	Small Small
Single cell protein	Animal feeds	0.3–0.5	Large
Vegetable oils	Foods, feeds	0.3–0.6	Large
Marine oils Waste treatment	Supplements Municipal, industrial	1–30 1 per kg algae	Small Large
Methane, H_2, liquid fuels	General uses	0.1–0.2	Large

* Market sizes ($ million): small <$10; medium $10–100; large >$100.

other zooplankton. How to design an optimal algal reactor – most designs we are familiar with today were originally described in this book: vertical and horizontal glass and plastic tubular reactors, circular ponds, covered troughs. The glaring exception was the mixed raceway pond design, now used almost universally in commercial systems (see Fig. 11.1). How to maintain desirable environmental conditions in the algal production reactors: pH, temperature, nutrients. How to harvest the algae economically. How to maximize productivity of the cultures (for example by using the 'flashing light' effect described by Kok, 1953).

One major issue not addressed in Burlew's book was the economics of algal production. Not even an analysis of inputs required for algal cultivation was presented. Such an analysis, based on the pilot plant work described in Burlew's book (consisting of a large sausage-shaped transparent plastic bladder), was presented by Fisher (1961), who concluded that the

Algae genus or type	Product content	Production system	Current status
Many	>5%	Tubular, indoors	Commercial
Red Cyanobacteria	1–5%	Tubular, indoors	Commercial Commercial
Cyanobacteria Other	0.1–1%	Indoor ponds, fermenters	Research
Dunaliella	5%	Lined ponds	Commercial
Greens, diatoms, etc.	0.5%	Unlined ponds Lined ponds	Research Commercial (?)
Greens, unknown	<1%	Fermenters	Research
Chlorella, Spirulina	100%	Lined ponds	Commercial
Porphyridium Greens, others	50%	Lined ponds	Research Commercial
Diatoms, Chrysophytes	100%	Lined ponds	Commercial Research
Chlamydomonas N-fixing species	100%	Indoor ponds Lined ponds	Commercial Research
Chlorella Cyanobacteria	10% 10%	Lined ponds Lined ponds	Research Conceptual
Green algae, others	100%	Unlined ponds	Research
Greens, others	30–50%	Unlined ponds	Research
Diatoms, others Greens, others	15–30% n.a.	Lined ponds Unlined ponds	Research Commercial
Cyanobacteria, greens, diatoms	30–50%	Unlined ponds	Research

costs of cooling the system made such a design unacceptable. Despite the limitations of the technology (the A. D. Little Co. pilot plant leaked and overheated, the cultures 'crashed' repeatedly, productivities were low and erratic and harvesting was a problem), the use of microalgal cultures as a future food source continued to be actively promoted.

However, not without critics: Odum (1971), for example, accused proponents of this technology of perpetrating a 'cruel illusion' on the unsuspecting public by claiming that microalgae could be a source of food for the hungry masses of the world. Based on an average yield of 12.5 g m^{-2} d^{-1} (extrapolating to 20 tons acre^{-1} yr^{-1}) for the pilot plant described in Burlew's book, and a cost of \$2.8 m^{-2} yr^{-1} reported by Fisher (1961), Odum calculated that it would take more energy to produce algae than could be derived from it. His calculation was extremely simplistic: he divided the Gross National Product by the total energy used in the US to estimate the energy input into this system! He goes on to castigate the scientists involved for duping the public with '. . . this fantastic means for putting public funds into research . . .'.

The arguments by critics against and proponents for microalgal biotechnology as a cure-all for most problems of mankind (see Goldman and Ryther, 1977 and Oswald and Benemann, 1977, for examples), continue in essentially unmodified form today. The sceptics allege inordinately high energy, water, fertilizer, and CO_2 requirements, major technical problems (species control, harvesting, etc.) and high capital and operating costs. The supporters point to high (sometimes almost fantastic) productivities and portray microalgal cultures as having the dual advantages of agriculture (use of sunlight) and fermentations (hydraulic systems and continuous production), while arguing that most problems are only apparent or could be solved by R & D. Most of these pro and con arguments are either superficial or incomplete. Only by analysing the engineering and biological fundamentals of microalgal biotechnology can we arrive at a more realistic view of the real potentials and limitations of this technology.

Enclosed reactors

The basic inputs into microalgal mass cultures are sunlight, water, CO_2, N, P, and other nutrients. The use of artificial lights is very expensive ($>\$10$ kg^{-1} of biomass for the electricity alone) and no good rationale for its use can be made in most cases, even for high value products where economics is not the overwhelming factor.

Some have suggested producing microalgae heterotrophically in conventional type aerobic fermenters. However, the relatively low growth and metabolic rates of microalgae (compared to bacteria and yeast) in the dark suggest that fermentations are unlikely to be a practical approach to microalgal biotechnology. Several private companies have active programmes underway to develop microalgal products (vitamins C and E, for example) through fermentation. However, no commercial successes have been reported. In any event, it would appear that genetic engineering approaches using bacteria or yeast, already used in fermentation processes, are more attractive than genetic or physiological manipulations of microalgae to achieve high productivities in fermenters.

Many researchers have proposed and developed 'closed' reactors, consisting of plastic or glass covered ponds or tubular reactors. Arguments for closed systems are that they help to prevent contaminations, help conserve water and CO_2, increase operating temperatures and overall exhibit higher productivities than open systems. However, none of these arguments is correct. Contamination is not prevented unless the reactors and all inputs are sterilized, which is economically impractical (except possibly for product values exceeding $\$100$ kg^{-1} algal biomass). Water use is not a sufficient cost factor to justify the additional capital and operating costs, and neither is CO_2, loss of which can be prevented by operating the system within prescribed limits of pH and alkalinity. As Fisher (1961) found, higher temperatures in closed reactors require expensive cooling. Covered systems, because of light attenuation, can be expected to have lower productivities than open ones. In most cases, the use of covered systems cannot be justified on economic grounds.

Thus, I predict that the use of indoor cultures using artificial light, of heterotrophic fermentations, of tubular reactors, or of covered ponds will be limited to relatively few, very high value products. Examples would include pharmaceutical products (algae were considered as a source of steroids in the 1950s and are currently under investigation as a source of anticancer agents, see Ch. 7), isotopically enriched and radioactively labelled compounds (many ^{13}C and ^{14}C containing compounds, for example, are already produced by algal cultures), deuterated lubricants (Delente, 1987), and phycofluors (phycobiliproteins used as fluorescent labels in diagnostics) (Glazer and Stryer, 1984). All these products have in common small-scale markets, from a few hundred grams to kilograms. Another use for laboratory reactors and covered ponds is the production of inoculum for large-scale production systems. Although these applications can be important from a commercial perspective, the future of microalgal biotechnology rests with using sunlight in low cost, open, pond systems.

Engineering designs of microalgal pond systems

With sunlight as the input, productivity and reactor scale are expressed in terms of areal units (not volumetric units, as in fermenters) and will be limited by the intensity of sunlight and conversion efficiencies of photosynthesis. Perforce this requires reactors with large surface areas (and high surface to volume ratios) if significant scale productions are aimed for. In turn, this forces significant cost restraints on the designs of microalgal production facilities (cost per unit area) even if the most favourable and optimistic conversion efficiencies for photosynthesis are assumed.

The engineering design of microalgal production systems which can meet the severe cost constraint of microalgal production can be specified in relative detail, even in the absence of specific information about the site of the plant, the actual species to be cultivated, or the product to be produced (see Ch. 10).

An algal pond must be shallow (10–30 cm), to minimize water handling and maximize cell concentration which, in turn, minimizes harvesting efforts. It must be mixed at a rate sufficient (>10 cm s^{-1}) to prevent deadspots and allow nutrients to be distributed throughout the pond (as well as periodically redistribute the algae in the water column) but not too fast (<30 cm s^{-1}), since power costs rise as a cube function of the linear mixing velocity and fast mixing could create excessive turbidity. Even at 30 cm s^{-1} mixing velocity the total cost of mixing is only about $0.01 kg^{-1}$. Microalgal production is not an energy intensive process, despite many allegations to the contrary. The pond must be provided with a source of CO_2, and operated within a pH range and at an alkalinity which would allow maximal algal productivity but would not result in excessive loss of CO_2 to the atmosphere. The pond bottom would need to be sealed so as to minimize percolation. These factors are to some degree interactive with each other and the size of the individual growth unit.

The basic microalgal pond design which has evolved is that of a paddle-wheel mixed raceway type pond (Fig. 11.1). I foresee that the use of this type of microalgal production system will continue in the future for most applications, although significant engineering advancements are still likely in areas such as hydraulics (particularly of larger ponds), nutrient and CO_2 supply, pond lining, and even mixing systems. The algal production reactor (pond) is however only one component, though the most visible, of an algal production facility. Harvesting the algal biomass and processing it to produce the desired product(s) are as important, and can be even more expensive, than the cultivation of the algae themselves.

Harvesting and processing of microalgae

Harvesting has been a preoccupation of many researchers in this field over the past few decades, including myself. Harvesting of the algal biomass present is a major technical (e.g. economic) problem. The problem is due to a combination of the small size of the microalgae (3–30 μm) and their low concentration in the pond medium (<500 mg l^{-1}). By comparison, the biomass concentration in most bacterial or yeast fermenters is about 100-times higher. Thus, the use of centrifuges, feasible in fermentations, cannot be used economically in most cases for microalgae, as the costs of centrifugation would be well above $1000 ton^{-1} of biomass (organic dry weight).

Several solutions exist to this problem: growing larger more easily harvested algae; operating the cultures under longer detention times to increase the algal biomass concentration; operating very shallow ponds (<10 cm), as density is an inverse function of depth; using chemical flocculating agents instead of centrifugation; or discovering novel harvesting technologies which can avoid these problems.

Briefly enlarging on the above alternatives: Large algal cells or colonies do not seem to be very competitive in outdoor cultures, though the reason for that is not clear. Lengthening the detention time of the cultures, beyond that resulting in maximal productivity (generally two to four days depending on insolation and temperature), results in a rapid decrease in productivity as culture density increases above the optimal. The reason for this is somewhat puzzling, as increasing respiration alone could not account for the large decline noted. Myers (1970) pointed to high chlorophyll levels in the most light-limited (longest detention time) cultures as one possible explanation. Very shallow ponds are not practical due to hydraulic limitations and wide temperature fluctuations, limited nutrient storage (particularly CO_2), and excessive O_2 build-up. Chemical flocculation has become the method of choice in the removal of algae from waste treatment ponds and other applications. However, the chemicals and the flocculation process are expensive (both capital and operating), being only marginally cheaper than centrifugation. Many different 'alternative' algal harvesting processes have been described in the literature, taking advantage of various physicochemical properties (adherence to microbubbles in dissolved air or electroflotation, for example) and biological characteristics (phototaxis, flotation) of the algae.

Algal harvesting is likely to remain an active area of research. However, I believe that harvesting need no longer be considered the limiting factor in algal biomass production, which it has been until rather recently. One reason is the realization that a 'universal' harvesting method is neither necessary nor desirable. Each algal species, and each production system, will require a harvesting process optimally adapted to both the requirements of the organism and of the processing steps (drying, extraction) that follow harvesting.

A more important reason is that recent experience has demonstrated that for most (if not all) algal species it is possible to develop an appropriate, and economic, harvesting system. In most of the cases that we have studied (with a variety of different microalgae, using both fresh and saltwater systems) the algae can be induced to flocculate and settle spontaneously, a process referred to as 'bioflocculation'. The use of novel, highly effective and low cost, flocculating agents is also dramatically changing the economics of chemical flocculation (J. C. Weissman, personal communication). In conclusion, harvesting is a problem that even if not completely solved, no longer looms as the major impediment to large-scale low cost algal mass culture.

After harvesting of the algal biomass, which should result in a 50- to 200-fold concentration, the algal slurry (5–15% dry weight) must be quickly processed, lest it spoil (in only a few hours in a hot climate). Processing is, at present, a major economic limitation to the production of low cost commodities (foods, feeds, fuels) and is a significant cost even in the case of higher value products (β-carotene, polysaccharides). It is difficult to discuss processing in general terms, since it is highly specific to each particular product and production system.

One major problem is the high water content of algal biomass, which makes sun drying (the method applicable to most crops) difficult and spray drying too expensive for low cost foods or animal feeds. Spray drying is the method of choice for higher value products (>$1000 ton^{-1}) where the entire biomass is desired; however, it can cause significant deterioration of some algal components, such as the pigments. If only a fraction of the biomass (e.g. protein, lipids, pigments, extracellular polysaccharides, etc.) is desired, a wet processing or solvent extraction of the biomass may be considered. At present the major limitations are the lack of experience with such processes and, thus, their uncertain economics. Processing is rapidly becoming a major area of R & D in microalgal biotechnology, and I believe that the development of economically viable processing technology will be a greater impediment to the commercialization of microalgal products than either cultivation or harvesting.

Economics of microalgal cultivation

Although current algal cultivation ponds are no more than about 0.5 ha in size, there are no obvious limitations to the size of individual growth units, and ponds up to 100 000 m^2 (10 ha) or even larger may be feasible. Ultimately, issues such as wind fetch, hydraulic slope, and mixing effectiveness (dispersion coefficient) will limit the scale of individual growth ponds. At any rate, from an economic viewpoint, a 10 ha pond approaches maximum economies of scale.

Besides the harvesting system used, the processing technology, total system scale and the size of the individual growth ponds, the cost of a microalgal production system depends on a number of site related factors: slope, site preparation costs, land costs, soil type, etc. Of these the most uncertain, at present, is the effect of soil type on system operation; with sandy soil a liner must be used to prevent excessive percolation. If a long-life, durable, plastic liner is used, costs of $100 000 ha^{-1}, just for the liner, can be estimated. However, for many soil types a liner may not be necessary and lower cost liners (clay, buried plastics) may be feasible. At present, experience is too limited with unlined production systems to be able to predict the effect of a 'dirt' pond on the long-term operational stability, specifically in regards to species control, of such a system. This is an area where more research is required.

Another major cost in microalgal production is the various 'support' systems: water and nutrients supply, treatment, storage and distribution; inoculum preparation and build-up; used media treatment, recycling and disposal; power supply and distribution; laboratories and maintenance operations; roads and fences; etc. To these must be added such capital costs as interest during construction, working capital, engineering fees, contingencies, and R & D. Operating costs include, besides the obvious inputs such as labour, water, CO_2 and other nutrients, indirect costs such as maintenance, interest, taxes, insurances, overheads, sales, promotions, etc. Indeed, when all costs are considered, the actual costs of building and operating the algal growth ponds themselves often is but a small fraction of the overall costs of producing a microalgal product.

The actual costs involved in the production of microalgal biomass have been analysed by myself and colleagues, taking into consideration the above listed factors (Benemann et al., 1987). Our analysis assumed a generally favourable site and a two-stage harvesting process ('bioflocculation' followed by centrifugation), but did not include any costs for processing the biomass produced (about 15% solids) to a final product. We estimated that a small system (about 10 ha) with a conventional plastic liner could produce algal biomass for somewhat above $3000 mt^{-1} (metric ton), while larger (100 ha) dirt bottom pond systems, could produce biomass at just under $1000 mt^{-1}. Costs drop to about $500 mt^{-1}, close to the target for commodity products, at the 1000 ha scale.

These estimates are based on relatively specific cost analysis for such system components as paddle-wheels, earthworks, water supplies, fertilizer use and costs, labour requirements, etc. They must be considered generic in nature, as only a product, site, and process specific cost estimate can provide a clear answer to the economic realities of a particular process. Such generic estimates do, however, serve the useful purpose of identifying major cost factors, specific future research required to improve the overall process, cost sensitivities to various critical assumptions (e.g. productivity at 70 mt ha^{-1} yr^{-1}) and the relative commercial viability of various processes and products.

One major cost identified by this analysis was CO_2 and its transfer to the ponds, amounting to about one-third the cost for the largest scale systems. Indeed, if microalgal biotechnology is to make a significant contribution in terms of food, feed, and fuel production, it must become independent of commercial sources of CO_2. The lowest cost and most plentiful source of CO_2 is air, and perhaps methods could be devised by which algae, just like higher plants, derive CO_2 from the atmosphere. However, even in the absence of such technology, there are many 'waste' and geological sources of CO_2 available at low or no cost.

Ultimately, the economics of microalgal production will depend on the validity of the basic assumptions underlying such cost estimates: the ability to reliably cultivate the microalgae ('species control'), and the productivity of the cultures. These issues are addressed next.

Species control in microalgal mass culture

The problem of algal species control has preoccupied me since I started working in the field of algal mass cultures. The problem is simple: is it possible to inoculate a desired strain of an algal species into an outdoor pond and maintain an essentially unialgal culture for any length of time? The current experience in this field would indicate that the answer is negative; only *Spirulina* and *Dunaliella* production systems have been developed as successful commercial enterprises over the past two decades. These algae are found in nature as unialgal cultures under conditions of high alkalinity (>15 g l^{-1} of $NaHCO_3$) and salinity (>100 g l^{-1}), respectively, which are also the conditions which are used in their cultivation. Productivities are generally low, less than half of what can be achieved with green algae or diatoms.

The green algae that have been successfully cultivated in commercial scale systems, or large pilot plants, have either been isolated from cultures that appeared (and became dominant) spontaneously (e.g. *Scenedesmus* in Germany or *Micractinium* in wastewater treatment in the US) or are maintained in the ponds through the production of a high volume of sterile inoculum, which is carefully built up under controlled conditions (*Chlorella* production in Japan and Taiwan). Most other efforts to maintain unialgal cultures in large-scale outdoor ponds have been unsuccessful. This is particularly true for most products listed in Table 11.1.

However, these failures can be readily ascribed (at least with hindsight) to fairly obvious shortcomings of the production system (no uniform mixing, poor designs, nutrient limitations), poor choice of location (low temperatures) and strains used (often laboratory adapted strains). I believe that by selecting 'wild' strains which grow well in outdoor ponds (which can serve as a selection device) and better managed production systems, it will be possible to operate outdoor ponds with a variety of both green algae and diatoms. Indeed, recent work by Weissman *et al.* (1988) has demonstrated that it is possible to grow a number of algal species in outdoor cultures on a sustained basis. In general, culture stabilities and productivities were good. As expected not all cultures tested could be successfully grown outdoors; some were sensitive to the temperature extremes experienced, others to the high O_2 present in the ponds during the daytime. However, there appears to be no lack of potentially suitable algal strains that can be grown in mass cultures.

Nevertheless, many of the commercially desirable algal species are unlikely to be able to be cultivated in outdoor systems, at least not the presently available strains. Two examples are *Porphyridium cruentum* which produces a valuable polysaccharide and *Botryococcus braunii*, which has attracted attention as a hydrocarbon producer. Present experience indicates that they cannot be grown in mass cultures because they contaminate very easily. The reason is that about half of their photosynthetic products are used for the biosynthesis of (respectively) polysaccharides and hydrocarbons, which reduces their relative growth rate compared to the more competitive species. Although in nature these products confer some (although presently unknown) competitive advantage on these species, in a mass culture system these products do not allow them to compete with invading species. This is a general problem with many of the species (and microalgal products) of greatest interest. One solution would be to select (or genetically engineer) strains which produce these products under metabolic control, allowing separation of biomass and product formation into different phases or stages of the cultivation process.

This general approach, of separating biomass and product formation, was recently demonstrated in my laboratory in experiments with lipid accumulating algae, which synthesize storage lipids in response to nutrient (nitrogen) limitation (Benemann and Tillett, 1987). In our experiments the algae were first grown under nitrogen sufficient conditions, resulting in maximal biomass production, and then with no or limited nitrogen, for a short period of time (a few days). Nitrogen limitation prevented continued cell division but not photosynthesis. In a few of the species tested, redirection of the cellular metabolism resulted in an accumulation of lipids, without loss of overall productivity.

I foresee that the problem of microalgal species control will remain a central one in the development of microalgal biotechnology. For many of the products and applications which are listed in Table 11.1, the major issue is not the economics of production but the actual ability to grow the desired organisms in outdoor mass cultures. One example is in the aquacultural husbandry of filter feeding bivalves, which requires the mass culture, on a reliable basis, of several different types of algal species. This has not yet been accomplished. For low value products, economic constraints require that the algal cultures be controlled with a minimum of inputs (e.g. inoculum size, herbivore losses, etc.). This will require long-term (10+ years) R & D to be satisfactorily demonstrated.

Theoretical limits of microalgal productivities

Assuming that the problems inherent in the control of microalgal cultures in 'open' (subject to contamination) ponds can be resolved, the next major issue is the actual biomass productivity that can be achieved with algal cultures operated with sunlight. Perhaps no question has generated more controversy in the past, a controversy that continues to this day. Warburg's experiments with *Chlorella* cultures (see reviews by Kok, 1960, and Pirt, 1986) already mentioned above, suggested that O_2 evolution (and CO_2 fixation) could proceed with less than the eight photons (quanta) per O_2 required by the Z scheme of photosynthesis. Although Warburg's experiments suffer from serious methodological problems and though the Z scheme is solidly established, reports of high rates of algal photosynthesis, apparently incompatible with the Z scheme, continue to appear. The most credible recent work is that of Pirt *et al.* (1980), who claimed almost 50% light conversion efficiency with algal cultures, and projected a productivity of 500 mt ha^{-1} yr^{-1} for areas with the highest insolation (over ten times that projected from the above described A. D. Little pilot plant work!).

Although Pirt's measurements and extrapolations have attracted some criticism, as there are significant methodological uncertainties in any research involving accurate measurements of quantum efficiencies, his recent review of the literature of algal productivities (under laboratory conditions), appears to support the view that algal biomass is produced in laboratory reactors with fewer quanta than the eight to ten predicted by the Z scheme (Pirt, 1986). Although I disagree with Pirt regarding the reliability of many of the measurements that he cites, my own review of the literature makes me conclude that many of the efficiencies reported are ambiguous in regards to the validity of the classic view of a minimum eight quantum requirement. However, neither Pirt (1986), nor other recent authors who question the Z scheme prediction (Osborne and Geider, 1987), propose a testable mechanistic alternative, which makes it more difficult to accept their views.

I propose one possible explanation for this discrepancy between theory and (apparent) practice. If the production of ATP during photosynthesis (an even more disputed number than the quantum requirement) is larger than generally thought, then it would be possible that ATP-driven reverse electron flow from PSII to a strong reductant (NADPH) could take place, reducing the quantum requirement proportionally. This would be the opposite of cyclic photophosphorylation. Although such an electron transport pathway has not yet been demonstrated, it should be recognized that neither has the mechanism of ferredoxin or NADPH reduction in photosynthetic bacteria been elucidated, which must work through reverse electron flow. The great plasticity of the photosynthetic apparatus (e.g. ratio of PSII to PSI reaction centres) suggests that such a mechanism may be possible. With this mechanism a quantum requirement which is not a whole number and can be below eight could be a real possibility. If this can be demonstrated, microalgal biotechnology will have made a major contribution to the whole field of photosynthesis and crop productivity.

Practical limits of microalgal productivity

By contrast with the laboratory experiments, most outdoor algal production systems report productivities in the range of 15–30 g m^{-2} d^{-1} with peak productivities approaching 50 g m^{-2} d^{-1} (Weissman et al., 1988). Optimistic projections from currently established outdoor productivities are, for favourable sites with little seasonal climatic variations, in the region of 100 mt ha^{-1} yr^{-1}. This corresponds to an efficiency of about 4% of total solar radiation and a quantum requirement of several-fold that of the Z scheme of photosynthesis. In the case of algae such as Spirulina and Dunaliella, productivities can be estimated at about half that. Why the large discrepancy between the outdoor productivities and those obtained from most laboratory work?

The laboratory work reported above was carried out under conditions of low light intensities, at or below that saturating for photosynthesis. In practice when dealing with sunlight, the major problem is not the maximal quantum efficiency but the fraction of absorbed photons that can actually be used in photosynthesis. The problem, first recognized by V. Bush (see Burlew, 1953) (but not addressed by Pirt, 1986), is that if algal photosynthesis saturates at low light intensities, but algal cells have a high light extinction coefficient (high pigment content), then more light will be absorbed than can be used in photosynthesis. Depending on the actual saturating and incident light intensities, the extinction coefficients and the densities of the cultures, the maximum efficiency (regardless of minimum quantum requirement) will be reduced by a factor of about two- to five-fold from that possible at low light levels.

The only solution to this problem, as already pointed out by Myers (1970), is to increase the light intensity at which photosynthesis saturates, which means to decrease the antenna and accessory pigment level of the cells (and thereby reduce also the extinction coefficient), without affecting the rest of the photosynthetic apparatus. In some algal strains grown under high light intensities such an adaptation is found. However, to prevent the cultures from becoming too dense (and thus reducing effective light intensity seen by each algal cell) they must be diluted frequently, which means fast growth rates. However, in a mass culture system cell densities must be high (high enough to absorb all the light) which means that growth rates must be (relatively) slow. As pointed out by Myers (1970) we must search for genetic or adaptive characteristics which provide maximum rates of photosynthesis per chlorophyll (pigment) unit at both high and low light intensities, even when algae are managed at low growth rates.

This is a high priority area of research, since it could increase productivities by several hundred percent. Both genetic and physiological approaches to this problem can be considered. One major problem remains, even if organisms are developed which are genetically able to control their antenna to reaction centre pigment ratios, namely the poor competitiveness of such strains in mass cultures. Any contaminating organism that had a higher pigment content would rapidly take over the culture because it would be able to absorb (and thus use) more light in light-limited zones of the culture (which accounts for most of it). This competitive advantage is so strong that any contamination can be expected to become dominant in a matter of a few days.

In essence the problem is lack of cooperation. Algal cultures are collections of individual cells competing with each other for sunlight (even though genetically identical). By maximizing each individual algal cell's competitive advantage under the light-limiting conditions found in algal mass culture ponds, the overall productivity of the system is reduced. In theory one way to overcome this problem is to structure the production system such that several non-mixing layers of algal culture operate simultaneously, with increasing antenna to reaction centre pigment ratios in the deeper cultures. This is the arrangement that higher plants strive for. For microalgal cultures that is not practical.

A more practical solution to this dilemma can probably be found by the judicious use of dual light and nutrient limitations of the algal cultures, as we recently demonstrated in my laboratory for the case of lipid production by nitrogen limited algae (Benemann and Tillett, 1987). Another alternative, one that is becoming popular in agriculture (at least at the experimental level) is to genetically incorporate herbicide resistance into the algal strain and operate the cultures in the presence of the herbicides. The currently available herbicides, effective in the ppb range, make such an approach feasible in principle.

In conclusion, it should be feasible to develop techniques which could overcome the major limitations on photosynthetic efficiencies imposed on algal cells by the current light saturation levels of photosynthesis in dense cultures. Solving the problems outlined above would allow the achievement of the high productivities forecast by, among others, Pirt, though for reasons that are fundamentally different than those advocated by those authors.

In principle it would be possible to apply similar arguments to higher plants; the much higher maintenance energy (respiration) and non-productive tissue components of higher plants suggest that microalgal cultures will maintain and even widen their current lead in terms of productivities of useful products. Indeed, even with current productivities, algal cultures are more productive by an order of magnitude (per unit area or even per unit water consumed) than conventional agricultural systems in the production of products such as proteins or vegetable oils. However, they are not yet economically competitive. Achieving higher productivities can help overcome some, though not all current economic limitations. I predict that algal cultures will become the most efficient and productive biological solar energy converting systems available to mankind. The major issue is the extent to which this technology can be applied.

Products from microalgae

The objective of microalgal biotechnology is to produce specific products which meet real consumer needs. Most of the current production of microalgae – in the 'health food' market – does not constitute, in my mind, a real product, as the consumer demand has been created without regard to any underlying value, efficacy, or need. Indeed, the history of microalgal production for the health food market has shown large swings in consumer demand making this a very high risk commercial activity (as evidenced by the several failed companies in *Spirulina* production). Thus I stress other potential products from microalgae (Table 11.1).

Two products, which are not 'health foods', are currently being produced in significant amounts from microalgae: 'linablue', a phycobiliprotein concentrate from *Spirulina* produced by the Dai Nippon Inks and Chemicals Co. at their plant in Southern California and sold in Japan as a colouring agent for ice-creams, yogurts, etc.; and β-carotene, a food colouring agent and a pro-vitamin A which is thought to be effective as an anticancer agent, produced at two plants in the US and two in Australia. In this regard I should mention that the Australian plants do not use the paddle-wheel mixed, raceway pond system I have discussed thus far. Rather they rely on harvesting algae from a more or less 'natural' environment (e.g. wind mixed ponds, no carbon dioxide enrichment). I predict that these plants will convert to the more conventional raceway production system. I also believe that the future is very bright for both of these products, with markets boosted by the increasing demand for natural food colouring agents and the recent findings that only natural β-carotene is likely to be effective as an anticancer agent. However, the current costs of production for both products must still be reduced considerably to exploit fully the commercial opportunities in this area. Considering that their commercial production started only recently, such cost reductions can be anticipated.

Microalgae, because of their high pigment content, are uniquely suited for the production of food colouring agents. One class of such pigments are the xanthophylls, which are presently used extensively in poultry feeds, which are often supplemented with marygold petal extract as a source of such pigments. Xanthophyll production by microalgae as a source of chicken feed pigmentation was already investigated in the 1950s, and has a significant potential. The economics of xanthophyll production are favoured by the fact that the whole alga would be sold as a feed, which would not require expensive extraction of the pigment. However, significant R & D is still required. I estimate that an economical process will require doubling to tripling the current xanthophyll content of most of the algae grown in mass culture. This could be accomplished in a number of ways: finding higher xanthophyll-containing algae (which can also be mass cultured), genetic strain selection, or physiological manipulations. Such a product could be developed in ten years.

Another relatively unique, and commercially quite interesting, market is the production of microalgae for the husbandry of filter feeding bivalves: clams, oysters, mussels, scallops, etc. Algal feeds have been produced as part of hatchery and seed operations for several decades, although the processes are very small scale (a few hundred litres), highly controlled (indoor tanks) and quite expensive (about $1000 kg^{-1} of dry weight algae). Growing algae on a larger scale to produce a market size animal, rather than small (<1 cm) seed, has been attempted by a number of groups for over twenty years, with consistent failures. These failures can be attributed to some fairly obvious shortcomings (poor choice of location, nutrient limitations). This area deserves additional work. I believe that this is also one of the near term (approximately 10 years) applications for the mass culture of microalgae.

A number of very high value specialty chemicals are or could be produced from microalgae. The 'phycofluors' for fluorescent labelling of diagnostic reagents were mentioned above. These are phycobiliproteins extracted from cyanobacteria and red algae. However, markets are minuscule (currently about 100 g yr^{-1}, with a market value of less than $1 million) and other sources (seaweeds) for these products exist. Isotopically labelled products have somewhat larger markets, but they are also measured in grams, not of significant interest for algal mass cultures. Although several companies are actively searching for pharmaceuticals among the microalgae, this is a gamble with very uncertain and long-term pay-offs. Even if successful, it is unlikely that it would lead to any significant needs for production of microalgae.

Lower value specialty chemicals have also been of some interest (see discussion of fermentative production of vitamin C and E above). Among these are several amino acids: proline from salt-stressed *Chlorella* and aspartic acid/arginine from cyanophycin granules, a storage compound of cyanobacteria, which is a copolymer of these two amino acids. Although it may be possible to design a (marginally) economic system for these products, it is likely that they would be produced as by-products from other processes (e.g. pigment extractions) rather than for their own sake.

Rather than small volume specialty chemicals, I believe that the real future of microalgal biotechnology lies in the larger scale production systems for lower value products: polysaccharides, vegetable oils, single cell proteins and protein isolates, fuels and fertilizers.

Some of these may not be far off in the future. One company has already sold an algal soil inoculum for almost ten years. Although there is no conclusive scientific proof that the algal inoculum (a culture of *Chlamydomonas mexicana*) actually does improve soil tilth at the application rates used. However, it is clear that microalgae play a significant role in soil ecology. Several companies are working on developing algal soil inoculum of nitrogen-fixing cyanobacteria. Of course, in rice production in developing countries nitrogen-fixing cyanobacteria already make substantial contributions to the nitrogen economy of that crop. Although I am not yet convinced (from the data I have seen thus far) that inoculation of rice fields with selected strains of algae has been proven effective, in principle it should be technically feasible. Although there are still significant problems to solve, my own research in this area has convinced me that the production, storage, transport, and soil (or rice paddy) inoculation of nitrogen-fixing cyanobacteria is both technically and economically feasible. Most important, I believe that the inoculated cyanobacteria can grow on the soil and contribute to soil fertility. Fertilizer production by a nitrogen-fixing cyanobacterial inoculum could become an effective and significant source of fertilizer in many agricultural applications.

The production of real commodity type products, (animal feeds, vegetable oils, fuels) would require significant advances in microalgal technology, as was discussed above. There are, however, no fundamental barriers to such enterprises, provided they are carried out at sites where all the requirements for algal mass culture (level land, water, CO_2) are available at minimal cost. The availability of all these resources at one site will limit the overall impact of this technology, although perhaps not as much as may be thought. Microalgae have the ability to utilize high salinity waters which are unsuitable for conventional agriculture. Thus both seawater and inland saline waters (of quite different composition than seawater) can be used. The presence of saline waters implies that competition for land with agriculture will be minimized. This leaves the availability of low cost ($<\$20$ ton^{-1}) CO_2 as the major resource limitation to the production of commodities from microalgal cultures (assuming that technological problems in cultivation, harvesting and processing discussed above can be solved).

One area of application, not yet discussed, is the use of algae in wastewater treatment. With animal wastes there is a potential for the use of algal cultures to simultaneously treat the waste waters and production of animal feeds. Although the direct use of algae as an animal feed would be limited by its high moisture content and the high cost of conventional drying processes, sun drying or mixing with dry feeds are possibilities. In the case of swine and hog production, a wet feed is feasible, suggesting that as a near term application of algal waste treatment/feed production.

Another important application is in the treatment of municipal and industrial wastewaters. Although 'oxidation' ponds are extensively used in a number of countries, including the US, as a low cost waste treatment alternative, these ponds cannot be properly considered algal production systems as they do not maximize algal output (indeed they try to minimize it, as it results in suspended solids and oxygen demand in the effluents). The 'high rate pond' (a shallow mixed raceway type pond) first proposed by Oswald (1963) for wastewater treatment maximizes both wastewater treatment and algal production. The major limitation to a more widespread application has been the cost of algal removal and the disposal costs of the algal sludge generated. Algal harvesting from high rate wastewater ponds by bioflocculation appears to be feasible (Benemann, *et al.*, 1980) and the algal biomass could be used for the generation of fuel, as potential contamination limits its use as animal feed.

Fuel production by microalgae can be considered the last frontier of microalgal biotechnology. Perhaps paradoxically, it is currently the major focus of R & D in this area both in the US and, to a lesser extent, Europe. The current emphasis in the US is on the production of oils (lipids) by microalgae as a future source of liquid fuels. Previously, gaseous fuels (methane, hydrogen) were emphasized. This will require long-term (25 years) research. Why such emphasis on fuel production if food or feed is likely to be of greater significance in the near term?

The answer is both technical and political; near term products do not need a long-term R & D commitment or support from the government, and current views are that liquid fuels will be a much more limiting economic factor in the foreseeable future than food or feed. By focusing on the long-term development of a specific algal biotechnology (such as oil production by microalgae), we can expect that most of the generic problems in microalgal biotechnology will be solved. Ultimately the market will decide which products provide the greatest value to both producers and consumers. Until then much work remains to be done.

Conclusions

Microalgal biotechnology at present is a limited and small component of the overall field of biotechnology, where advances in industrial fermentations, therapeutic proteins, genetically engineered plants and animals, and diagnostics have attracted major commercial interests and are resulting in many novel products. Microalgal biotechnology is another example of how the basic research in biological sciences over the past few decades is now resulting in significant new commercial activities. I will not hazard a guess of how significant, in terms of actual dollars, microalgal biotechnology will be in 10 or 25 years. However, even a pessimistic forecast would need to conclude, on the strength of what has already been accomplished, that the applications of microalgal biotechnology will repay the investments in basic research many fold. Thus I take this opportunity to acknowledge the debt we owe to the many scientists who pioneered this field, only some of whom I have mentioned above, for pointing the way and allowing us the opportunity to see their dreams realized. To quote Myers (1970): '. . . improvements in engineering practice based upon careful attention to laboratory data may still lead to practical achievements in mass culture'. Indeed they already have, and following the lead of these pioneers assures our continuing progress.

References

Becker E W 1985 *Production and use of microalgae, Ergebnisse der Limnologie* **20**

Benemann J R, Koopman B L, Weissman J C, Eisenberg D E, Goebel R P 1980 Development of microalgae waste water treatment and harvesting technologies in California. In Shelef G and Soeder C (eds) *Algae Biomass*, Elsevier Biomedical Press, Amsterdam, Holland, pp 457–96

Benemann J R, Tillett D M 1987 *The Effects of Fluctuating Environments on High Yielding Microalgae.* Final Report Solar Energy Research Institute, Golden, Colorado

Benemann J R, Tillett D M, Weissman J C, 1987 Microalgae biotechnology. *Trends Biotechnol.* **5**: 47–53

Borowitzka L J, Borowitzka M A (eds) 1988 *Microalgal Biotechnology* Cambridge University Press, Cambridge

Bruce J R, Knight M and Parke M W 1940 The rearing of oyster larvae on an algal diet. *J. Mar. Biol. Assoc. UK* **24**: 337–74

Burlew J S (ed) 1953 *Algal Culture – From Laboratory to Pilot Plant*, Carnegie Institution of Washington, DC

Delente J J 1987 Perdeuterated chemicals from D_2O-grown microalgae. *Trends Biotechnol.* **5**: 159–60

Fisher A W 1961 Engineering for algae culture. In Daniels F, Duffie J F (eds) *Solar Energy Research*, University of Wisconsin Press, Madison, pp 185–9

Glazer A, Stryer L 1984 Phycofluors. *Trends Biochem. Sci.* **8**: 423–7

Goldman J C, Ryther J H 1977 Mass production of algae: bioengineering aspects. In Mitsui A, Miyachi S, San Pietro A, Tomura S (eds) *Biological Solar Energy Conversion*, Academic Press, New York, pp 367–78

Harder R, von Witsch H 1982 Ueber Massenkultur von Diatomeen. *Ber. Deutsche Bot. Ges.* **60**: 146–52

Kok B 1953 Experiments on photosynthesis by *Chlorella* in flashing light. In Burlew J S (ed) *Algal Culture – From Laboratory to Pilot Plant*, Carnegie Institution of Washington, pp 63–75

Kok B 1960 Efficiency of photosynthesis. In Ruhland W (ed) *Encyclopedia of Plant Physiology*, Springer, Berlin pp 566–633

Lembi C A, Waaland J R (eds) 1988 *Algae and Human Affairs*, Cambridge University Press, Cambridge

Myers J 1970 Genetic and adaptive physiological characteristics observed in the chlorellas. In *Prediction and Measurement of Photosynthetic Productivity*, Wageningen Centre Agricultural Publishing & Documentation, 447–54

Odum E 1971 *Environment, Power, and Society*, Wiley Interscience, New York

Osborne B A, Geider R J 1987 The minimum photon requirement for photosynthesis. *New Phytol.* **106**: 631–44

Oswald W J 1963 The high-rate pond in waste disposal. *Dev. Ind. Microbiol.* **4**: 112–25

Oswald W J, Benemann J R 1977 A critical analysis of bioconversion with algae. In Mitsui A, Miyachi S, San Pietro A, Tomura S (eds) *Biological Solar Energy Conversion*, Academic Press, New York, pp 379–96

Pirt S J 1986 The thermodynamic efficiency (quantum demand) and dynamics of photosynthetic growth. *New Phytol.* **102**: 3–37

Pirt S J, Lee Y-K, Richmond A, Watts-Pirt M 1980 The photosynthetic efficiency of *Chlorella* biomass growth with reference to solar energy utilization. *J. Chem. Technol. Biotechnol.* **30**: 25–34

Richmond A (ed) 1986 *Handbook of Algal Mass Culture*, CRC Press, Boca Raton, Florida

Shelef G, Soeder C 1980 (eds) *Algae Biomass*, Elsevier Biomedical Press, Amsterdam, Holland

Weissman J C, Goebel R P 1988 Photobioreactor Design: Mixing, carbon utilization, and oxygen accumulation. *Biotech. Bioeng.* **31**: 336–44

Index